THE ILLUSTRATED HANDBOOK OF AVIATION AND AEROSPACE FACTS

BY JOE CHRISTY

TAB BOOKS Inc.
BLUE RIDGE SUMMIT, PA. 17214

Acknowledgment

I am indebted to TAB Aviation Editor Steve Mesner for his interest and cheerful contributions to this work, and to TAB Editorial Assistant Lorri Willard for her careful and tireless indexing.

Other TAB Books by the Author

No. 2252 *How To Install and Finish Synthetic Aircraft Fabrics*
No. 2258 *The Complete Guide to Single-Engine Beechcrafts—2nd Edition*
No. 2265 *WWII: US Navy & Japanese Combat Planes*
No. 2268 *The Complete Guide to Single-Engine Cessnas—3rd Edition*
No. 2275 *WWII: Luftwaffe Combat Planes & Aces*
No. 2291 *Refinishing Metal Aircraft*
No. 2298 *Low-Cost Private Flying*
No. 2322 *Racing Planes & Pilots*
No. 2325 *The Private Pilot's Handy Reference Manual*
No. 2327 *American Air Power: The First 75 Years*
No. 2347 *Engines for Homebuilt Aircraft and Ultralights*
No. 2357 *Build Your Own Low-Cost Hangar*
No. 2367 *Your Pilot's License—3rd Edition*
No. 2377 *Aircraft Construction, Repair & Inspection*
No. 2387 *High Adventure—The First 75 Years of Civil Aviation*

FIRST EDITION

FIRST PRINTING

Copyright © 1984 by TAB BOOKS Inc.
Printed in the United States of America

Reproduction or publication of the content in any manner, without express permission of the publisher, is prohibited. No liability is assumed with respect to the use of the information herein.

Library of Congress Cataloging in Publication Data

Christy, Joe.
The illustrated handbook of aviation and aerospace facts.

    Includes index.
    1. Aeronautics—History. 2. Astronautics—History.
I. Title.
TL515.C52   1984   387.7'0973   84-8714
ISBN 0-8306-2397-3 (pbk.)

Cover illustration by James H. Farmer.

Contents

Introduction		v
Prelude		vii
Section I U.S. Civil Aviation		1

1 Those Magnificent Men (and Women!)   3
French pioneers—International air meets—U.S. air meets—First female pilots—First transcontinental flight—Stinson Sisters—Early exhibition pilots

2 Builders and Barnstormers   13
Curtiss—Martin—Wright-Martin—Boeing—Lockheed—Sloan and Standard—Burgess-Dunn—Fairchild—Vought—Gallaudet and Consolidated—Dayton-Wright—Ryan—Cessna—Travel Air—Early air mail

3 Power Paces Progress   46
Curtiss OX series—Hispano-Suiza—Kirkham—Lawrance—Whirlwinds—Pratt & Whitney—Le Blond—Kinner—Menasco—Warner—Velie—Szekely—Camine—Jacobs—Siemens-Halske—Alexson—Cirrus—Ranger—Continental—Lycoming—Morehouse—Allison—Rolls-Royce—Rotaries

4 The Wings Grow Stronger   75
Lindbergh—Eaglerock—Pitcairn—Buhl-Verville—Fokker—Ford-Stout—Stinson—Cessna—Beech—Stearman—Great Lakes—Piper—Taylorcraft—Monocoupe—Luscombe—Aeronca—First dusters—Technological advances—NACA

5 The First Airlines   106
Kelly Bill—Contract Air Mail routes—Colonial—National Air Transport—Western Air Express—Varney Air Lines—Ford-Stout Air Lines—Pacific Air Transport—Northwest Airways—Florida Airways—Clifford

Ball—Capital Airlines—Air Commerce Act—Morrow Board—Air mail subsidies—Boeing Air Transport—United Air Lines—North American—AVCO—Walter F. Brown—SAFE-Way—American Airways—Ludington—Black Committee—Air Corps flies the mail—Black-McKeller Act—Civil Aeronautics Act of 1938—Pan Am

6 General Aviation During the '30s — 143
Technological advances—Private airplanes—American Eagle—Porterfield—Spartan—WACO—Civilian Pilot Training Program

7 American Air Racing — 154
National Air Races of the '30s—Travel Air Mystery—Purses—Thompson Trophy races—Greve Trophy races—Postwar racing—Goodyear races—Modern air racing—Unlimiteds—Formula I—Sport Biplanes

8 The Homebuilts — 175
Heath—FAA responds—The builders—Materials—The EAA—Oshkosh Fly-In—International Aerobatic Club—Akro pilots—Championship aerobatics—Aresti Aerocryptographics

9 New Problems, New Opportunities — 192
General Aviation during the '40s and '50s—Piper—T-Craft—Culver—Mooney—Ercoupe—Fairchild—Aeronco-Citabria—Cessna—CallAir—Luscombe—Temco—G.I. Bill—FAA—Modern general aviation—Ultralights and ARVs—Rutan—Airline deregulation—Controllers strike—Slots and flow control—Labor, the airlines, and the Supreme Court

Section II U.S. Military Aviation — 227

10 The First U.S. Military Airplanes — 229
Wright machine of 1909—First Navy airplane—Marine Corps Aviation formed—With Pershing in Mexico—Military air, 1914—First aircraft armament—WWI "firsts"—U.S. airpower in WWI—Frank Luke—Billy Mitchell—Navy/Marine Air, WWI—Aircraft production and procurement

11 The Tumultuous '20s — 265
Foreign policy—Navy Bureau of Aeronautics—Military air races—Metal propellers—McCook Field—Airplanes vs. battleships—First aircraft carrier—First world flight—Mitchell trial—Morrow Board—Lexington and Saratoga commissioned

12 The Uncertain '30s — 313
U.S. military posture—Navy dirigibles—Navy blimps—Army balloons—Toward war—Japan invades China—Hitler invades Poland—U.S. Neutrality Acts—Battle of Britain—Army and Navy aircraft to Iceland—Navy combat aircraft 1931-1940

13 World War II — 350
Negotiated peace?—America is attacked—Battle for the Philippines—Coral Sea and Solomons—Doolittle raid—Battle of Midway—Guadalcanal—North Africa—Palm Sunday Massacre—Italian campaign—Cross-channel invasion—Victory in Europe—China-Burma-India—Flying Tigers—The Aleutians—Victory in the Pacific—Superfortresses against Japan

14 The Era of Confrontation — 413
National Security Act—SAC formed—Boeing B-52 genesis—Berlin airlift—Turbojet development—Speed of sound exceeded—Korean War—Vietnam War

15 American into Space — 442
Dr. Robert Goddard—Germans build on Goddard's work—First U.S. rockets—Apollo series—Moon landing—The space shuttles—Unmanned space vehicles—Extraterrestrials?

Index — 455

Introduction

My purpose in preparing this book was to compile a general reference on American aviation, both civil and military, that offers the maximum amount of useful data in an interesting fashion. Toward that end, I eliminated subscripts, footnotes, and asides wherever possible, but that left a few details that demand explanation.

I use the word "airplane" throughout the book, but "airplane" is an American word that first appeared in a 1916 aviation magazine. Prior to that time, "flying machine" and the British "aeroplane" were the most common generic terms. Throughout the '20s and '30s, pilots often used the word "ship." Some of the gray eagle types still do.

Until World War II, aircraft propulsion systems were "motors." Then the military taught a couple of million young men that "motors" generate electricity, while aircraft powerplants are "engines," which explains why the three-engine Ford and Fokker transports of the late '20s were "tri-motors."

"Airlines" is sometimes spelled herein as two words, sometimes as one. That is because the air carriers themselves do it that way. E.g., United Air Lines, American Airlines.

In the military section, we begin with the Aviation Section of the Signal Corps which, by popular usage, became the "Air Service" during WWI and, officially, the U.S. Army Air Service in 1920, the U.S. Army Air Corps in 1926, the U.S. Army Air Forces in 1940, and today's independent U.S. Air Force in 1947. We employ herein the term appropriate to the period in time under discussion.

In May 1942 the USAAF changed the designation of pursuit squadron (PS) and pursuit group (PG) to fighter squadron (FS) and fighter group (FG).

The Army had "scout" planes early in WWI, which soon became "pursuit" planes, a term that remained in use until WWII when popular usage dictated the change mentioned above. However, the Army Air Forces continued to employ the "P" designation for its fighter aircraft until the new U.S. Air Force changed that in 1948. Therefore, for example, the Mustang was a P-51 until 1948, when it became an F-51.

The U.S. Navy used the term "fighter" for

such airplanes from the time it ordered its first machine of that type, the Curtiss HA "Dunkirk Fighter" of 1918.

The Air Corps/Air Force has always been relatively straightforward about aircraft designations, employing simply one or more letters followed by a number. For example, the Curtiss P-40 was the 40th fighter (pursuit) allocated a designation, the Boeing RB-47 the reconnaissance ("R") variant of the Air Force's 47th bomber.

The Navy's system until 1962 was somewhat more complex. One or more letters designating aircraft type or function were followed by a letter denoting the aircraft's manufacturer, hence AD for Attack, Douglas; TBF for Torpedo Bomber, Grumman; SBD for Scout Bomber, Douglas, etc. When more than one aircraft of the same type was procured from the same manufacturer, a number was placed between the letters. Thus the F6F was the sixth fighter design from Grumman, the SB2C was Curtiss' second scout bomber, and so on. The system can sometimes cause confusion to those not familiar with it. For example, during WWII, the Navy made extensive use of two "F4" aircraft—Grumman's F4F Wildcat and Vought's F4U Corsair. (To further complicate matters, the Corsair was also license-built as the FG [first fighter from Goodyear] and the F3A [third fighter from Brewster].)

The Department of Defense decided to consolidate and unify the Army/Air Force and Navy/Marine designation systems in 1962, beginning everything back at the number one. Thus the Navy's F4H Phantom, F8U Crusader, and AD Skyraider became, respectively, the F-4, F-8, and A-1. The Air Force's F-111 was the highest number assigned in the fighter series, and the F-110—the USAF version of the Phantom—became the F-4, the same as the Navy's. Except for the Northrop F-5 and the unassigned F-13, existing aircraft filled the F-for-fighter numbers until the all-new F-14 of the early '70s. (The F-numbers are up to 20 at this writing.) The B-for-bomber numbers are currently stuck at one, the B-1 being the only new bomber contracted since 1962.

Another point deserving of clarification may be our use of the "Me" prefix for WWII Messerschmitts. Some aviation writers today use the prefix "Bf" for these aircraft because most were built by Bayerisch-fleugzeugwerke. However, these machines were called Me109s, Me110s, etc., by airmen of both sides during WWII, and therefore we prefer to stick with historical rather than technical fact. The gifted Willy Messerschmitt was chief of design at Bayerisch-fleugzeugwerke.

Finally, about the national insignia on our military airplanes: The white star on blue circular background with red center was officially adopted 19 May 1917.

From January 1918 to August 1919, concentric circles, with red on the outside, then blue, with a white center were used. Then the military services returned to the white star on circular blue background with red center.

On combat aircraft, the star was added to each side of the fuselage in February 1941, and the star was removed from the upper surface of the right (starboard) wing and from the lower surface of the left (port) wing.

Immediately after Pearl Harbor the Navy returned to two wing stars, upper and lower surfaces, and retained them until 1943.

The red center was removed from the star at the end of May 1942.

At the end of June 1943, white rectangles were added to each side of the blue circular field, and the entire emblem outlined in red. In September 1943 a blue border replaced the red one.

In January 1947 a red stripe, centered in the horizontal bars, brought the national insignia to its present form—or at least its present shape. In the late '70s it was discovered that the red stripe in the "star and bar" was not only visually conspicuous but added measurably to an aircraft's infrared "signature," making it more vulnerable to heat-seeking missiles, so some combat types such as the A-10, F-16, and F-18 now carry low-visibility national insignia of black or medium gray.

Prelude

Man's first successful aerial vehicle was a hot air balloon launched at Annonay, France, by Joseph and Etienne Montgolfier on 5 June 1783. Having observed that smoke rises, the Montgolfier brothers decided that smoke contained a mysterious gas that was lighter than air. They constructed a 35-foot balloon (*"balon"*) of linen-reinforced paper under which they built a fire fueled by damp straw and sheep's wool. Their balloon filled with the dense smoke and rose majestically into the sky.

The brothers followed their triumph with another ascent on 19 September 1783, sending aloft as passengers a duck, a sheep, and a rooster to establish that the upper atmosphere was safe to breathe. Then, the following month, French scientist Pilatre de Rozier and the Marquis d'Arlandes made a flight over Paris in a Montgolfier balloon. Man's conquest of the air had begun.

During the next 100 years, lighter-than-air vehicles—employing either heated air or hydrogen as a lifting agent—became as commonplace as their limited usefulness would allow, evolving at last into Count Ferdinand von Zeppelin's giant rigid airships, the first of which flew in Germany on 2 July 1900, more than three years before the Wright brothers took to the air in the world's first successful *heavier-than-air self-propelled* flying machine.

The invention of the airplane was preceded by a great deal more wishful thinking than scientific inquiry. A host of experimenters and dreamers proposed—and even built—a number of flying machines during the nineteenth century. None was successful, although several made small contributions to the quest. Britain's Sir George Cayley and Germany's Otto Lilienthal were the most significant of the serious experimenters who preceded the Wright brothers. America's Octave Chanute, a retired structural engineer, built several multi-wing hang gliders in 1896 and 1897 that sailed up to 350 feet along the south shore of Lake Michigan. But Chanute's best machine was essentially a refinement of an earlier Lilienthal design, and Chanute's only direct contribution to manned flight was the Pratt truss system for bracing biplane wings.

Lilienthal flew his first monoplane hang glider in 1891, later switching to biplane configurations. Flying from an artificial hill near Berlin and the Stollner Hills near Rhinow, he achieved glides up to 750 feet in length before he was killed in one of his machines on 9 August 1896.

Lilienthal was a long way from achieving controlled, powered flight when he died, and his calculations of lift produced by a cambered wing surface were in error, but photos of him in flight were published around the world and his work inspired many others—including the Wrights.

The extent to which Lilienthal was influenced by the work of Sir George Cayley is a matter of conjecture, but Cayley's paper *On Aerial Navigation*, published in 1810, and the successful flights of his model gliders establish that this remarkable Englishman was the first to propose the proper configuration for a fixed-wing, heavier-than-air flying machine. Cayley suggested a curved upper wing surface for increased lift, wing dihedral for lateral stability, and his models featured cruciform tails. In the early nineteenth century he was so far ahead of everyone else in this field that it may be said that nothing more of significance was added to Man's meager store of knowledge on this subject until Lilienthal began his experiments 80 years later.

Lilienthal's flights in his hang gliders actually produced almost no new scientific data; his carefully recorded figures on the lifting properties of cambered wing surfaces were found to be in error by the Wrights after they began their experiments. But Lilienthal's influence was great, at least partly because the halftone process for reproducing photographs in newspapers came into use just in time to picture Lilienthal's accomplishments and prove to the world that man *could* sail through the air on artificial wings.

Another Englishman, Percy Pilcher, built an improved monoplane glider following Lilienthal's death, but Pilcher worked in isolation and died in the crash of his machine in 1899 without influencing the invention of the airplane. And perhaps we should mention that in Russia in 1884 (sometimes mistakenly reported as 1882), a steam-powered airplane built by Alexander Mozhaisky and piloted by I.N. Golubev was thrust into the air from a ski-jump type ramp for a few seconds, but proved incapable of supporting itself in flight.

The most controversial of the early experimenters was Professor Samuel P. Langley, a respected astronomer and secretary of the Smithsonian Institution. Langley became seriously interested in mechanical flight about 1886, and ten years later built a 25-pound steam-powered model of hickory and silk that flew three-quarters of a mile—an event witnessed by Dr. Alexander Graham Bell, who also believed that man-carrying flying machines were possible, and who reported Langley's success to President McKinley.

Langley's model achieved lateral stability in flight by means of dihedral in each of its two wings, which were positioned in tandem—an arrangement used by British experimenters Thomas Walker in 1831 and D.S. Brown in 1874, the latter having tested model gliders so configured. Langley's model also possessed cambered upper wing surfaces (first used by Sir George Cayley in 1809), although Langley's airfoil shape was very inefficient, with its deepest point much too far from the wing's leading edge.

Apparently influenced by Dr. Bell's enthusiastic account of Langley's work, President McKinley obtained an Army appropriation in the amount of $50,000 to finance construction of a full-size, man-carrying version of the "Aerodrome," as Langley called his model.

Langley hired 23-year-old Charles Manley, an engineering student at Cornell University, and designed a quarter-scale model of a full-sized Aerodrome. This flew in 1901 with a small gasoline engine.

The full-sized Aerodrome was completed in July 1903, and taken to its launch track atop a houseboat in the Potomac River. Manley had built it according to the 63-year-old Langley's instructions, and also designed and built its engine after it became clear that the Balzar engine contracted by Langley was much too heavy and produced only a fraction of the power expected. (Some historians have referred to the Aerodrome's powerplant as the

The Langley Aerodrome on its catapult in the Potomac River. It failed to fly in two attempts, suffering structural failure each time. (NASM)

"Manley-Balzar" engine, apparently in the belief that Manley merely modified Balzar's creation. Manley did expend considerable effort attempting to get more power from it, but finally gave up and designed a truly remarkable five-cylinder radial that was water-cooled and produced 52.4 hp at 950 rpm; it weighed 207.5 pounds including coolant. This engine proved its reliability in three separate ten-hour test runs, and its weight-to-horsepower ratio was not equalled until the appearance of the Liberty engine in 1917.)

On 7 October 1903, the Aerodrome was ready for test. Manley climbed aboard, gave the signal for release of the catapult spring, swooshed down the launching rail—and fell into the Potomac. Only slightly injured, he swam to safety. A photo taken just as the machine left its launching device showed its forward wing twisted grotesquely out of shape.

On 8 December 1903 Manley had the repaired Aerodrome back atop the houseboat for another try.

The Langley machine plunges into the Potomac just nine days before the Wright brothers made the world's first successful airplane flight. (NASM)

ix

But this time the machine's rear wing failed as it left its launching track, and Manley narrowly escaped drowning when he became entangled in the wreckage. Nine days later the Wrights flew at Kitty Hawk, and Langley abandoned his project. He died in 1906, a thoroughly honorable man, but the Aerodrome—or a reasonable facsimile thereof—would be revived eight years later and used by others to perpetrate a shameful hoax designed to discredit the Wright brothers—about which, more momentarily.

Orville and Wilbur Wright possessed a significant advantage over most of those who had preceded them in the quest of manned, mechanical flight: *They approached the question as airmen*. Except for Lilienthal, all the others had mostly proceeded with preconceived ideas as to what a flying machine should look like. They produced "eyeball" designs, and few bothered with any serious investigation of the dynamics of flight, while none gave any real thought as to how a flying machine would be controlled if it *should* succeed in getting off the ground and achieve sustained flight. Most appeared to believe that an airplane could be steered like a boat, and that wing dihedral would provide the necessary lateral stability.

The Wrights, however, were concerned with *controlled* flight from the outset, and they approached their task suspicious of every "discovery" and each bit of data that had been passed along by experimenters before them. Thus, Wilbur discovered very early that the best information available on the lifting properties of wings with various upper surface curves—painstakingly recorded by Lilienthal—was in error. The Wrights did adopt the Pratt truss system for wing bracing, but their ultimate success was exclusively theirs. They constructed a small wind tunnel that allowed them to witness the shift in the center of pressure above the wing as its angle of attack was changed, and which

Wilbur Wright. (National Archives)

Orville Wright. (National Archives)

aided in the selection of a reasonably efficient airfoil shape for their wings. Eventually the brothers would put a similar, deliberate effort into propeller research, and their marvelously reasoned solution to the problem of roll control and turning flight—wing warping—was the answer to a major aerodynamic question that others had not yet asked. The addition of a hinged rudder, interconnected to the wing-warp controls, solved the problem of adverse yaw in turning flight.

The Wrights built their first aircraft in August 1899. It was a biplane kite with a wingspan of five feet, and was flown tethered to a stake, its primary purpose being to test the wing warping system. This system warped downward the outer trailing edges of the wings on one side while simultaneously warping upward the trailing edges on the opposite side. Hinged ailerons (first used by the French experimenter Esnault-Pelterie in 1904) would soon supplant the Wrights' wing warping system, but the principle was the same.

Following tests with their kite, the Wrights built three biplane gliders, one each year 1900-1902 inclusive, and accomplished more than a thousand glides from the sand hills at Kitty Hawk, North Carolina.

At the conclusion of the 1902 tests the brothers were confident that they were ready to build a powered machine, and the Wright Flyer I (Flyers II and III would follow in 1904-1905), fitted with a 12-hp gasoline engine designed by the Wrights, was ready to fly late in 1903.

Wilbur won the coin toss for the privilege of making the first flight in the new machine on 14 December, but scarcely got off the ground before overcontrolling with the forward-mounted elevator

The 1902 Wright glider; flight control theories proven, the Wrights needed only to add a proper propulsion system—which required that they design and build their own engine and propellers. (NASM)

The age of manned flight begins, 17 December 1903. On his first attempt that day, Orville flew 120 feet in twelve seconds. Three more flights were made, with the brothers alternating at the controls. The fourth flight, at noon, with Wilbur piloting, lasted 59 seconds and was measured at 852 feet.

and nosed into the sand. Damage was slight, and at 10:35 a.m. on Thursday, 17 December, Orville took his turn. He flew 120 feet in 12 seconds into a 20-22-mph wind. During the next hour and a half, three more flights were made with the brothers alternating at the controls. On the final flight Wilbur flew 852 feet while remaining aloft for 59 seconds. Those flights, witnessed by five local people, were the first in the history of the world in which a powered heavier-than-air vehicle had taken off under its own power, achieved sustained, controlled flight, and then landed at a point as high as that from which it had taken off.

During the next two years the brothers regularly flew from Huffman Prairie, about eight miles east of their home in Dayton, Ohio. They made some 40 flights in 1905, the longest being for a distance of 34 miles. The Wrights avoided publicity awaiting patent protection and the expected sale of their invention to the U.S. and British governments. Satisfied that their Flyer III was a practical aircraft, the brothers did not fly again until the summer of 1908. Orville demonstrated a two-place Flyer at Ft. Myer (near Washington, D.C.) in September that year, the U.S. Army at last having been forced to take the Wrights seriously as reports persisted of successful flights by experimenters in France. (Accurate sketches of the Flyer III had been published in France in 1906, and although no one there understood the Wrights' system of lateral control, Henri Farman had, in 1907, remained aloft for a minute and 14 seconds and completed a wide, skidded circle. The Farman machine had no means of lateral control except for the yaw effect of its rudder.) At Ft. Myer, one of Orville's propellers failed in flight, damaged the airframe, and sent his machine crashing to the ground out of control. Lt. Thomas Selfridge, riding as a passenger, was killed, Orville seriously injured. But a month earlier, Wilbur had begun a series of demonstrations in France that captivated all Europe, and despite Orville's accident, the world had suddenly become

Wilbur with passenger near Auvors, France, in 1908. After witnessing one of Wilbur's many flights in France, Maj. B.F.S. Baden-Powell, Secretary of Britain's (hopeful) Aeronautical Society, said, "That Wilbur Wright is in possession of a power which controls the fate of nations is beyond dispute."

aware that *man could fly*.

Meanwhile, Alexander Graham Bell had brought together and financed a small group of men who were interested, as was Dr. Bell, in manned flight. That was the "Aerial Experiment Association," led by motorcycle racer/builder Glenn Hammond Curtiss. This group produced four flying machines based upon what could be learned about the Wright Flyers. These machines appeared between March and December 1908. The first barely flew and crashed on its second test. The next one managed to remain airborne for slightly more than a thousand feet during the best of its five attempts before crashing, while the third aircraft built by the A.E.A. members, the "June Bug," made a number of flights, the longest of which was two miles.

The June Bug was said to be Curtiss' design; it was fitted with a 30-hp water-cooled V-8 designed by Curtiss and engine builder Charles Kirkham and featured wingtip ailerons. Its best flight did not compare with the flights made three years earlier by the Wrights in their Flyer III, but the June Bug led Glenn Curtiss into the aircraft manufacturing business, and within two years Curtiss would be the Wrights' principal competitor—as well as their primary target in a series of bitter court confrontations resulting from Curtiss' alleged infringement on Wright patents.

The courts held in favor of the Wrights in every case that went to trial, but Curtiss managed to grow and prosper—primarily due to profits generated by the Curtiss exhibition teams—and held the Wrights at bay with appeals and other legal maneuvers until, at last facing a showdown in 1914, Curtiss resorted to a consciousless ploy that will forever shadow his very real contributions to aviation.

It occurred to Curtiss that if the Langley Aerodrome could be flown, then he could argue in court that the Wright patents were not based on an original invention.

Curtiss was sympathetically received by the man who had succeeded Langley as secretary of the Smithsonian, Dr. Charles D. Walcott, and Walcott not only loaned the Langley machine to Curtiss, but also sent along one Dr. Albert P. Zahm of the Smithsonian to document Curtiss' effort. It should be noted that Dr. Zahm had earlier been a Curtiss witness in court.

There were, however, other interested parties—particularly Griffith Brewer of Britain's Royal Aeronautical Society, to whom history is indebted for the true facts concerning the Curtiss caper.

Curtiss made no effort to fly the original Aerodrome. By 1914 a lot of progress had been made in airplane design. Curtiss was building flying boats by then, and had at least one able and experienced airframe designer on his staff (Douglas Thomas, designer of the famed WWI Curtiss JN-4 Jenny), and it was obvious to Curtiss that the Aerodrome was incapable of flight without significant modification.

Curtiss constructed new wings for the Aerodrome—wings of different planform, different camber, and differently braced. The Penaud-type tail (fixed surfaces with limited and very slow movement as a unit, and similar to that which Cayley had employed on his models 100 years earlier) was modified to operate with a Curtiss control system; the propellers were changed, the pilot's seat repositioned, and the machine was mounted on floats. There was still no means of lateral control except the stability provided by the dihedral angle at which the four wing panels were attached.

This hybrid could not sustain itself in flight; its longest hop was of five seconds duration. There-

Glen Curtiss (1) is pictured with Chance Vought (third from right) in 1911. Vought would also build airplanes and help found the company that is today known as United Technologies. (U.S. Navy)

fore, Curtiss removed the Manley engine and its twin pusher propellers and installed a 90-hp Curtiss V-8 with a single tractor propeller in the nose of the machine. Then, on 1 October 1914, he accomplished six takeoffs that resulted in flights of 20 to 65 seconds duration. It was then up to the Smithsonian's Dr. Zahm to legitimize the hoax.

Zahm's affidavit read in part: "The machine was the same machine in construction and operation as the original structure. The frame was the same; the engine was the same; the horizontal and vertical rudder located at the rear was the same; the vertical rudder under the machine was substantially the same; the propellers were the same and the wings were identical in construction with the original machine, except, as I have heretofore stated, they were perhaps a little more roughly built and a little heavier."

Actually, Zahm wrote this affidavit before Curtiss made the engine change. It therefore did contain *one* truth when it was written. We should note that the original Aerodrome had a total weight of 830 pounds. The machine flown by Curtiss weighed 1520 pounds.

It is not clear who returned the Aerodrome to its original configuration, but it was placed on display in the Smithsonian Institution just as Charles Manley had built it in 1903 along with a plaque that read: "The First Man-Carrying Aeroplane in the History of the World Capable of Sustained Free Flight . . ."

That is why the Wright brothers' Flyer remained in Great Britain's Science Museum until 1948. By that time, pressure from many individuals (as well as the possibility of a Congressional investigation) brought a formal apology to the Wrights from the Smithsonian, and the Wright Flyer claimed its rightful place there on 17 December 1948, the 45th anniversary of its historic first flights. Unfortunately, Orville had died the previous January at age 77. Wilbur had been spared the hurt of the whole shameful mess, having died of typhoid fever in 1912 at age 45.

The Wrights' lawsuits against Curtiss were never settled in court. The WWI patent pool effectively placed them on "hold" until 1919, and by that

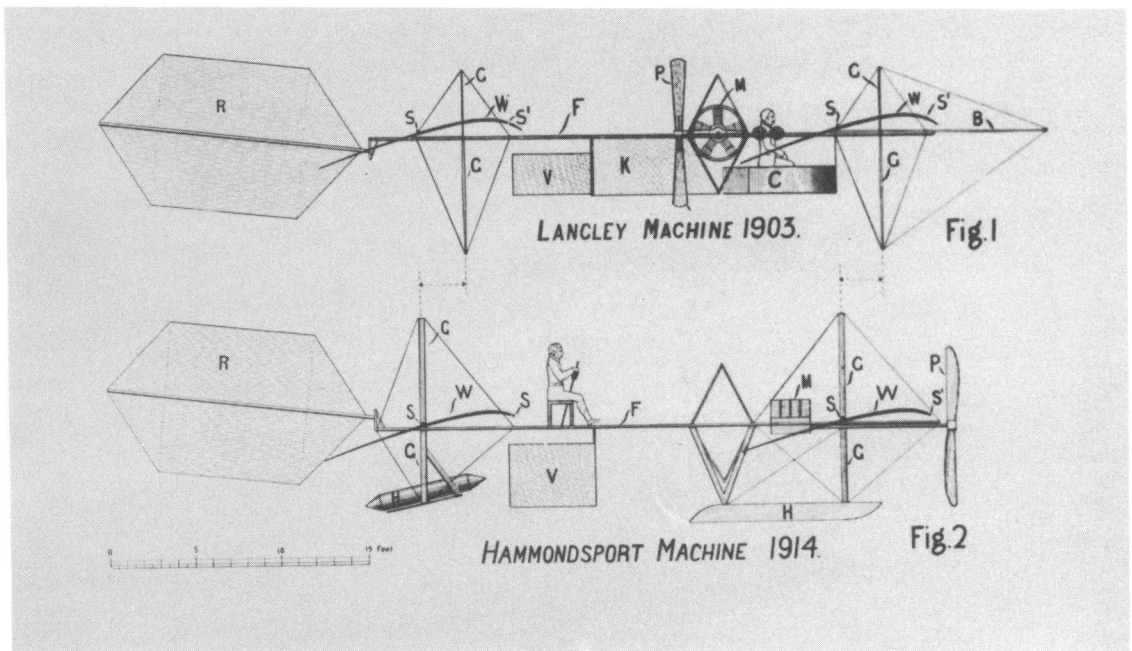

Comparison of the original Langley Aerodrome and the Curtiss-modified machine of eleven years later reveals the differences. (NASM)

Curtiss making one of several short hops 17 December 1914 in an effort to prove that the Langley Aerodrome was capable of flight, and therefore that the Wright patents were not based on an original invention. Curtiss' ploy fooled no one. (NASM)

time neither Orville nor Glenn Curtiss controlled the corporations that bore their names, and neither board of directors was willing to pursue the matter. Both men lived to see the merger of those companies in 1929, which formed the (then) giant Curtiss-Wright Aeronautical Corporation, but Orville had long since retired to his mansion in Dayton, and Curtiss was a Florida real estate developer. Curtiss died a year later of complications following an appendectomy.

We will encounter Curtiss-Wright in chapters to follow.

Section 1
U.S. Civil Aviation

Chapter 1

Those Magnificent Men (and Women!)

The first commercial use of the airplane was exhibition flying. Prior to WWI there was no other way to make money with this exciting new vehicle. Engines were too unreliable; payloads too small. Man could fly, but just barely.

That was enough, however, to justify promotion of several great aerial expositions to celebrate the achievement of this age-old dream. The first was a week-long event held on the Betheny Plain near Rheims, France, in August 1909. By that time France was already beginning to lead the world in development of the airplane. Except for advances made by Glenn Curtiss (which included off-water flying), almost every innovation, every improvement to the flying machine between 1908 and 1914 originated in France. The French were ready and eager for the air age; the rest of the world had to be dragged into it.

The roseate aeronautical climate in France was apparently due to several factors. Ballooning had been born there more than a century earlier, but a more direct influence was the work of several Frenchmen who had been experimenting with flying machines since 1903, spurred by published accounts in France of the Wrights' successful gliders. These accounts included detailed drawings and photos of the Wrights' 1902 glider in flight.

Among these French aviation pioneers were Ernest Archdeacon, Robert Esnault-Pelterie, Gabriel and Henri Voisin, Leon Levavasseur, and Louis Bleriot. Englishman Henri Farman and Brazilian Alberto Santos-Dumont added their achievements to the French aviation scene.

The diminutive Santos-Dumont, heir to a coffee fortune, began with a small, cylindrical gas bag fitted with a tiny gasoline engine; in this device he flitted among the rooftops of Paris much to the delight of Parisians. He turned to heavier-than-air machines in 1906 and, after several failures, produced a 200-pound monoplane of 24 hp, the "Demoiselle," in 1909. The Demoiselle's best flight was of about 12 minutes duration. Other machines in both America and France were more advanced by then, but Santos-Dumont's contribution was sig-

3

nificant because he was a colorful figure and his was such a class act that he provided much impetus to aircraft development in Europe at just the proper time.

International Air Meets

The 1909 Rheims Meet attracted all of the top aviators and the best aircraft designs produced up to that time. A French aviator, Maurice Tissandier, flew a copy of the latest Wright machine. The Wrights did not attend. Glenn Curtiss was there, however, with his specially designed Rheims Racer.*

Curtiss won the Gordon Bennett cup race with a speed of 47 mph, the *Prix de la Vitesse* at 46⅝ mph, and placed second in the *Tour de Piste*. There were protests over his misrepresentation of his engine's power, but Curtiss' reputation as an airplane builder was markedly enhanced.

The success of the Rheims Meet prompted a number of similar events, the next one being an air spectacular promoted by actor Dick Ferris and held at Dominguez Field near Los Angeles in mid-January 1910. The ubiquitous Glenn Curtiss established a speed record with a passenger at 55 mph, and leaped into the air with a takeoff run of 98 feet in 6.4 seconds. But France's Louis Paulhan, flying a Farman biplane, took home top honors and most of the money for his 75-mile round trip to Santa Anita and return with an elapsed time of one hour and 58 minutes. Paulhan also established a new altitude record of 4165 feet. It was indicative of the progress being made in aircraft development. The previous altitude mark 508 of feet had been made at Rheims only five months earlier.

The next great aerial meet was at Boston Harbor during the first two weeks of September 1910. Prizes totalled $100,000, the principal one being for a race around Boston Light. This was won by Enland's Claude Graham-White in a Farman, barely nosing out Glenn Curtiss. Graham-White flew the 5½ miles in six minutes.

The Boston Harbor Meet was the largest to date, but it was surpassed by the greatest aerial tournament prior to WWI, which was held at New York's Belmont Park, October 22-31, 1910. Described as the top society sporting event of the decade, the Belmont meet attracted not only the best pilots from both sides of the Atlantic, but what passed as the "jet set" of that era as well.

The Bennett Cup race at Belmont generated great excitement as Graham-White was hard pressed to beat out such aeronauts as America's John Moisant, flying a Bleriot; England's Hubert Latham, piloting an Antoinette; Alec Oligvie in a Wright Model C; and Walter Brookins' Baby Wright. Graham-White averaged 61 mph for the 62-mile course. Brookins crashed the Baby Wright when he turned to fly to the aid of Alfred Leblanc, who had wrecked his machine after averaging 66.2 mph in the speed dash. John Moisant won the race around the Statue of Liberty from Belmont Park, but was disqualified on a technicality and the $10,000 prize went to France's Count Jaques de Lesseps (Tenth son of the Suez Canal's builder). Ralph Johnstone claimed an important record for America when he established a new altitude mark of 9714 feet.

At Belmont the girls first appeared as pilots. Tiny Mlle. Helene Dutrieux was enthusiastically applauded by spectators for her flights, although she was not allowed to compete with the men. Other early "aviatrices," as they were called, were Americans Blanche Scott, Ruth Law, John Moisant's pretty young sister Mathilde, and a petite West Coast drama critic, Harriet Quimby. Katherine Stinson would follow in 1912, and her younger sister Marjorie in 1914.

Daring Birdmen

The success of the first great air meets, with their attendant publicity, correctly suggested that ready-made audiences awaited throughout the U.S. for living proof that Man could actually fly. There-

---

*Curtiss himself referred to this machine as the "Rheims Racer" or the "Rheims Machine." It was also called the "Golden Flyer," and there is evidence that it was the same airplane, fitted with a 50-hp engine, that some researchers have called the "Gold Bug," and which Curtiss sold to the Aeronautic Society of New York in the spring of 1909, powered with a 30-hp air-cooled V-8. It was loaned back to Curtiss in July 1909, and Curtiss appears to have built new wings for it. He registered his machine at Rheims at 30-hp, but it was found to have a new 50-hp engine.

fore, the Wrights, Curtiss, and Moisant organized exhibition teams and toured the country for the edification of the public and the enrichment of themselves. These teams appeared in all the large cities, while freelancers performed individually at county fairs and in smaller towns. The best of the freelancers were represented by booking agent Bill Pickins as the "Lincoln Beachey Flyers," Beachey being perhaps the most daring and soon the most popular of this nomadic fraternity.

Beachey had begun as a dirigible pilot at age 19 in 1905. He flew a sausage-shaped gas bag from which was suspended a light framework containing engine and aeronaut. But when the spectators at the Dominguez meet showed far more interest in winged aircraft than in Beachey's and Roy Knabenshue's "rubber cows," Beachey went to Curtiss, learned to fly, and by 1911 began giving airplane exhibitions booked by Pickins. His craft was a Curtiss pusher fitted with a water-cooled V-8 engine —an engine that would evolve into the famed OX-5 of WWI.

The Curtiss airplanes of this period had at least two advantages: good engines, and pilot's controls that were more natural to use than the Wrights' control systems. (Almost every manufacturer of flying machines employed a different control system through 1911.) The Wrights used two control sticks—one moving forward and backward to activate the elevators (pitch control), and one that when moved forward put down-warp on the right wings and when pulled back put down-warp on the left wings for banking (roll control). At the top of this latter stick was a lever that controlled the rudder for yaw control. Earlier, the Wrights had used a shoulder yoke that effected wing-warp when the pilot leaned in the direction of the desired bank.

The early Curtiss machines used a wheel atop a vertical column that raised and lowered the elevators when pulled back or pushed forward (pitch), and which controlled the rudder when turned right or left. Ailerons were activated by a shoulder yoke as in the earlier Wright machines.

Again, it was the French who eventually produced the more logical system in use today, which incorporated foot-operated pedals for rudder control and a wheel that activated both elevators and ailerons. For many years this was referred to as the "Dep" control system, presumably because it was introduced on the Deperdussion machines in 1912. The designer of Deperdussions was Louis Becherau. Most historians have passed Becherau by, but he designed exceptionally clean mid-wing monoplanes, one of which was the first airplane to exceed 100 mph (in 1912), and raised the mark to 126.59 on 29 September 1913. Later, Becherau would design the SPAD of WWI ("SPAD" for the company that built it, *Societe Pour Aviation et Derives*).

Although the Wrights had unlocked the secrets of controlled manned flight, it remained for others to build upon those discoveries. By 1912, the pupils could teach the teachers a few things.

However, some of the most famous of the early birdmen preferred the Wright machines, perhaps at least partly because they had learned to fly with the Wright control system. Calbraith P. Rodgers was one of them.

Rodgers may or may not have been a truly competent pilot, but one thing is certain: He had more than his share of determination. He was the first to fly coast-to-coast across the United States. Sponsored by the makers of Vin Fizz, a popular grape drink of the day, and in pursuit of a $50,000 prize offered by William Randolph Hearst to the first aviator to complete such a journey within 30 days, Rodgers left New York on 17 September 1911. He crashed 19 times and required 49 days for the trip, but refused to give up. He finally arrived in Pasadena on 5 November with one leg in a cast and only a rudder and a single wing strut remaining of his original Wright pusher. He had made 69 stops—many unscheduled due to weather and mechanical trouble—and actually spent 82 hours and four minutes in the air, covering 3220 miles. His average speed in the air was slightly over 39 mph.

In mid-1912 (by which time Calbraith Rodgers had died in a crash), one of America's pioneer women pilots took to the air. Katherine Stinson was not yet 17 years old when she soloed a Wright Model B at Max Lillie's flying school on Chicago's

Cicero Field. She required 3½ hours of dual instruction. (At $90 per hour, one tended to learn fast.) Katherine's indulgent mother, Emma, found money for a down payment on a used Model B Wright, and little Katy was soon booked for as many exhibition flights as she cared to make. Represented by Bill Pickins, she received $500 for each appearance—an impressive sum indeed in 1912 dollars. Not that she did anything spectacular by today's standards; at that time it was enough for most people to see a a flying machine rise off the ground and sail overhead, and to witness such a feat performed by an unassuming young woman was indeed a memorable sight.

In 1914, sister Marjorie followed in Katherine's propwash, attending the Wright school at Dayton, Ohio. Another Model B Wright was purchased for Marjorie's use, then she too began touring the county fair circuit. Brother Eddie attended the Wright school a year later, after having spent most of 1914 living in a hangar at Cicero Field with the Laird brothers, Matty and Charles, and Buck Weaver. These four would all significantly impact American aviation in years to come.

In 1915, Lincoln Beachey was killed while performing at San Francisco's Panama Pacific Exposition. Katherine bought the wreckage of his airplane to obtain its 80-hp Gnome rotary engine for a new biplane being built for her by Elmer Partridge and "Pop" Keller. She performed a loop in this machine, becoming the fourth American pilot to master this maneuver, and then included a "loop-the-loop" in her subsequent appearances. (The first inside loop in history was performed by France's Adolphe Pegoud in 1913. Beachey was the first American to do it, followed by DeLloyd Thompson, Art Smith, Katherine, and, in 1916, Matty Laird. Jimmy Dolittle performed the first outside loop in 1927.) Matty Laird appears to be the first pilot to perform a loop with a passenger, that passenger being Marjorie Stinson.

Emil Matthew "Matty" Laird was 20 years old in 1916, and was flying his third homebuilt airplane as a successful exhibition pilot, He, too, was represented by Pickins, and flew night exhibitions with battery-actuated railroad fuses fixed beneath his wings. By 1916 the public expected more than a few circles of the field, or even a loop or two. Laird's mechanic was George "Buck" Weaver, who would name and help organize the WACO company several years later.

Matty Laird, as many other early birdmen, not only designed and built his own airplanes but had taught himself to fly as well. Clyde Cessna did the same, at about the same time, as did Allan Lockheed, Giuseppe Bellanca, and Glenn Martin. Laird built his first airplane, a shoulder-wing configuration, at age 17 when he was employed as a $4.50 per week office boy. He had almost no money to spend on the project, but somehow managed to scrounge most of the material for the machine. A sympathetic high school instructor allowed him to use the school's woodworking tools; propeller maker Ole Flotorp (whose shop was just across the street from Cicero Field) provided a prop at minimum cost, and Laird's engine—part motorcycle, part automobile—was purchased for $40 on a time-payment arrangement. It would normally run for about 30 minutes before heat expansion induced valve failure due to a lack of lubrication. But the crude monoplane flew, and Laird crashed but once teaching himself to fly.

Up to that point, Laird's experience was not unusual; many, whose names have been lost to us, who merely wanted to fly, did much the same. Laird, however, would go on to build some of the finest commercial biplanes ever produced, and the Laird factory would be the first established in Wichita, Kansas—the spawning ground for so much of America's private aircraft industry.

The Henry Farman biplane of 1909 was powered with a 50-hp Gnome engine and had a speed of 37 mph. It was much in evidence at the Rheims meet.

One of the earliest aerial photos shows a spare wheel lashed to the aircraft's skid and was taken over Rheims, France, 1909.

Mlle. Helene Dutrieux flew at Belmont, L.I., but was not allowed to compete with the men.

Marjorie Stinson learned to fly at the Wright brothers' school in Dayton in 1914 and flew exhibitions in a Model B Wright pusher. (Marjorie Stinson)

Lincoln Beachey began his career as an exhibition pilot in 1905 at age 19. He flew Baldwin dirigibles with Roy Knabenshue's troupe before turning to airplanes in 1910 to become, because of his daring, the best-known of the early birdmen. Beachey died when his monoplane shed its wings during a performance at the San Francisco World's Fair, 14 March 1915. (NASM)

The Bleriot XI of 1909 had a speed of 36 mph, was powered with a 25-hp, three-cylinder Anzani engine, and was commercially available for about $2000. In such a machine, France's Louis Bleriot flew the English Channel 25 July 1909. Many Bleriots were sold and/or copied in America. Clyde Cessna built his first airplane after inspecting a Bleriot in Oklahoma City.

Calibraith Rodgers (R) made the first transcontinental flight 17 September to 10 December 1911. He crashed 19 times and arrived in California from New York with only a wing strut and rudder of his original aircraft, A Wright Model B. His sponsor was the manufacturer of "Vin Fizz," a popular soft drink. (NASM)

Crashes during the birdman era often required the services of friends armed with baseball bats to discourage souvenir hunters.

A.L. Longren of Topeka, Kansas, built this copy of a Curtiss pusher and toured the county fairs on the southern plains in 1912-1913. (Joe Durham)

Matty Laird's first airplane, built in 1913 when Laird was 17 years old, was "a loose formation of discarded material arranged in airplane form." A friend wrecked it. (E.M. Laird)

Katherine Stinson with her 1916 airplane, which was built for her by "Pop" Keller and Elmer Partridge. Engine was the 80-hp Gnome rotary salvaged from Lincoln Beachey's airplane.

Laird exhibition flight at St. Paul, Minnesota, 7 November 1916, by that time including a loop, crowded Presidential election news to a secondary position on front page of the Dispatch. (E.M. Laird)

Poster announcing Laird visit to La Grange, Indiana, 6 October 1916 reveals the hype designed to attract crowds to see the "death cheating aviator." Promoter Bill Pickens had also booked Beachey, the Stinson sisters, and other top exhibition pilots of the time.

Chapter 2

Builders and Barnstormers

There were as many as twelve aircraft manufacturers in the United States at the end of 1916, if we include the companies that were more shop than factory, producing three or four airplanes per year. Little more than 200 flying machines had been commercially produced since 1903, of which about half had gone to the U.S. Army and Navy.

Curtiss-Wright

Curtiss Aeroplane and Motor Corporation was easily the largest airplane maker in America at that time, although the recently formed Wright-Martin Aircraft Corporation was well-financed and positioned for the rush of orders anticipated as a result of the war in Europe—that is to say, the expectation that America would soon be drawn into that war. This was the third Curtiss company. The first was formed in March 1909, upon dissolution of the Aerial Experiment Association. It was the short-lived Herring-Curtiss Company, a partnership with one Augustus M. Herring, a self-important type formerly associated with Octave Chanute in gliding experiments. Herring and Curtiss parted abruptly after the Rheims Meet, apparently because of a disagreement over Curtiss' winnings there. Then Curtiss winged it on his own until 1916, by which time the need for expansion capital was clear. Meanwhile, Curtiss had brought the seaplane and flying boat to a practical stage of development and, in January 1911, had established a flying school on North Island, across the bay from San Diego, where he trained both Army and Navy pilots. Therefore, confronted with the probability of large orders from the military, Curtiss agreed, early in 1916, to a proposal from John North Willys to form a public corporation.

Willys, of Willys Car Company (later, Willys-Overland), brought in Clement Melville Keys, an investment banker who consolidated Curtiss' holdings and marketed a stock issue. Willys emerged as the chief minority stockholder of Curtiss Aeroplane and Motor Corporation, which established plants in Garden City and Buffalo, New York. After the United States entered WWI on 6

13

April 1917, the Buffalo plant in particular was greatly expanded, aided by an $8 million advance from the government. Willys was president of the new company; Curtiss was board chairman as well as president of the Engineering Division at Garden City. Keys was a vice-president.

During WWI, Curtiss produced about 5000 JN-4 Jenny trainers for the Army (along with the Navy version of the Jenny, the N-9), thousands of Curtis OX-5 engines—a water-cooled V-8 of 90 hp that evolved from earlier Curtiss V-8s—and a number of flying boats.

At war's end, Willys took his profits and sold his stock to C.M. Keys, correctly assuming that there would be little money to be made in aircraft manufacturing in the foreseeable future. Keys would reorganize in 1923 as the Curtiss Aeroplane and Motor Company (formerly, "corporation"), and remain in control for another ten years, engineering its merger with Wright Aeronautical Corporation in August 1929 to form Curtiss-Wright Aeronautical Corporation. Both Glenn Curtiss and Orville Wright lived to see the day that the names of these two old enemies were permanently linked together.

Within a few months, other av-properties were brought into C.M. Keys' Curtiss-Wright family, most purchased with C-W stock: the Travel Air Company, Keystone Aircraft, Moth Aircraft, the Curtiss-Robertson Company of St. Louis, Caproni, and New York Air Terminals, Inc.

A year earlier, Keys had organized North American Aviation Company. Originally capitalized at $25 million and backed by General Motors, Bankamerica-Blair, and Hayden, Stone & Company, North American was intended as a holding company, its mission being the acquisition of promising aviation properties. The aviation boom that followed Lindbergh's flight to Paris in 1927 found aviation enterprises as attractive to investors as the computer industry would be 30 years later. With fewer legal constraints at that time than are extant today on stock trading and corporate mergers, it was a time for empire building. Thus North American soon controlled Eastern Air Transport (later, Eastern Air Lines), Sperry Gyroscope, Ford Instruments, and Transcontinental Air Transport (TAT, a coast-to-coast passenger airline which, merged with Western Air Express, became Transcontinental and Western Air, or TWA—today, Trans-World). By way of an earlier GM investment, Keys had a tenuous hold on National Air Transport, a New York-to-Chicago air mail operation. These corporate marriages meant that the Curtiss-Keys forces not only manufactured airplanes and airplane components but also possessed their own markets.

Martin

Wright Aeronautical Corporation, which Keys merged with Curtiss Aeroplane and Motor to form the $200+ million base for his aviation dominion, was chartered in 1919. Its true beginnings, however, go back to 1908, when Glenn Luther Martin, a Ford and Maxwell dealer in Santa Ana, California, decided to build a copy of Curtiss' June Bug. The Martin machine was completed in mid-1909, at first fitted with a 12-hp Ford engine. When that powerplant proved inadequate, a 30-hp Elbridge marine engine was installed and Martin set about the task of teaching himself to fly. That took another year; then he flew some exhibition dates and began construction of two additional airplanes in a vacant cannery.

Martin moved to Los Angeles in 1912 with the intention of building and selling airplanes, hopefully to the military. He had progressed to a fairly advanced tractor biplane when, in 1914, Grover C. Loening, then an aeronautical engineer with the Signal Corps, was sent to the Army's training field at North Island. The fatality rate in the Curtiss and Wright pushers was unacceptable and Loening was ordered to do something about it. A recent class of 14 student pilots had suffered eight fatal crashes. (From 1909 through 1914, Army fliers averaged one death for approximately each 100 hours of flying time.)

Loening quickly determined that the airplanes at North Island were little more than death traps. He went directly to Martin, and the order for 17 Martin Model TT training planes that resulted established Martin as an aircraft manufacturer.

Wright-Martin

A year later, Martin sold out to a syndicate headed by W.B. Thompson of the Simplex Automobile Company and including Richard Hoyt of Hayden, Stone & Company, Thomas Chadbourne and Harvey Gibson of Manufacturers Trust, and Albert Wiggen of Chase National Bank. This group also purchased the Wright patents, along with some hardware, from Orville Wright and formed the Wright-Martin Aircraft Company.

During WWI Wright-Martin, managed by the Goethals Engineering Company, operated plants at New Brunswick, New Jersey and Long Island City, New York, during which period the company produced 51 airplanes and 5816 Hispano-Suiza aircraft engines. The "Hisso" engine, built under a license agreement with a French firm, was rated at 150 hp and was an excellent (for that time) water-cooled V-8 with a cast block. Originally designed in Spain by Swiss engineer Marc Birkigt, it would later be built in 180-hp and (after the war) even more powerful versions.

Key enginemen at Wright-Martin were George Mead and Andrew Willgoos. The Army's engine inspector at Wright-Martin was Captain Frederick Brant Rentschler, a Princeton graduate who had learned the machinists' trade in his father's shops in Hamilton, Ohio.

World War I ended on 11 November 1918 and, early in 1919, with orders from the military reduced to a trickle, Wright-Martin was liquidated and its New Jersey plant sold to the Mack Truck Company. However, Fred Rentschler convinced Wright-Martin's major stockholders to provide a residual $3 million for the formation of a new company that would continue to profit from their possession of the Wright name and patents, and their position as aircraft engine builders. Thus, Wright Aeronautical Corporation was born. Rentschler held onto Mead and Willgoos, the men who, along with Sam Heron and Edward T. Jones, would father the great radial aircraft engines, the Whirlwinds and Wasps, as we shall presently see.

Boeing

In the meantime, up in the Pacific Northwest, 34-year-old William E. Boeing, heir to a lumbering fortune, and Conrad Westervelt, a naval officer assigned to the Seattle shipyards, had formed a partnership to build an airplane or two. Their design work started late in 1915 in a makeshift shop on the shore of Lake Union. The partners were most impressed by the Martin TT, and after buying a Martin for study, produced two "B&Ws" that looked very much like Martins.

It isn't clear just when this partnership was dissolved. Cmdr. Westervelt was sent east by the Navy before the first B&W made its maiden flight, but it appears that Westervelt remained a partner while this enterprise operated through 1916 as the Pacific Aero Products Company. It became the Boeing Airplane Company, a corporation, on 30 April 1917, when Boeing moved to the Heath Shipyard on the Duwamish River in Seattle. By that time, he had developed the Boeing Model C trainer, essentially a refined Martin TA seaplane, and the Navy liked it well enough to place an order for 50 of them.

During WWI, Boeing also built Curtiss H5-2L flying boats for the Navy. With the cessation of hostilities, Boeing held his work force together by building bedroom furniture and a device known as the Hickman sea sled. Then, from 1920 until 1927 the company lived on airplane orders from the military, except for a few flying boats built for an early air mail route and the carrying of passengers.

Lockheed

The Loughead (Lockheed) brothers—Allan, Malcolm, and half-brother Victor—grew up in Niles, California, the sons of Flora Haines Loughead, a writer for the *San Francisco Chronicle*. Allan and Malcolm worked as automobile mechanics between 1904 and 1910, by which time they were 21 and 23 years of age, respectively. In the meantime, Victor, two years the eldest, had become an engineer and was associated with one James E. Plew of Chicago, who owned a Curtiss pusher. Through Victor, Allan obtained a job with Plew, taught himself to fly the Curtiss, and flew exhibition dates in the Midwest until wrecking the airplane at Hoopeston, Illinois, in 1911.

Allan and wife Dorothy returned to San Francisco where he and brother Malcolm, with limited financial backing from a taxi executive named Mamlock, designed and built a three-place floatplane fitted with an 80-hp Curtiss V-8 engine. They called it the "Model G" to disguise the fact that it was their first effort. The Loughead Model G first flew on 15 June 1913, but the brothers' need to meet family responsibilities kept the biplane mostly in storage for two years until the 1915 Panama Pacific Exposition offered an opportunity for some profitable flying. Allan cleared more than $4000 hopping passengers in the Model G, which was a significant stake in 1915.

The brothers took their airplane to Santa Barbara the following year and offered a sightseeing service that was modestly successful. Then, in 1916, backed by machine shop operator Burton Rodman, and with 20 year-old John K. Northrop aiding with design, the Lougheads produced a large twin-engine flying boat that they called the Model F-1. Anticipating, as did so many others, that the United States would soon be drawn into WWI, the Lougheads and their associates hoped to build such aircraft for the Navy.

The F-1 launched the Loughead Aircraft Manufacturing Company, but the big craft was not completed until March 1918. Although the Navy was impressed by the F-1 (particularly so because Carl Christofferson flew it over water from Santa Barbara to North Island for the Navy's evaluation), the Loughead company received an order instead for two Curtiss flying boats.

With termination of the war, the Loughead group decided that there would be a strong market for an inexpensive, single-place sportplane, especially among the thousands of war-trained pilots returning to civilian life. In Chicago, Matty Laird had come to the same conclusion; so had a group in Lorain, Ohio, that included Buck Weaver and Charlie Meyers and which worked under the name Weaver Aircraft Company (WACO). (This same dream would surface again at the end of WWII. It proved to be without substance both times.)

The Loughead sportplane was designed by John "Jack" Northrop and was unique in several ways. The fuselage, constructed of spruce strips, was formed in two halves in a concrete mold. Spruce rings inside the mated halves provided additional strength. This was not the first example of stressed-skin construction—it had been employed in France as early as 1910—but it was highly unusual and it not only established the construction principle that made possible the famous wooden Lockheeds to come (the Vegas, Orions, etc.), but also presaged the development of the significant Northrop Alpha, Beta, Gamma series in the early '30s.

It was a commendable effort, but no buyers appeared for the Model S-1. (No WACO "Cooties" were sold, either. Laird found a customer for one of his motorcycle-powered machines, also called a "Model S.")

That was the end of the Loughead Aircraft Manufacturing Company of Santa Barbara. The company was liquidated in 1921. But five years later, Allan and Jack Northrop would get together again to manufacture airplanes. Backed by tile manufacturer F. S. Keeler, they formed the Lockheed Aircraft Company (both Allan and Malcolm, tired of having their name mispronounced, decided to spell it as it should be pronounced).

The new company was incorporated in December 1926, and work on Northrop's new design began in a rented building at the corner of Sycamore and Romaine Streets in Hollywood. Keeler was president of Lockheed Aircraft by virtue of the $25,000 he had invested. Allan was vice-president and general manager; Ken Jay, a former Army Air Service pilot and accountant for Keeler, was secretary-treasurer.

Jack Northrop chose the name "Vega" for his new creation, after one of the brightest stars in the firmament. It first flew on the fourth of July 1927, and its debut coincided with the beginning of an exciting new era in American aviation.

Other Early Builders

Among the dozen or so minor aircraft builders in existence in 1916, some—perhaps most—obtained orders from the Army and/or Navy for airplanes during WWI. Some were unable to deliver a single machine; some built a handful, and a few produced a hundred or more. Almost none remained in business after the war, and none had any lasting influence on American aviation.

Nevertheless, all represented sincere efforts, and if most were forced to work outside the mainstream of aircraft development, they were nonetheless true pioneers. They included a group in Redwood City, California, organized under the name of United States Aircraft Company that evolved, in 1915, from the shop of California early birdman Simon "Chris" Christofferson. United States Aircraft was well-managed by non-aviation people, and delivered 100 JN-4 Jennies to the Army during WWI.

Sloan Aircraft of Plainview, New Jersey, which produced in small numbers a Jenny-type trainer fitted with distinctive round radiators (in the early '20s, Matty Laird purchased a number of those radiators in the surplus market and used them on the first Laird Commercials), was acquired by Japanese interests, renamed Standard Aircraft Corporation, and moved to Elizabeth, New Jersey. Standard would survive into 1919 building JR-1Bs for the Post Office Department's new aerial mail routes. Standard's designer, Charles H. Day, would foster formation of the New Standard Aircraft Corporation in Paterson, New Jersey, in 1928 with some new civilian biplane designs.

Burgess-Dunn of Marblehead, Massachusetts, was put together by W. Starling Burgess, who had worked briefly with Augustus Herring in 1910, and Maj. John W. Dunn, an Englishman who designed tailless airplanes with sweptback biplane wings. About 15 Burgess-Dunns were built for the Army and Navy in 1914 and 1915. The company was absorbed by the newly-formed Curtiss Aeroplane and Motor Corporation a year later.

The first airplanes to carry the Fairchild name were a couple built in 1911-12 by Walter L. Fairchild and flown at Mineola, Long Island. These were shoulder-wing monoplanes that resembled the French Antoinettes, and are noteworthy because they employed steel tubing in their structure.

Other early builders included Sturtevant, Aeromarine, Benoist, Lawson-Willard-Fowler, Gallaudet, and Chance Vought. Chance Vought was an Ivy League-type whose fascination with flying machines led to a friendship with Glenn Curtiss as early as 1909. Vought was able to form his own firm, the Chance Vought Aeroplane Company, in 1917. He designed exclusively for the military, and we will return to him later.

Edson F. Gallaudet organized an engineering company in 1910 and built airplanes in Connecticut and Rhode Island for several years. The Gallaudet Scout and Bullet were the best-known models. The Bullet, an unusual monoplane, had a pusher propeller in the tail driven by an engine mounted in the nose. A tubular steel spar in the wing permitted changing the angle of incidence in flight. The Gallaudet company assets were purchased by Consolidated Aircraft in 1923. Ruben Fleet's Consolidated Aircraft would eventually become the Convair Division of General Dynamics.

After the United States entered WWI, a number of aviation manufacturing firms were hastily formed to seek military contracts. One of the largest was another "Wright" company, Dayton-Wright, put together by prominent men whose patriotism was exceeded only by their greed. It seems clear now that Orville Wright was duped into lending the use of his name to this company. He was supposed to function as a consultant, but Dayton-Wright's use for Orville was limited to their need for the name of an instantly recognizable airplane builder on their letterheads, because none of the company's principals had ever been so engaged.

Dayton-Wright

Dayton-Wright was organized by Edward A. Deeds of the National Cash Register Company, and Charles F. Kettering, inventor of the automobile electric self-starter. They had previously formed Dayton Engineering Laboratories Company (DELCO) to build Kettering's starter, and in 1916, when DELCO was absorbed by United Motors, a subsidiary of General Motors Corporation, that resulted in a union which possessed a lot of muscle. Officials of United Motors included Howard E. Coffin of Hudson Motor Car Company, Henry M. Leland of Cadillac, and Jesse G. Vincent of Packard.

Dayton-Wright was assured of plenty of business because it was important to the United Motors group. Jesse Vincent designed the engine selected as the "Standardized Aircraft Motor of the United States" by President Wilson's Aircraft Production Board. The Chairman of the Aircraft Production Board was Howard Coffin. Edward Deeds was a board member, as was a Detroit automobile lawyer by the name of Emmons. Vincent's 400-hp engine, popularly called the "Liberty," could be fitted to the DeHavilland DH-4 warplane, a British design. Dayton-Wright built the DH-4 in large numbers. At the end of the war, Dayton-Wright did not close with the cancellation of all warplane contracts, but produced several experimental planes for the Army before its hard assets were sold to Consolidated Aircraft in 1923.

The automobile interests controlling United Motors managed to keep the Liberty engine in production until 1922, producing some 22,000 of them altogether. The Army was forced to use them throughout the '20s, although much better engines were available. We will encounter Mr. Deeds and Mr. Coffin again, which is why the story of Dayton-Wright, although it exclusively served (read that "exploited") the needs of the military, is told here. Both Deeds and Coffin would later have a noteworthy influence on civil aviation.

Barnstormers

Although the barnstormers of the '20s have often been characterized as daredevils and irresponsible nomads, most took no chances that were not forced upon them by circumstance. It was a matter of survival—a bent airplane earned no money. Some accidents were inevitable because aircraft maintenance was almost exclusively of the shade-tree variety, and performed by the pilot himself. The only airports belonged to the military, along with a few used by the aerial mail service. There was no law requiring pilots or aircraft to be licensed, and no airworthiness certificates to worry about. Each pilot determined the airworthiness of his plane—a judgment based on common sense though often tempered by economic considerations. Spruce wing spars were commonly repaired with the yellow pine that orange crates were made of in those days. (Perhaps that explains the origin of the term "crate" so often applied to airplanes in a dubious state of repair.)

Most accidents produced no fatalities and often no serious injuries due to low wing loadings and slow speeds. But we will never really know how dangerous the barnstormers were because there was no federal agency to regulate or keep statistics on them. They possessed a freedom that is hard to imagine today—a freedom that nurtured some colorful and often remarkable people.

Charles W. Meyers was a barnstormer. The son of a newspaperman, he grew up on Long Island and built a couple of hang gliders in 1913 at the age of 17. After high school, he worked for the Aeromarine company in Keyport, New Jersey, where he learned aircraft construction and caged a few hours' dual instruction with the company test pilot. In 1916, Meyers joined the Royal Flying Corps and spent WWI in Canada and England as a flight instructor.

"I was demobilized in April 1919," Meyers remembered during an interview shortly before his death in 1972, "So Buck Weaver, who had worked with Matty Laird before the war, joined me and we started down through Ohio barnstorming in a couple of Canucks—Canadian Jennies—we bought in Montreal. You know, the U.S. Government didn't immediately release its Jennies to the surplus market. All you could buy in the U.S. at that time were Standards, with the understanding that you'd take the fire-prone Hall-Scott engines out of them.

"There was quite a bit of money to be made barnstorming then. At the start, we got $20 for a three-minute ride; $25 for a stunt ride, which included a climb to 2500 feet from which we'd do a loop and spin, recovering at about 500. And there were plenty of customers. So Buck and I bought a third airplane and sent for E.P. Lott, a fellow I'd worked with at Aeromarine. He'd been an Air Service mechanic during the war. We taught him to fly and he worked with us that season. Lott ended up flying for United some years later."

WACO

In mid-summer, Meyers, Weaver, and Lott were joined by Elwood "Sam" Junkin and Clayton Bruckner, a pair who had worked together in aircraft construction since 1914. The group spent the winter in Lorain, Ohio, having decided to design and build a marketable "Flivver plane," an inexpensive, single-place sportplane for those thousands of war pilots who would surely want to continue flying. From this tenuous dream the name WACO emerged, partly for Weaver Aircraft Company, and also because Buck had served as an Air Service instructor at Waco, Texas, during the war. (While the city is called "Way-ko" in Texas, the Army pilots pronounced it "Wah-ko," and the long line of WACO biplanes that followed were so called.)

The first WACO, a rather crude parasol monoplane powered with a two-cylinder air-cooled Lawrance engine obtained on the war surplus market, crashed on its third test flight, and Weaver narrowly escaped serious injury. The next two WACOs, tiny biplanes, were equally unsuccessful and the group went to a larger "Jenny replacement" design, a single-bay three-place biplane fitted with an OX-5 engine of 90 hp. This WACO appears to have been well-conceived, and it eventually found a buyer in a slow market.

The fourth WACO—Model Number 3—was built at Medina, Ohio, and the WACO 5 followed at Alliance before the company settled at Troy, Ohio, to produce the WACO 6. However, not until the WACO 9 appeared in April 1925 did sales take off. Thirty Model 9s were sold within four months of its introduction.

Meanwhile, both Buck Weaver and Charlie Meyers had long since departed WACO—Weaver to Chicago, where he took a job as pilot for wealthy seed merchant Charles Dickinson, and Meyers into the Carolinas to barnstorm. Charlie's creative urge recurrently surfaced and he modified or built an airplane each winter, the rest of the time hopping passengers, delivering newspapers by air to rural customers, serving as the first aerial deputy sheriff, and otherwise seizing every opportunity to make a dollar with his airplane.

During the winter of 1925-26, Meyers designed and built a small biplane in a North Carolina barn. It weighed but 305 pounds empty and was powered with a 32-hp Bristol Cherub engine, but the Meyers Midget cruised at 100 mph and attracted considerable attention at the 1926 National Air Races in Philadelphia. Charlie hoped to market the Midget in kit form, but crashed it in a railroad yard when its engine failed at low altitude. That was his signal to return to WACO. Barnstorming had become steadily less profitable, and Charlie knew that civil aviation was entering a new era.

Meyers was warmly welcomed at WACO (actually, the Advance Aircraft Company after Weaver's departure, although its products were always known as WACOs. The company at last became the WACO Aircraft Corporation in June 1929). Designer Sam Junkin had died after bringing forth the WACO 9, and Clayton Bruckner, a machine shop genius but without aircraft design or marketing talents, needed Meyers as much as Meyers needed a job.

Advance Aircraft sold 164 WACO 9s in 1926—a sizable share of the total new airplane market that year of 650 units. The price of the OX-5 WACO 9 was $2500 at the factory.

Meanwhile, Charlie Meyers designed a follow-on, the WACO 10, and 435 of the 1000 civil aircraft built in 1927 were WACOs. A major innovation was the oleo shock-absorbing landing gear strut on the Model 10. Meyers then designed new wings for Model 10 to produce the WACO Taperwing before leaving to join the new Great Lakes Aircraft Corporation in Cleveland. But WACO would maintain its position as a major builder of private airplanes—both open-cockpit and cabin biplanes—until WWII.

Swallow

Few people—including aviation people—in Wichita, Kansas, today can identify William A. "Billy" Burke or Jacob Melvin Moellendick. That seems strange, because these two men brought the first airplane factory to Wichita—a factory that clearly spawned others and started an industry that largely sustains this metropolis on the Plains to this day. True, Burke left the scene rather early, and the few who remember Moellendick say that he was like a bulldog chasing a locomotive: plenty of guts, but not much judgment. Nevertheless, their contribution to general aviation cannot be denied.

Moellendick and Burke became acquainted in Okmulgee, Oklahoma, before WWI. Burke was a successful exhibition pilot, flying a pusher biplane built by A.K. Longren in Topeka, Kansas, as well as a Buick-Franklin automobile dealer. Moellendick worked as an oil driller in the Oklahoma oilfields, and made a practice of accepting part interest in each wildcat venture in lieu of wages. Early in 1919, Jake Moellendick was working in the new Wichita discovery and owned "a piece of a hole" that came in a gusher. Jake was in the money.

Meantime, Billy Burke, anticipating the end of the war, had written to Matty Laird in September 1918 to offer $2500 for Matty's prewar exhibition plane. Burke had seen the airplane at Ashburn Field in Chicago (which replaced Cicero Field) in 1916 and admired it.

Laird would not sell his "Boneshaker" (so named because of the vibration imparted to the airframe by its Anzani engine), but he did sell Burke the Laird Model S, a small biplane powered with a 50-hp Gnome rotary engine, which Matty hoped to produce for the presumed postwar boom in private flying. Burke took delivery of this machine in the late spring of 1919, and Matty thereupon painted a sign on his hangar proclaiming it the E.M. Laird Company of Chicago. That business would not exist for long, however, because no one else showed up to buy a Model S, and because events were conspiring elsewhere to put Matty in the airplane manufacturing business in Wichita.

During the fall of 1919, a trio of ex-Army pilots, headed by Joe Witt, organized a barnstorming and air taxi service in Wichita. They had two Canucks, a Jenny, and a 40-acre field on the northeast edge of town not far from the Jones Motor Car plant. They had the financial backing of several Wichita businessmen, including hotelman George Siedhoff, contractor J.H. Turner, and Jake Moellendick. Jake was a natural target for Witt and his partners as they went looking for capital, because Jake was highly visible around town: He was loud, generous, and possessed a decided affinity for good whiskey. He also liked long shots. Jake not only bought in, but became Wichita Aircraft Company's majority stockholder.

Aware that this made him the boss, Jake immediately started bossing. He didn't know much about airplanes, but he knew a man who did—his old friend Billy Burke down in Okmulgee, Oklahoma. Jake offered Burke the position of manager and chief pilot with Wichita Aircraft and, after Burke convinced Jake that better airplanes were needed, dispatched Billy to the Chicago Aircraft Show to seek them.

The 1919 Chicago Aircraft Show offered but one new airplane—the two-place Curtiss Oriole, which Burke quickly determined to be overloaded with anyone aboard besides the pilot—so Billy went to Ashburn Field to see Matty Laird.

Matty had assessed the changing aviation scene just as Burke had done. Exhibition flying, even with stunts and nighttime pyrotechnics, was on the way out. People were now ready to ride in "one of them contraptions" and pay for the privilege, and Matty planned a three-place biplane to be powered with the OX-5 engine as the machine that would earn the most money for the least operating cost.

By the time Matty had explained his new design, Burke had forgotten the Wichita Aircraft Company. With mounting excitement, he proposed that he and Moellendick put up the cash to match Laird's plans, equipment, and know-how for a three-way partnership and build the new Laird "Commercial" in Wichita in quantity.

Jake Moellendick proved equally enthusiastic, and the E.M. Laird Company of Wichita was formed with Moellendick and Burke each contributing $15,000.

The first Laird Commercial took shape in a woodworking shop behind the Wichita Forum, and early bird pilot Lester Bishop, who a little later flew the aerial mail, pitched in to help.

"Matty and Charles Laird worked on the fuselage of the new ship," Bishop recalled, "while Walt

Weber and I worked on the wings. The drawings were made by Matty with a common pencil on plain paper as work progressed. He carried most of the plans in his head. He had made some of the metal fittings in Chicago, and the rest were made by hand by Weber. So, there in Wichita, America's first commercial airplane was born. It was a very fine airplane . . . "

Work started on the plane early in February 1920, and it was completed during the second week in April. During this time, the E.M. Laird Company bought out the faltering Wichita Aircraft Company, obtaining an airfield and four T-hangars.

On the afternoon of 11 April 1920 the new airplane was taken to the Laird company's new field, assembled, and test flown by Matty. When he landed, a spectator exclaimed that it flew "just like a swallow," whereupon the partners agreed that that was a proper name for it.

The Laird Swallow was a success. The first company advertisement, appearing in the June 1920 issue of *Aviation* magazine, described it as the "first commercial airplane; capable of carrying a pilot and two passengers with fuel enough for 225 miles at full speed." This announcement was greeted by orders for eleven airplanes, and freewheeling Jake Moellendick planned a production schedule of 20 machines for the rest of the year.

List price of the Laird Swallow was $6500. Available at extra cost were dual controls, long-range fuel tank, and "streamline wires." It had a span of 36 feet, was 23 feet 4 inches in length, weighed 1075 pounds empty, and its useful load was 675 pounds. It cruised at 85 mph, and landed at 38 mph. The Swallow's wing-load safety factor was assured by a test common at the time: The plane was suspended inverted by its four wing-attach fittings at the fuselage, then 6000 pounds of sandbags were heaped upon the wings. If nothing broke—*bending* was okay—it was presumed that the airplane's structural integrity had been established, and that concluded the static load testing. Flight testing was equally simple, although spins were usually included (tailspins were commonly taught to beginning flight students during the '20s and '30s).

In mid-June, Moellendick and Burke decided that they would start an airline service between Wichita and Kansas City and asked Matty to design a five-passenger cabin plane. The Laird Transport was completed early in July 1921, but the Moellendick airline never materialized. Fitted with two 90-hp OX-5s, it was underpowered. Modified, with a single 400-hp Liberty in its nose, its performance improved, but company pilot Walter Beech wrecked it on a cross-country trip to Tulsa.

Beech had been barnstorming out of Arkansas City, Kansas, with Errett Williams and Pete Hill when he decided that the Laird factory in Wichita was in business to stay. Jake Moellendick hired Beech as a salesman-pilot in mid-1921. Walter was 30 years old and long removed from the Tennessee farm where he had grown up. His natural mechanical ability had taken him to Europe before the war as a truck manufacturer's representative. Walter's interest in flying machines reportedly began with a homebuilt glider in 1906; in 1914 he rebuilt and, according to some sources, briefly flew a Curtiss pusher.

Walter had spent WWI at Rich Field, Waco, Texas, where he was in charge of the engine overhaul shop with the rank of master signal electrician (Sergeant). We may presume that Beech, Burke, and Weaver were acquaintances there.

Other Swallow employees at that time included Walter Weber, Homer Clark, Walter "Pop" Strobel, Bill Snook, and the Stearman brothers, Waverly and Lloyd.

Lloyd Stearman, who was four years younger than Laird, came from Harper, Kansas, and had been an architectural student at Kansas State University when the U.S. entered WWI. He enlisted as a Naval Aviation student aviator and learned to fly, but the war ended before he got overseas. Lloyd had been working in Wichita for an architectural firm when the first Swallow took to the air, and he at once applied to Laird for a job. Matty put him to work on fuselage assembly.

Throughout 1921 Laird Swallows sold at the rate of about one per month. Early in 1922, Beech sold a Swallow to the Sterling Oil Company, then made a deal with the Nourse Oil Company to fly their representatives throughout the central Plains. Nourse officials were so pleased with the reception their people enjoyed arriving by air that the company purchased three Swallows and hired pilots to fly them.

In October, Beech flew a Laird Swallow to victory in the "On to Detroit" race that was a part of the 1922 Pulitzer Trophy program at Selfridge Field near the Auto Capitol. Another Swallow captured second place.

The Pulitzer race itself was strictly a military affair. The civilian airplanes in attendance, in addition to the two Laird Swallows, consisted of surplus Jennies, Standards, two Thomas Morse Scouts, and a Curtiss Oriole. And that was a fair representation of the private aircraft available at the time. The E.M. Laird Company of Wichita and Advance Aircraft of Alliance, Ohio, were the only two airframe makers of consequence offering new airplanes to the general public except for Curtiss, which offered a small flying boat, the Seagull, in addition to the Oriole. Neither could be considered successful.

Actually, there were more than a dozen planemakers in the United States by the fall of 1922, but most catered solely to the military, and the others were barely afloat.

Ryan

T. Claude Ryan started Ryan Flying Company, later Ryan Airlines, that year, although scheduled flights between his home base in San Diego, and Los Angeles, were not begun until 1925. In the meantime, with the help of Hawley Bowlus, Ryan modified surplus Standards into four-passenger cabin biplanes. The first pure Ryan design appeared in 1926.

Tubal Claude Ryan was reared in Parsons, Kansas, where his father operated a steam laundry. The Ryan family moved to Orange, California, in 1912 when Claude was fourteen. He learned to fly in the U.S. Army Air Service in 1920, flew forest fire patrol in DH-4s during 1921, and then bought a surplus Jenny in San Diego to go into business for himself.

The rebuilding of war surplus Standards was also the business of Lincoln Aircraft Company of Lincoln, Nebraska, which began operation in 1920 with Ray Page, Horace Wild, and Vincent Burnelli as organizers. They marketed the Lincoln-Standard, and supplemented that tenuous activity with the operation of a flying school of sorts.

Both the Grover Loening and Thomas Morse plants, established in 1917, were still active in 1922, but neither offered civilian designs. Later, Loening would produce a few amphibians for private owners. Thomas Morse never entered the civil market.

Douglas

Donald W. Douglas, who had worked for Martin, along with Lawrence D. Bell, in 1918-19 organized his first company ("Davis-Douglas") in 1920 at age 28 when he rented office space in the back of a Santa Monica barbershop. The "Davis" was David R. Davis who in effect financed the company with an order for an airplane. Davis was a young man of some wealth who wanted to be the first to fly coast-to-coast nonstop, and the Davis-Douglas "Cloudster" probably would have enabled him to claim that distinction had its Liberty engine not failed over El Paso. We should note that the big biplane was probably the first airplane capable of taking off with a load greater than its own empty weight. It was acquired by Claude Ryan and his financial angel, B. Franklin Mahoney, in 1925 to start Ryan Airlines, later changed to Los Angeles-San Diego Air Lines. Douglas, in the meantime, had built three Douglas DT torpedo planes for the Navy, followed, in 1924, by the four "World Cruisers" for the Army, two of which successfully completed the first round-the-world flight that year. Douglas would not make a significant contribution to civil aviation until the first of his famed DC series of airliners appeared in mid-1933. Assets of the original Davis-Douglas company would be purchased by the new Douglas Aircraft Company 30 November 1928.

So the Laird Swallow had what market there was for new private airplanes pretty much to itself into 1923. Through September of that year a total of

43 Swallows had been built, one of which Walter Beech sold to barnstormer Clyde V. Cessna a month earlier.

Cessna

Cessna's background was similar to that of Matty Laird, except that Clyde was considerably the older man. Cessna was a 31-year-old farm implement mechanic living in Enid, Oklahoma, when, in 1910, a trio of early birdmen staged an exhibition in Oklahoma City. Clyde attended and was smitten. He made sketches of the Queen Monoplane (a Bleriot copy marketed by Willis McCormick in New York; Grover Loening was McCormick's engineer) flown by one of the aeronauts and, back in Enid, ordered a bare fuselage from McCormick. Cessna designed and built his own wings, tail, and landing gear, depending upon memory and the sketches he had made in Oklahoma City. He converted a four-cylinder 60-hp Elbridge marine engine to power his machine. The craft was ready for its first flight in May 1911.

Like so many other air pioneers, Clyde had to teach himself to fly and test his homemade airplane simultaneously. The result was predictable. But on his thirteenth attempt, Cessna got off the ground and back down again without doing serious violence to either his machine or himself. He thereupon quit his job and took off on an exhibition tour.

An accident shortly afterwards wrecked the airplane and put Clyde in the hospital for a few weeks. By then, however, he had flown several successful exhibitions—the first, at Cherokee, Oklahoma, paid $300—and he had contracts in hand for other appearances.

During the winter of 1911-12, Cessna took his wife and two children, Eldon and Wanda, to the Cessna family homestead near Rago, Kansas, and built a new plane fitted with an Anzani engine. This craft, the "Silver Wings," prophetically employed the leaf of a buggy spring in its landing gear.

For the next four years Cessna worked the county fairs in season and built a new and improved airplane each winter. Late in 1916, he built three airplanes in the plant of the Jones Motor Car Company at Wichita when the makers of the Jones "Light Six" offered free use of their facilities in exchange for the presumed advertising value of this activity. One of these craft, the Comet, had an enclosed cockpit and was capable of almost 100 mph.

Cessna was not accepted for military service in WWI (he was 37 years old). He spent the war years on his farm and returned to barnstorming in 1919. Rather than build still another airplane, he purchased the Swallow in August 1923.

The sale to Cessna was one of the last before Matty Laird pulled out of the company and returned to Chicago to build airplanes without partners. Moellendick had proven increasingly difficult. Jake's aggressive nature and fondness for good whiskey had previously caused Billy Burke to leave the company, and Matty's departure was precipitated by what Laird considered to be Moellendick's mismanagement.

Jake reorganized under the name of Swallow Airplane Manufacturing Company and promoted Lloyd Stearman to the position of chief engineer. Stearman designed the New Swallow early in 1924—a clean, single-bay biplane that maintained Swallow's position in the market. By October, however, Moellendick faced another crisis when Lloyd Stearman and Walter Beech resigned.

Travel Air

According to John Nevill, top aviation writer of the '20s, Lloyd and Walter quit because Jake refused to switch from spruce to steel tubing in the Swallow's fuselage. That may well have been the case, for it is not recorded that anyone ever won an argument with Moellendick. It is possible, however, that Beech and Stearman had concluded that it was time they did for themselves what they had been doing for Jake. Stearman had plans for a new biplane that Walter liked very much, and there were positive signs that the market for new private airplanes was improving. Whatever the true reason, Stearman and Beech did leave Swallow in October 1924 and joined with Clyde Cessna to start a new aircraft manufacturing company in Wichita.

Cessna put up $6200 and Beech $5700; Stearman's equity was represented by $200 and his drawings of the new airplane. The new company began as a partnership in a small rented workshop at 471 West First Street in downtown Wichita. But by the time the first airplane was completed, in mid-January 1925, it was clear that more capital was needed, so the partners decided to incorporate. The *Wichita Eagle* of 26 January announced the new shareholders as William Snook and Walter Innes, Jr. The newspaper also noted that Stearman was designing a five-place cabin monoplane to be offered along with the Travel Air biplane. Mr. Innes was credited with providing the new company with its name, Travel Air Manufacturing Company. Innes was a local pilot; Snook a businessman.

Travel Air's position was further bolstered when, in mid-summer, financier Daniel C. Sayre ordered three airplanes, gave Beech $10,000 up front (the Travel Air fitted with an OX-5 engine, later called the Model 2000, was originally priced at $3500), and then purchased Travel Air stock. Sayre's confidence in the company certainly influenced several local businessmen who then bought Travel Air stock.

An early Travel Air employee was Miss Olive Ann Mellor, a local business college graduate hired as secretary and office manager. Olive Ann would become Walter's wife in 1930 and, after Walter's death in 1950, would take his place as chief executive officer in Beech Aircraft Corporation, a position she would hold until her retirement more than 30 years later following acquisition of Beech Aircraft by the Raytheon Company.

In the meantime, up in Chicago, Laird had moved into his new factory at Ashburn Field (4500 West 83rd Street) and his new Laird Commercial biplane, offered with an OX-5 or Curtiss C-6 of 160 hp, was finding a few buyers. It was priced at $4500. By way of comparison, the Super Swallow, fitted with a 150-hp Hispano-Suiza, was selling for $2750, while the WACO 9 had a list price of $2225.

By Aerial Post

The world's first official flight of regular mail took place on 17 August 1859 from Lafayette, Indiana, when John Wise lifted off in his balloon *Jupiter* with 123 letters and 23 circulars bound for New York. Due to uncooperative winds, Wise got only as far as Crawfordsville, Indiana—30 miles away—where he placed the U.S. Mail sack aboard the New Albany and Salem Railroad train.

A unique semi-regular aerial mail service was established during the Franco-Prussian War in 1870-71, when a total of 67 balloons—55 of them carrying mail—were sent aloft from inside the besieged city of Paris to sail over the encircling German Army into unoccupied France. More than two million letters were posted by balloon with a limited number of replies returned by carrier pigeon. Five of the balloons fell into enemy hands.

Mail was carried along with passengers on the world's first scheduled airline, although the schedules were sometimes a bit uncertain. The *Deutsche Luftschiffahrts Aktien Gesellschaft* (DELAG) had air terminals in Frankfurt, Berlin, Hamburg, Potsdam, and Dresden, and began service 22 June 1910 with the rigid dirigible *Deutschland*. Three additional Zeppelin-type dirigibles were added to the line during the next three years, the *Schwaben, Viktoria Luise,* and *Sachsen*. These four great airships carried more than 37,000 passengers without injury to any before the beginning of WWI ended the service. As much as 1000 pounds of mail was carried on some flights.

In America, numerous one-time mail flights were made prior to WWI, but all were stunts. During a 1911 air meet at Garden City, Long Island, Earle Ovington flew authorized mail over a three-mile "route" to Mineola on an irregular schedule. Similar stunts were part of other early air meets, including the third big aerial exposition at Dominguez Field in 1912. Glenn Martin was one of the pilots flying from Dominguez to Compton, four miles away.

The true beginning of regularly scheduled U.S. Air Mail was on 15 May 1918. On that day mail flights were begun between Washington, D.C., and New York City, with U.S. Army pilots and planes flying under the direction of the U.S. Post Office Department.

As early as 1910, U.S. Representative Morris Sheppard of Texas had introduced a bill in Congress that would provide authorization and money to start an aerial mail service. The measure failed. Later, as a senator, Sheppard was back with a similar bill. This time, with the enthusiastic support of Postmaster General Albert Burleson, a majority of Sheppard's fellow legislators voted in favor of the bill, which authorized $100,000 for each of the fiscal years 1917-1918 to begin such a service.

May 15th was an ideal spring day in the Nation's Capitol. The Army's Maj. Reuben H. Fleet, responsible for the pilots and airplanes to be used in getting the aerial mail service started, flew into the polo grounds on the banks of the Potomac River about 9:30 A.M. in the Curtiss JN-6H (a Jenny powered with a 150-hp Hispano-Suiza engine) that Lt. George Leroy Boyle would pilot to Philadelphia with the mail.

By 10:30 A.M., President and Mrs. Wilson were on hand for the historic takeoff, along with many government dignitaries including Assistant Secretary of the Navy Franklin D. Roosevelt. There followed, however, an embarrassing interlude when the Jenny's engine could not be started. At last, someone remembered to check the fuel and that problem was solved.

Lt. Boyle took off, circled the field—and flew away in the wrong direction.

Two hours later, hopelessly lost, Boyle landed in a plowed field near Waldorf, Maryland, to seek directions but smashed his propeller when his plane went up on its nose in the soft earth. The mail was quietly put aboard a train 24 miles from its origin.

The other segments of the route were flown successfully that day. Lt. Torrey Webb flew from Long Island's Belmont Park to Philadelphia, where Lt. James Edgerton waited to take the westbound mail on to Washington, D.C. Meanwhile, Lt. Paul Culver was also at Philadelphia, expecting to take Boyle's load on to New York. He took 200 letters to New York.

The Army pilots had the service established within two months and the Post Office Department was ready to take over. Captain Benjamin B. Lipsner was separated from the Army Air Service to become the first superintendent of the air mail (Lipsner was not a pilot, but a good administrator—a "mother-hen" type), and the first civilian air mail pilots were Max Miller, Edward Gardner, Maurice Newton, Edward Langley, and Robert Shank, plus a reserve pilot by the name of Boldenweck. Except for Newton, all had been Air Service instructors.

The Post Office Department, in the persons of Air Mail Superintendent Lipsner and Second Assistant Postmaster General Otto Praeger, had to replace the Army's Jennies, and selected specially designed Standard JR-1Bs. These were a little faster than the JN-6Hs (66 mph cruise), and also Hisso-powered.

On 12 August 1918 the civilian pilots began flying the mail between New York and Washington D.C., employing the same shuttle system used by their predecessors and with equal reliability. In anticipation of increased mail loads, a pair of Curtiss R-4Ls were placed in service on the New York-Philadelphia segment during the first month of operations. The R-4L was somewhat larger than the Standard JR-1B, and fitted with the 400-hp Liberty engine. During the first year of operation, 1208 air mail flights were completed. There were 90 forced landings—53 due to weather, and 37 because of engine failure. Air mail postage revenues for the year totalled $162,000, and while Supt. Lipsner listed total operating expenses as $143,000 ($64.80 per hour of flying time), that was apparently "government style" bookkeeping and did not include prorated costs of mail trucks and other essential Post Office support. The price of an air mail stamp, 24¢ during the time the Army pilots and planes were employed, was lowered to 16¢ when the Post Office Department took over the routes.

In May 1919, the Chicago-Cleveland air mail route was added, and six weeks later service was begun between New York and Cleveland over the Alleghenies. By the time the air mail was two years old it had reached Omaha, and on 8 September 1920 the Post Office could offer regular transcontinental service between San Francisco and New York, although it was not air mail all the way. The planes flew only during daylight hours, and the mail was transferred to trains overnight. That did not greatly speed the mail, but airfields were established coast-to-coast along the route, the pilots gained experience over some difficult terrain, and useful lessons were learned in aircraft maintenance.

Meanwhile, late in 1918 the Post Office Department had filed requisitions for 100 DH-4s and 100 extra Liberty engines in anticipation of their appearance on the surplus lists at war's end. This equipment, still in original crates, was obtained early in 1919, and the DH-4s were placed in service as rapidly as they could be modified as cargo carriers and their Liberty engines extensively altered in an attempt to increase their reliability. Some of these airplanes were reworked at the Post Office Department's aircraft repair depot on Maywood Field, Chicago. Others were rebuilt under contract by several civilian contractors.

At least 20 different aircraft types were tried by the Post Office, including a cargo version of the Martin bomber, a tri-motor Caproni, and the Junkers-Larson JL-6, but the deHavilland DH-4Bs remained the mainstays as long as the Post Office operated the routes.

Postal authorities recognized from the beginning that the air mail must fly at night if the service was to achieve a marked savings in time over surface mail, especially on the transcontinental route. The 16¢ air mail stamp had been discontinued on 18 July 1919, and letters marked "Air Mail" were then accepted at the regular 2¢ per ounce rate charged for First Class stamps. When that failed to fill the cargo compartments of the airplanes, the mail sacks were filled with First Class mail. The 2¢ per ounce rate for air mail would remain until August 1923, when the transcontinental route was lighted and regularly scheduled flights were dispatched with mail on a 24-hour per day basis. At that time, three new air mail stamps were issued in values of 8¢, 16¢, and 24¢. The U.S. was divided into three air mail zones and the cost of air mail became 8¢ per ounce per zone.

Jack Knights' Epic Flight

The lighting of the airways did not come about without great effort. There was skepticism in the Congress concerning the practicability of the air mail. It wasn't as dependable as the trains, wasn't much faster, and had operated at a deficit after the first year. Senator Warren Harding, who believed that the air mail service was a waste of money, was elected President in November 1920, and would be inaugurated the following March. Therefore, fearful that Harding would seek to terminate the service once in office, Postmaster General Burleson and his assistant, Otto Praeger, determined that the mail must be flown, without interruption, coast-to-coast. Such a demonstration should have a dramatic effect on the public and the Congress.

Mid-winter was not the best time for such a test. There were of course no radio aids to air navigation then, and the pilots served as their own weathercasters. But at 6 A.M. on Washington's Birthday, 1921, two mail planes took off from Hazelhurst Field, L.I., New York, headed west, and two left Marina Field, San Francisco, at 4 A.M. (Pacific Time) flying east. Pairs of relay planes would be waiting at all of the regularly scheduled stops in between.

However, a blizzard had moved out of the Northern Plains into the Midwest and the westbound mail planes got only as far as Chicago before being grounded by the blinding snowstorm. In the meantime, one of the eastbound planes crashed at Elko, Nevada, although the other reached North Platte, Nebraska, to land in the snow at 7:50 P.M. Flown by Frank Yeager, it was the only hope remaining for an uninterrupted transit of the air mail between the East and West Coasts.

The pilot scheduled to take the mail eastbound from North Platte was James H. "Jack" Knight. He waited impatiently as the tailskid was repaired on the big biplane and then, at 10:44 P.M., was off into the winter night.

Knight flew the 276 miles to Omaha navigating by dead reckoning, aided by bonfires set by public-spirited citizens at Lexington, Kearney, and Central City. He reached Omaha at 1:10 A.M. and, as he warmed himself by the big pot-bellied stove in the operations office, learned of the storm ahead that had apparently doomed the air mail service's trailblazing effort. The backside of the storm was somewhere between Omaha and Chicago. Knight had never flown the route. He stuffed newspapers inside his fur-lined flying suit, filled his coffee thermos, and returned to his airplane.

Knight lifted into the blackness at 2 A.M. and stayed low in an attempt to follow the Rock Island Railroad tracks into Des Moines. The clouds were low and he encountered brief snow flurries, but the rococo dome of the Iowa Capitol at last appeared through the murk. There was too much snow to risk a landing, so Knight continued to Iowa City only to discover that the ground personnel there had obviously gone home to bed—there were no lights to mark the airfield. But certain disaster was averted when a night watchman heard Knight's engine and raced to the center of the field to light a flare. Knight had but ten minutes of fuel remaining when he managed to land in the blowing snow with only that single beacon to guide him.

The night watchman helped refuel the airplane and then Knight was off again. The snow gave way to a Mississippi Valley fog that forced him higher, but at last a gray dawn revealed holes in the undercast. Soon afterwards, smoke from Chicago's industries puddled in the mists ahead to mark his destination. The big wheels of the DH-4 sank into the snow of Checkerboard Field at 8:40 A.M. The mail had been flown all the way from San Francisco in slightly over 26 hours—and Jack Knight had flown almost 800 miles of that distance through the night to make it possible.

Pilot J. D. Webster departed Chicago bound for Cleveland with the mail at 9 A.M. and Ernest Allison flew the route from there to New York, arriving at 4:30 p.m. to complete the bold demonstration. Seven pilots had spanned the 2660 air miles between San Francisco and New York in 33 hours and 20 mintues with the United States Mail (averaging 80 mph including time on the ground). The fastest transcontinental trains required 108 hours for the same journey.

The feat was headlined in newspapers throughout the country, and President-elect Harding, who recognized a popular issue when he saw one, praised the air mail service and later supported the lighting of the airways.

The first transcontinental airway was lighted by 1 July 1924 and, following a 30-day test, the air mail began scheduled 24-hour operation. A total of 289 flashing beacons marked the route, including those at emergency fields and in between. On clear nights a pilot could always see several beacons ahead on his course. The lighted emergency fields averaged about 30 miles apart.

Post Office Department pilots received a maximum base pay of $3600 per year, plus 5¢ to 7¢ per mile flown. That would equate to perhaps ten times as much today. Each pilot flew five to six hours per day, two to three days per week. The maximum number of pilots employed at any one time was 55.

Once the airway was lighted, and the mail flown through the night coast-to-coast on regular schedule, the Post Office Department was ready to relinquish the routes to private contractors. The Air Mail Act of 1925 (H.R. 7064), popularly known as the Kelly Bill, gave the postmaster general that authority, and thereby set in motion a most intriguing period in the development of America's airline industry.

U.S. Civil Aircraft Production 1912-1941

Year	Production	Year	Production
1912	29	1927	1,374
1913	29	1928	3,127
1914	34	1929	5,516
1915	152	1930	2,690
1916	269	1931	1,988
1917	135	1932	803
1918	29	1933	858
1919	98	1934	1,178
1920	72	1935	1,251
1921	48	1936	1,869
1922	37*	1937	2,824
1923	56*	1938	1,823
1924	60	1939	3,661
1925	342	1940	6,785
1926	654	1941	6,844

*Includes remanufactured DH-4s

The Curtiss Jenny powered with a Hispano-Suiza engine was a JN6-H as above. The photographer mistakenly captioned this as a JN-4. Top speed was 80 mph with the "Hisso" engine of 150 hp.

Flight of the first flying boat was achieved 10 January 1912 by Glenn Curtiss. Curtiss developed the flying boat concept quickly and sold a number of them to the Navy during WWI.

Katherine Stinson learned to fly at Chicago's Cicero Field in July 1912 and soon commanded high fees as an exhibition pilot. She is pictured here in Japan during the winter of 1916-1917. (Charles Meyers)

Aeromarine biplane built at Keyport, New Jersey, 1914. Aeromarine concentrated on the military market.

The first Lockheed was the 1913 Model G, a three-place seaplane designed and built by Allan and Malcolm Loughead in a San Francisco garage. Engine was an 80-hp six-cylinder Kirkham. (Lockheed-California).

The second Lockheed was the F-1, which was completed in 1918. Powered with a pair of 160-hp Hall-Scott engines, it was the world's largest flying boat with a span of 74 feet and an advertised capacity of ten passengers. (Lockheed-California)

Loughead Aircraft Manufacturing Company received a Navy contract for two Curtiss-designed HS2L flying boats in 1917. Delivered early in 1919, these single-engine pushers were fitted with Liberty V-12s of 400 hp. The brothers employed up to 85 workers during WWI. (Lockheed-California)

John K. "Jack" Northrop, self-taught aircraft design genius, was co-founder of the second Lockheed company. Northrop conceived the Vega—and many other advanced designs to follow with three different Northrop companies.

The 1920 Loughead S-1, a Northrop design, was a single-place sportplane intended for a market that failed to materialize. Its lower wing could be rotated in flight to act as a brake or flap. Both upper and lower wings folded for storage. The S-1's two-cylinder 24-hp engine was a Loughead development, and its molded fuselage inspired the Vegas and other wooden Lockheeds that followed. (Lockheed-California)

Introduced in mid-1927, the five-place Vega served early airlines, and established many speed and distance records. Fitted with a 450-hp Wasp engine, the new NACA cowling, and wheel pants, it had a top speed of 185 mph. (Lockheed-California)

Air Express NC-514E was the 65th Lockheed built; it went to the New York, Rio, and Buenos Aires Line in Argentina early in 1930. Price was $19,885.

E.M. "Matty" Laird in his second homebuilt airplane, 1914. Matty's engine was put together from scrounged parts. A few county fair bookings provided money for a better engine and materials for an improved flying machine. (E.M. Laird).

The 1915 Verville, an aerodynamically clean design for its time. Alfred Verville, who began as an engineer with Curtiss in 1914, designed experimental aircraft at the Air Service's McCook Field during the early '20s; these possessed such features as full cantilever wings, retractable landing gear, and monocoque fuselage construction. (NASM)

Victor Carlstrom (R) established inter-city records in 1919 in a Curtiss R-4 powered with a Curtiss VX engine of 180 hp. With back to camera is Matty Laird, then using a crutch as the result of a crash while testing a Katherine Stinson airplane in San Antonio while Katy was in Japan. Second from left is Charles Kirkham, Curtiss engine designer-builder from 1907 until 1919. Exhibition pilot Art Smith is third from left. Photo was taken at Chicago's Ashburn Field. (E.M. Laird)

The barnstormers of the '20s flew war surplus Jennies, Standards, and Canucks, the latter being the Canadian-built Jenny. Some Thomas-Morse Scouts were also available, and were often converted to two-placers and fitted with OX-5 engines for barnstorming. This Chaplin machine is a JN-4 Jenny. (Francis Dean)

The barnstormers attracted crowds with aerobatics and wing-walkers. The object of the airshow was to gather potential customers for $20 three-minute rides around the pasture. Competition steadily eroded this bonanza and the price of a ride was down to $3 by the mid-'20s.

Barnstormer Albert C. Reed (L) of the TLR Flying Circus and happy passengers with Reed's OX-5 Standard "somewhere in Kansas" in 1921. This three-plane operation also boasted two Jennies. The other pilots were Beeler Blevins and Tex LaGrone. (Earl Reed)

Charles "Speed" Holman performs minor maintenance on his Curtiss Jenny. As many other famous pilots of that era, Holman barnstormed before becoming nationally known. Later, he would turn to racing and would be chief pilot for Northwest Airways (Northwest Orient). He was killed 17 May 1931 performing aerobatics in his Laird Speedwing at a benefit airshow in Omaha. (Noel Allard)

The first WACO, fitted with a two-cylinder Lawrance engine, was built at Loraine, Ohio, in 1919 by company founders Buck Weaver, Sam Junkin, and Clayton Bruckner. (Charles Meyers)

WACO number two was constructed from the crashed remains of the first, and was an attempt to market a low-cost personal airplane for the anticipated boom in private flying following WWI that did not materialize. That same dream resulted in the Navion, Sea Bee, and Aeronca Sedan following WWII and proved equally false. (Charles Meyers)

The first Laird Swallow under construction in Wichita, February 1920. Lloyd Stearman and Walter Beech were early Swallow employees. (E.M. Laird)

Billy Burke, an Air Service instructor at Rich Field, Waco, Texas during WWI, was largely responsible for establishing Wichita, Kansas as an aircraft manufacturing center. (Charles Meyers)

Lloyd C. Stearman (L) designed the New Swallow in 1924. Matty Laird's brother, Charles (center), followed with the Super Swallow. Unlovable Jake Moellendick, who put up most of the money for Wichita's first airplane factory, is at right. (Lloyd Stearman)

This Laird Swallow ended up atop the Laird company's office at 49th and Hillside in Wichita, blown there by a storm in 1921. (E.M. Laird)

Walter Beech offered $3 airplane rides to his employees at Travel Air on paydays. His secretary and office manager, Olive Ann Mellor, sold tickets. (Beech)

This twin-engine Burnelli was planned as an airliner in 1924—before there was a viable passenger airline in the nation. Vincent Burnelli continued to believe in his airfoil-shaped fuselage concept for another 30 years, and two transports were built by Canadair so configured after WWII.

The Ryan M-2 Bluebird of 1926. Hawley Bowlus, Walter Locke, and Jack Northrop worked on this design, which evolved from the open-cockpit Ryan M-1 mail planes flown by Vernon Gorst's Pacific Air Transport.

Clyde Cessna in his 1912 "Silver Wings" in which he toured the county fair circuit as an exhibition pilot. Average fee for a takeoff, several circles around the fair site, and landing near the crowd was $300 to $500. Many fair-goers doubted that a flight would be made, and cheered when such a machine actually left the ground. (Cessna)

Cessna's "Comet" of 1916 was powered by an Anzani engine and was capable of 120 mph. (Cessna)

The seven-place Laird transport was powered with two OX-5 engines; it was another premature effort to promote an airline not yet in existence. Buck Weaver, a founder of WACO, is in the cockpit. Walter Beech wrecked this airplane following an engine failure. (Matty Laird)

Herb Harkom and his 1929 Travel Air 4000 at Beech Field, 14 July 1966. The 4000 series was available with a number of different engines, ranging from 125 hp to 300 hp. Some installations were made on Group Two Approvals to the original Approved Type Certificate. With a J6-7 Whirlwind, the price at factory was $9100. (Beech)

The Curtiss HA, powered with a Kirkham K-12 engine, was an unsuccessful Navy fighter that was pressed into service with the new U.S. Aerial Mail experiment in 1919. (NASM)

The deHavilland DH-4 became the Post Office Department's standard mail plane during the early '20s after being rebuilt in the air mail maintenance shop at Chicago. The fuselages were plywood-covered. (American Airlines)

Chapter 3

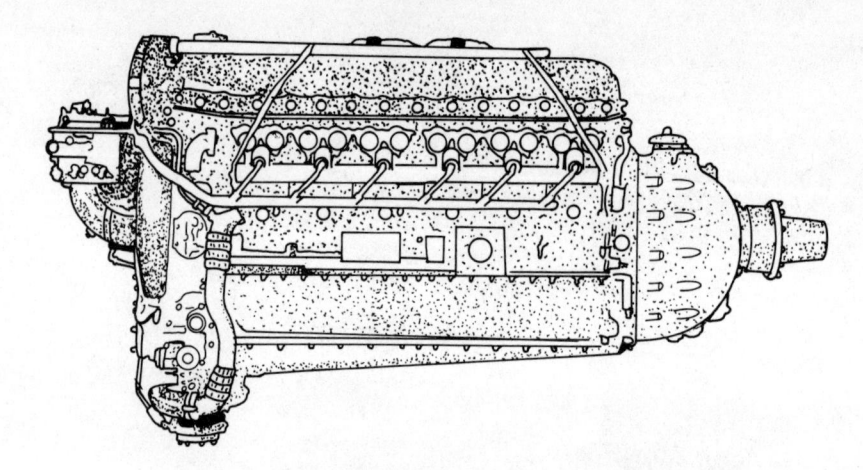

Power Paces Progress

Once, during development of the Pratt & Whitney R-1830 Twin Wasp engine, engineer W. A. Parkins, who had been having his problems, was forced to stop and explain the engine's workings to an important visitor. As Parkins concluded, the VIP smiled brightly and observed. "Actually, you people are simply trying to contain and control fire, aren't you?"

"Yes, sir," the harried engineer replied, "and that's simply all the Devil has to do in Hell, as I understand it!"

Perhaps that heartfelt rejoinder pretty well sums up the frustrations of aircraft enginemen from the time of the Wright brothers. The special requirements of successful aircraft powerplants—minimum weight, maximum reliability, minimum fuel consumption, maximum power—leave little room for their designers to maneuver. Nevertheless, the aircraft industry progresses only as the means of powering aerial vehicles progresses.

The first successful airplane engine was, of course, that which powered the Wright brothers' Flyer. It had four cylinders, in-line, water-cooled, and it operated laying on its side. It weighed 179 pounds and produced 15.76 hp at 1200 rpm, although it could maintain that output for only about three minutes, after which it stabilized at 12 hp. Carburetion was achieved by metering gasoline into a hot section attached to the water jacket throught which air was drawn. The resulting fuel/air mix was then directed to the combustion chambers. There were no spark plugs; the spark for combustion was made by two contact points inside the combustion chamber. Dry batteries were used to start the engine, after which it ran on a single magneto.

The Wrights designed the engine, and their longtime assistant, Charles Taylor, built it, possessing in the Wrights' Dayton shop only a lathe and drill press driven by belts from a one-cylinder gas engine. Taylor completed the job in six weeks and the engine was ready for test in January 1903. Propeller and drive train design was begun the following month, and it is apparent from a study of the

Wright Papers ("Notebook H") that considerably more effort went into propeller design than in the engine itself.

The Wrights designed and built some half-dozen improved engines prior to their demonstrations in France and at Ft. Myer in 1908, culminating in a four-cylinder powerplant that operated in an upright position and produced 30 hp. Except for the Manley engine, which was fitted to Prof. Langley's unsuccessful Aerodrome and never had a chance to fly, the Wright engines were as good as any used by their contemporaries before 1909. By then, however, Curtiss in America and several Frenchmen were flying better engines—particularly the Antoinette, Renault, and Gnome rotary, each of which generated 50 hp.

Curtiss

The Curtiss engines, primarily the work of machinist Charles Kirkham, evolved from the Curtiss motorcycle engines. The first Curtiss airplane, the 1908 June Bug financed by the Aerial Experiment Association, was powered by a Curtiss air-cooled V-8 of 30 hp. Curtiss' 50-hp water-cooled V-8, which powered his Rheims Machine in 1909, was an excellent engine and the parent of the Curtiss/Kirkham V-type engines that followed. The Curtiss Model L of 70 hp appeared in 1911, and it became the Model O when the water jacket was changed from copper to Monel. In 1913 Henry Klecker redesigned the rocker arm assembly of the Model O, which raised the horsepower to 90, and the Model OX resulted. Five minor improvements during the next year made it the OX-5.

The OX-5 was, as we've said, a water-cooled V-8. Its rated 90 hp was achieved at 1400 rpm, and it weighed 390 pounds dry. Approximately 15,000 OX-5s were built during WWI and a great many of those were available in the surplus market in the early '20s, often for as little as $50. The OX-5's weaknesses were its valve system (which wore out valve guides in less than 100 hours), the fact that it was water-cooled (and required that much attention be accorded its plumbing), and the single magneto (which had to be maintained in perfect adjustment).

In the late '20s an engineer by the name of Tank designed air-cooled cylinders for the OX-5 that employed Buick valves. This was a popular modification; it eliminated the plumbing problems and increased output to 115 hp.

Other Curtiss V-8s produced during WWI included the 100-hp Model OXX, the OXX-6 which featured dual ignition, the 160-hp Model VX, and the Model V-2 and V-3 of 200 hp. The general configurations of all were very similar.

The Curtiss D-12

The Curtiss D-12 grew from a design by Charles Kirkham, who had designed and built engines for Curtiss that dated back to Curtiss' motorcycle days. Kirkham formally joined the Curtiss firm as chief motor engineer in 1914, and after perfecting—if that is the proper word for it—the OX-5, and further work on the VX engine, began work on a 12-cylinder, V-type *enbloc* engine originally planned for 300 hp.

The enbloc, or monoblock, V-type engine was pioneered by a young Swiss engineer, Marc Birkigt, who had since 1905 designed some very successful automobile engines for the Spanish firm of Hispano-Suiza. At the beginning of WWI in 1914, Birkigt designed a monoblock V-8 of 150 hp for aircraft use. This engine was ready for production late in 1915 and so intrigued Kirkham that he determined to employ the monoblock principle in the new Curtiss engine.

The usual water-cooled engine of that time consisted mainly of a very stiff and relatively heavy crankcase upon which steel cylinders were bolted individually. Each cylinder was surrounded by its own cooling mantle, while the valve mechanism in the head was, in most cases, left exposed. Birkigt's Hispano-Suiza engine departed radically from that type of construction. Its cylinders were formed from an aluminum single-block casting into which were screwed four forged steel barrels that were threaded on the outside for their entire length. These barrels were closed at the top, forming a compartment that served as a combustion chamber. A very light crankcase was attached to the cylinder blocks; thus, the Hisso could be described as having the crankcase hung on the cylinders instead of

47

having the cylinders on the crankcase. The valves—two per cylinder and mounted vertically in the cylinder head—were actuated by a single overhead camshaft that operated the valve stems directly without the interposition of either pushrods or rockers. The entire valve mechanism was enclosed in an oiltight cover that fitted closely over the cylinder block. It was an excellent engine for that day, and it was eventually licensed for manufacture in a number of countries. In the United States, Wright-Martin acquired the rights to it.

Kirkham's first attempt to build a monoblock engine ended in failure. It was a V-12 called the Curtiss AB, and was running in test in April 1917, concurrent with America's entry into WWI. At 725 pounds, the AB was fairly heavy; its valves and reduction gearing were troublesome and, in any case, Kirkham came to believe that he had aimed too low at 300 hp. He scrapped the AB and immediately designed a V-12 that would produce 400 hp. The new engine was at first called the D-1200 (1145 cubic inches) and weighed between 625 and 660 pounds dry, depending upon which of the Curtiss and McCook Field records one is willing to accept. This powerplant was tested late in 1917, by which time it was officially called the K-12.

Although the K-12 failed to complete a 50-hour test run at McCook, the Army report was favorable and encouraged development. However, by then the Aircraft Production Board had ordered mass production of the Liberty engine—a V-12 of 400 hp with individual cylinders bolted to its crankcase—as the "standard" aircraft engine for the duration of the war, and the K-12 was without a market. Actually, the Navy liked the K-12 and offered a contract for 750 of them, but the Aircraft Production Board refused to authorize the necessary priority of materials needed for such production. As mentioned earlier, this wartime board was dominated by men who had a vested interest in the Liberty.

K-12 development did continue, however, and Curtiss airplane engineers designed a special two-place "Battleplane" around it in order to exploit the K-12's small frontal area and low weight-to-power ratio. This aircraft was built in both triplane and biplane configurations, the former for Navy evaluation, the latter for the Army Air Service. As a triplane it was known as the Curtiss 18-T Wasp; it achieved a climb rate of 2000 fpm and a top speed of 162 mph. The Army's version, the Hornet, crashed early in its test program and was abandoned.

The K-12 had its teething troubles, and 15 had been built when Kirkham, unwilling to accept advice or help in solving its problems, resigned from Curtiss early in 1919 to be replaced by Finlay R. Porter, former chief motor engineer at McCook Field.

Porter, aided by Arthur Nutt, redesigned the K-12, changing Kirkham's single casting of the block and crankcase into separate castings and providing a new exhaust system, along with a new crankshaft. This engine, redesignated the C-12, was ready for test in January 1920, but company president John N. Willys had sold his stock in Curtiss Aeroplane and Motor a month earlier to Clement Keys, and Porter left the firm with Willys. That left young (26-year-old) Arthur Nutt to face the fact that the reduction gear problem remained in the C-12. Nutt solved that by eliminating these gears altogether, which resulted in the direct-drive Model CD-12.

When the CD-12 proved to have main bearing weaknesses, Nutt, with Keys' backing, gave the engine still another complete redesign. The resulting Model D-12 could therefore be said to be an Arthur Nutt design—at least, to the extent that any engine is ever the design of *one* person.

The Curtiss D-12 (military designation V-1150) weighed 680 pounds dry, was rated at 460 hp at 2300 rpm burning 73 octane fuel, and had a normal fuel consumption of 38 gph. Its recommended time between overhaul was 200 hours. A total of 1192 D-12s were produced from 1922 to 1932 at an average price of $9100.

The D-12, which was built through the D-12F supercharged version, was produced almost exclusively for the U.S. military and the export military markets, powering such aircraft as the Curtiss PW-8, Curtiss Falcons, Boeing PW-9, the Hawk series of biplane fighters through 1928, and others.

The D-12 was followed by the V-1550 and V-1570 Conqueror. The Conqueror was made in both geared and direct-drive versions. Its compression ratios varied from 5.9:1 to 8.3:1 as fuels improved and more efficient cooling developed. Originally run at 1550 cubic inches displacement and 575 hp, the Conqueror was soon upped to 1570 cubic inches and 600 hp at 2400 rpm, at which output it consumed 56 gallons of fuel per hour. This engine was developed by Arthur Nutt (who had long since risen to chief engineer of the Curtiss company's Motor Division) as a direct descendent of the D-12, its crankcase, crankshaft, and connecting rods being almost identical to those of the D-12, though its crankshaft diameter was increased from three inches to three and a half. Conqueror cylinders were cast *enbloc* in two banks of six as were D-12 cylinders, but differed from the D-12 in that the cylinder sleeve in the D-12 was closed with its valves seating directly in the closed end, while the Conqueror sleeve was open at both ends, screwed into the aluminum block, and its larger valves seated on aluminum-bronze inserts. The frontal area of the Conqueror was about two inches narrower than that of the D-12 due to redesign of the overhead camshaft drives. The Conqueror weighed 750 pounds dry and was produced from 1925 through 1932; it was fitted to the Hawk P-6 series fighters, Curtiss B-2 bomber, Curtiss A-8, Consolidated A-11, and others. The Conqueror was the last liquid-cooled engine produced by Curtiss. Curtiss Aeroplane and Motor was merged with Wright Aeronautical in 1929; thereafter Curtiss-Wright concentrated on development of the Wright air-cooled radials.

The Wright Whirlwinds

The WWI Wright-Martin company was licensed to build the Hispano-Suiza engine in 1916, and produced more than 5000 of them before war's end, by which time the "Hisso" was available in 150 and 180-hp versions, with a 300-hp model in development. These were known, respectively, as the Wright Models A, E, and H.

After Wright-Martin was liquidated in 1919 and Wright Aeronautical born in its place, with Frederick B. Rentschler as president, Wright continued to supply Hissos (commonly called "Wright-Hissos") to the military while its best enginemen, Andrew Willgoos and George Mead, attempted to produce an air-cooled static radial engine of 1453 cubic inches displacement known as the Wright Model R-1.

The air-cooled engine, because of its potentially superior power-to-weight ratio and relative ease of maintenance, was an attractive concept for aircraft use, and the static radial configuration was the logical approach ("static" as opposed to the rotary radials of WWI). The British had been working on static radials in the 300 to 400-hp range since early in WWI. A.B.C. Motors, Ltd. produced the Wasp and Dragonfly seven and 14-cylinder, 180 and 350-hp engines respectively. These were not successful, but Cosmos Engineering Company engineer A. H. Roy Fedden began design of a 400-hp radial in 1918 that held much promise. Cosmos was liquidated at war's end and its assets—including Roy Fedden—were acquired by the Bristol Aeroplane Company. Bristol called Fedden's engine the Jupiter, and had it in production by 1921. Several models of the Jupiter were eventually offered in the 400 to 500-hp range and this engine dominated the British military aircraft market by 1927.

The Jupiter benefited from research into air-cooled cylinders performed by Samuel Heron at the Royal Aircraft Factory beginning in 1915. Heron also designed an air-cooled radial called the RAF-8, but left the government facility in January 1917 to join the Siddeley Deasy Company (later, Armstrong-Siddeley), where he continued work on the RAF-8 engine, by then called the Jaguar.

Heron left Armstrong-Siddeley in mid-1917 following an argument with his bosses over the Jaguar cylinder head design; he subsequently sailed for the United States, where he was employed by the Army Air Service at McCook Field to do research into air-cooled cylinders for aircraft use.

The Jaguar engine, rated at 400 hp, appeared in 1922 and enjoyed a great deal of success until the Jupiter took most of its markets in the mid-'20s. Then the American built Wrights and Pratt & Whitneys would prove superior to the Jupiter series—which was a tad ironic, because the worldwide acceptance of these American air-cooled radials was largely owed to the genius of Sam Heron.

The Wright Whirlwinds evolved from a 200-hp, nine-cylinder radial designed by Charles Lanier Lawrance in 1921. Lawrance (that's the way he spelled it) worked for pioneer Italian aircraft engine designer Alessandro Anzani in Paris prior to WWI. Anzani was building a variety of air-cooled engines—some of three cylinders in a fan configuration, and some twin-row radials of four to 20 cylinders. When the U.S. entered WWI. Lawrance returned to the United States and set up shop in a New York City loft where he produced some two and three-cylinder air-cooled engines that powered the Breese Penguin Trainer. (The Penguin was an idea borrowed from the French. These machines resembled the 1910 Bleriot with clipped wings. They did not fly, but allowed student pilots to taxi about the training field to gain some feel of the controls.)

In 1920 Lawrance designed the R-1, a nine-cylinder radial of 150 hp that powered the Elias TA-1 Trainer, three of which were purchased by the Air Service. The Navy, in the person of Lt. Cmdr. Kraus, chief of aircraft engine design in the Bureau of Engineering, liked the Lawrance R-1 and asked Lawrance to design a similar engine of 200 hp for the small fighters that would be needed for the Navy's first aircraft carrier. The Lawrance Model J resulted, and Lawrance received a Navy development contract for it in February 1921.

The first Model J failed its Navy test in May 1921, but the Navy, with uncommitted funds available as its fiscal year neared an end, gave Lawrance an order for 50 production versions of the prototype, the J-1s, rather than return the unspent money to the U.S. Treasury (a practice not uncommon then as now). Then, following the successful 50-hour test run in January 1922, J-1 production began—albeit at a slow pace, due to Lawrance's very limited facilities. Eventually 250 J-1s were manufactured.

The Navy used the not-altogether-satisfactory J-1 in its TS-1 fighters, which were carried aboard the *USS Langley,* America's first aircraft carrier, commissioned in March 1922. The Army fitted a J-1 to an experimental pursuit (fighter), the Dayton-Wright XPS-1, a parasol monoplane with a retractable landing gear.

Meanwhile, the newly-established Bureau of Aeronautics (BuAer) under Rear Adm. William A. Moffett took the position that the Navy should not depend upon a single and/or limited source for its essential hardware, particularly if a national emergency should arise. Therefore, Moffett's aircraft engineman, Lt. Cmdr. Bruce Leighton (Naval Aviator No. 40) pushed both Wright and Curtiss to develop air-cooled radial engines for the Navy's airplanes. Such engines, once perfected, would be much easier to maintain, especially aboard ship. Curtiss, with a large investment in its D-12, paid lip service to the idea but took no action on it at that time. Fred Rentschler at Wright had tried to build such an engine for the Air Service in 1919 without success and, although his top engineers—Mead and Willgoos—had learned much from that exercise, Rentschler was not interested in trying another, much preferring to sell his several Hisso models. Whereupon Leighton informed Rentschler that the Navy would buy no more Hissos or Hisso parts.

Lt. Cmdr. Leighton may or may not have been bluffing, but in the end Rentschler decided not to risk the loss of half of his market and agreed to attempt development of the kind of engine the Navy wanted.

We cannot be sure of exactly what happened at that point because Rentschler later told the story one way, while a couple of his close associates told it another. Perhaps it was Rentschler's idea to buy out the tiny Lawrance Aero Engines Company and develop the J-1; perhaps it was done at Leighton's insistence. In either event, that is what happened. Rentschler's board of directors voted to pay a reported half-million dollars for the Lawrance company and offer Charles Lawrance a vice-presidency at Wright Aeronautical.

The Navy was pleased with the deal, which was completed in May 1923. Then George Mead and Andy Willgoos were charged with the task of improving the J-1's reliability factor. This resulted in some modification of the J-1 and its redesignation as the J-3 in the fall of 1923 (the J-2 was a Lawrance design that was not built), and the Navy then standardized on the J-3 for its 200-hp needs.

In 1924, another improvement to the J-3 cylinder design produced the Wright J-4, which the company called the Whirlwind and it represented the next-to-last step before appearance of the famed J-5 Whirlwind, the airplane engine that touched off a transportation revolution.

Pratt & Whitney

Again, though Frederick Rentschler told it a different way (his board of directors would not authorize sufficient money for research), it seems clear in retrospect that, when Rentschler resigned his position at Wright on 21 September 1924, he had every reason to believe that Mead and Willgoos had learned their air-cooled engine lessons well and were ready to produce a truly great aircraft engine of that type. In addition to their unsuccessful 1919 effort, they had worked on an improved version of it redesigned at McCook Field, as well as a Navy air-cooled radial design of 400 hp, and they had begun design of the 450-hp Wright P-2 (shortly before Rentschler left the company) that would later lead directly to the successful Wright Cyclone. Therefore, one may be forgiven for suspecting that Rentschler's departure from Wright—and his subsequent formation of Pratt & Whitney Aircraft Corporation to produce, in record time, a fabulously successful air-cooled radial engine of 400 hp—was a carefully calculated move, intended to reap the rewards for himself that were just a step away from realization at Wright at that time. This, especially, in view of the fact that Mead, Willgoos, and other key people at Wright assured Rentschler that they would join him when he was ready to re-enter the engine business.

Meanwhile, Sam Heron, working with Air Service Lt. Edward T. Jones at McCook Field, had developed several air-cooled cylinders for radial airplane engines that would, at last, give such powerplants extreme reliability. These cylinders, along with another Heron invention, the sodium-filled exhaust valve (which evenly dissipated heat and practically eliminated warped valves), were breakthroughs known to Rentschler at that time.

In any event, it is clear that Rentschler had something of great potential value to sell, because he soon had a factory, an initial quarter-million dollars up-front money, and an agreement for an additional $1,000,000 when he was ready to begin production of his new engine. True, he also had the proper background: good family, degree from Princeton, brother Gordon with the National City Bank of New York City, training as a moulder and machinist in the family's Hoover-Owens-Rentschler Tool Company back in Hamilton, Ohio, prior to WWI, and eight years' experience in the airplane engine manufacturing business. But none of that would have mattered much had the 37-year-old Reintschler not possessed a first-class business proposition to offer potential backers.

He went to a longtime friend of the Rentschler family, James K. Cullen, president of Niles-Bement-Pond, which owned the Pratt & Whitney Tool Company of Hartford, Connecticut. (we must include the fact, however painful, that Mr. Edward A. Deeds of the boondoggling WWI Aircraft Production Board was also an officer in Niles-Bement-Pond.) Before the Civil War, Francis Pratt and Amos Whitney were machinists in Colt's Pistol Factory. In 1860, they teamed up to open their own machine tool shop. Pratt & Whitney Tool had accumulated large cash reserves at the end of WWI, but had delivered so much manufacturing machinery that their markets were temporarily saturated and their own plant idle. Pratt & Whitney's needs and Rentschler's were complementary.

Assured of a factory and ample financing, Rentschler called on Rear Adm. Moffett and was promised a $90,000 engine development contract from the Navy. Then he told Mead and Willgoos that it was time for them to resign their positions at Wright, and put them to work in Willgoos' garage designing the new Pratt & Whitney engine, employing Heron's Type M cylinder with the rocker box cast integrally with the head, and Heron's sodium-filled exhaust valves. The engine would displace 1344 cubic inches, should weigh slightly under 650 pounds, and produce 400 hp at 1800 rpm. Mrs. Rentschler suggested that it be called the "Wasp."

With Wasp blueprints accumulating, Rentschler raided Wright Aeronautical again, obtaining a shop superintendent, production engineer, and factory manager in the persons of John Borrup, Charles Marks, and Donald Brown, respectively.

Pratt & Whitney Aircraft Company was incorporated on 22 July 1925, with Rentschler as president and Mead vice-president of engineering. Half of the original stock issue was retained by P&W Tool, the other half going to Mead and Rentschler. On 1 April 1926, with the Wasp largely proven in a series of ground tests, Rentschler decided that the experimental period was over. The company's total debt to the tool company was $202,713.29. At that point, the tool company provided production funds that brought its total money advanced to $1,030,413.70, which was secured by 7500 shares of seven percent preferred stock with a par value of $750,000, plus $280,413.70 it held in six percent notes.

Seven years later, during Senate hearings investigating Pratt & Whitney Aircraft (among others), it was revealed that 5500 shares of P&W Aircraft common stock had been issued in July 1925, with 1375 shares going to Rentschler, a similar amount going to Mead, and the remaining half—2750 shares of common—going to P&W Tool. Rentschler and Mead had paid 20¢ per share for their stock at that time, or $275 each.

Interestingly, in December 1926, Rentschler sold 110 shares of his stock to Charles W. Deeds, the 30 year-old son of Edward A. Deeds. Charles Deeds was then listed as treasurer of P&W Aircraft. That left Rentschler with 1265 shares, which had cost him $253.

In November 1928, P&W Aircraft split its common stock 79 for one, which left Rentschler with 101,200 shares. A little later, United Aircraft & Transport Corporation absorbed P&W Aircraft in a merger that involved another exchange of stock, and Rentschler emerged with 219,604 shares in the holding company. In 1933, when United Aircraft was quoted at around 97, the paper profit on Rentschler's original investment amounted to more than $21,000,000. Between 1927 and 1933 he was said to have received more than $1,500,000 in salaries and bonuses alone.*

Well, anyway, it was a start.

Radial Development

In the meantime, Charles Lawrance moved up to the presidency at Wright Aeronautical, and brought in Sam Heron and Edward T. Jones from McCook Field to replace Mead and Willgoos. Employing Heron's Type K cylinder and salt-filled exhaust valves, Heron and Jones reworked the J-4 Whirlwind and emerged with the J-5 in 1926. Therefore, the 220-hp Whirlwind and the 420-hp Wasp appeared at about the same time. Both were extremely successful, in both military and civil applications, and not only for their efficiency and unusual reliability, but also because they became available at a most propitious time. Both the Wasp and the J-5 were in production late in 1927, and several other highly important developments had come together by that time to provide all the necessary ingredients for an unprecedented boom in civil aviation.

The spark that touched off that boom was the solo nonstop flight from New York to Paris by Charles A. Lindbergh on 20-21 May 1927 in a single-engine Ryan monoplane. It was an accomplishment that captured the fancy of people everywhere, particularly in the Western World, as few other acts by an individual have done. Few took note of the fact that the key to Lindbergh's success was the Wright J-5 Whirlwind in his airplane, which functioned perfectly for 33½ hours.

Earlier, as the first J-5 at Wright and the first Wasp at P&W were undergoing tests prior to production, Heron and Jones at Wright were following up with the 525-hp Cyclone (R-1750), and the Mead-Willgoos team at P&W was preparing the

---

*The Special Committee on the Investigation of the Air Mail and Ocean Mail Contracts, United States Senate, 73rd Congress, 2nd Session, P 2196; "Testimony of F. B. Rentschler," p1801. Also, "Testimony of Charles W. Deeds," September 26, 1933, p 1695.

525-hp Hornet (R-1690). Then Wright introduced its new J-6 series offering Whirlwinds of five, seven, and nine cylinders rated at 140, 225, and 300 hp, respectively (military designations, R-540, R-760, and R-975, respectively; the military liked this system because it described engine configuration—"R" for radial, and cubic inch displacement). The Wright J-6-9 eventually replaced the nine-cylinder 225-hp J-5 Whirlwind.

All of these engines possessed good growth potential. In the early '30s the Wright R-975, for example, had grown to 420 hp; its direct competitor at P&W, the R-985 Wasp Junior that also began life at 300 hp, was later upped to 450 hp, while the first Wasp, the R-1340, ended up at 600 hp. The Wright R-1750 Cyclone, after several years' development, pointed the way to a much improved engine which, with a forged crankcase and increase to 1823 cubic inch displacement, became the R-1820 of 700 hp. The P&W R-1690 Hornet was up to 850 hp in the late '30s, after starting out at 525 hp.

By that time, the R-1830 Twin Wasp was in production producing 1200 hp, and the R-2800 Double Wasp of 2000 hp was in development. At Wright, the R-3350 was also coming along, similarly rated, while late models of the R-1820 Cyclone had, by 1939, achieved a rating of 1200 hp.

Among the developments that made these advances possible were better fuels (a field in which Sam Heron played a major role), improved metals, and controllable-pitch propellers. (We will take a look at the evolution of the propeller momentarily.)

Other aircraft engine builders emerged in the wake of the "Lindbergh Boom" to seek a share of the low-horsepower market. These included Le Blond of Cincinnati, Kinner in Glendale, California, Menasco in Los Angeles, and Warner in Detroit. Jacobs, Continental, Lycoming, and Ranger would soon follow.

Le Blond

The Le Blond Aircraft Engine Company was formed in 1928 from the defunct Detroit Aircraft Engine Company. It was a subsidiary of the well-established Le Blond Machine Tool Company and Glenn D. Angle, formerly with the Detroit company, remained in charge of development and production. The air-cooled Le Blond radials—in three, five, and seven-cylinder models of 40, 60, and 90 hp respectively, were only marginally successful. This engine at Detroit had been known as the "Air Cat."

Kinner

The Kinner Airplane and Motor Corporation was incorporated in 1919, but its five-cylinder air-cooled radial, the K-5 of 100 hp, had no market until it was fitted to the Simplex Red Arrow—a mid-wing, open-cockpit monoplane—in 1928. Then, in 1929, the Alexander Eaglerock, Swallow TP-K, Fleet 2, and American Eagle civil biplanes all appeared with the Kinner engine. Its best known application would come years later in WWII as the powerplant in the Ryan PT-22 trainer, that version of the Kinner being the 160-hp R-540 (the Wright R-540 was no longer in Air Force inventory by then). Its designers were W.B. and C.M. Kinner.

Menasco

Al Menasco formed Menasco Motor Company in 1926 to market a quantity of French-built Salmson Z-9 engines purchased as surplus from the U.S. Government. These were nine-cylinder air-cooled radials of 250 hp that Menasco gave new cylinders, crankshaft, and other significant mods. The entire stock of 50 engines was contracted to the O.W. Timm Airplane Corporation of Glendale for that firm's Timm Biplane, a single-engine, seven-place cabin craft of hefty dimensions that found few buyers.

The most successful Menascos were the in-line inverted air-cooled engines of four and six cylinders. These were the C4 of 363 cubic inches, normally rated at 125 hp and known as the Pirate, and the 544-cubic inch Buccaneer or B6 Model rated at 200 hp. With gear-driven superchargers the Pirate became the C4S and produced 150 hp, while the Buccaneer as the B6S was rated at 290 hp. These powerplants achieved fame during the '30s when fitted to a number of successful racing planes.

The Pirate also provided power for some WACOs during the '30s, as well as the popular Ryan ST.

Warner

The original Warner Scarab, a seven-cylinder air-cooled radial of 110 hp, made its first public appearance at the 1932 National Air Races. The Scarabs were later built in 125, 145 and 185-hp versions, and were fitted to many private airplanes of the '30s, including the Monocoupe, Gee Bee Sportster E, and the Ryan SC. The Warner Scarabs were produced in Detroit.

Velie

The Velie Motors Corporation in Moline, Illinois, was an old and well-established manufacturer of automobile engines, and it, too, sprang into the low-horsepower airplane engine market as the boom in private flying became apparent late in 1927. The Velie M-5 and L-9 were five and nine-cylinder radials of 70 and 160 hp respectively. The M-5 powered the Monocoupe 113, Star Cavalier, Monoprep, and Nicholas-Beasley NB3, among others.

Szekely

The O.E. Szekely Corporation of Holland, Michigan, began production of three air-cooled radials in 1928, offering engines of three, five, and seven-cylinder configurations, producing 40, 70, and 110 hp respectively. The 40-hp version—variously rated between 35 and 45 hp, depending upon propeller/rpm combination—was the only one manufactured in quantity, powering the American Eagle Eaglet, the Curtiss-Wright Junior, and the Buhl Pup—or Bull Pup.

Caminez

The Fairchild-Caminez was a most interesting air-cooled radial. It was produced in limited quantities by the Fairchild-Caminez Engine Corporation, formed in 1925 as a subsidiary of the Fairchild Airplane Manufacturing Company at Farmingdale, Long-Island. This engine was the brainchild of Harold Caminez, at one time in charge of engine development at the Army's McCook Field. The four-cylinder Caminez had no crankshaft; instead, it possessed a huge, two-lobed cam, and the pistons ran directly on this cam by means of roller bearings in the piston skirts. The advantage of this arrangement, with the four-stroke cycle, was that each piston completed a power stroke with each revolution of the shaft speed, resulting in a high power output per cubic inch of displacement at a low propeller speed, the shaft being one half that of a normal crank engine. The four-cylinder Caminez engine was rated at 135 hp at 1000 rpm. Charlie Meyers flew this engine fitted to a WACO 10, and reported that it vibrated excessively. The Caminez was also installed in several other commercial biplanes during the late '20s.

Jacobs

The Jacobs air-cooled radials that would power WWII primary trainers and utility craft (Beech Stagger-wings, UC-43, Howard UC-70, WACO UC-72, etc.), prewar WACOs, Model 17 Staggerwings, and the postwar Cessna 190/195 series—among others—evolved from the 1929 A.C.E. Model La.1, a seven-cylinder 150-hp development of the even earlier Jacobs & Fisher engine that greatly resembled the German Siemens & Halske. The 225-hp Jacobs was the R-755, the R-830 was rated at 285 hp, and the 300-hp version was the R-915.

Siemens & Halske

Siemens & Halske built engines for the German Air Force during WWI, and some later versions were imported into the U.S. during the '20s. These engines were air-cooled radials of five, seven, and nine cylinders covering a power range from 60 to 600 hp, although the imports were almost all in the 95 to 113-hp class. The original importer was K.G. Frank, who called the engines "Yankee-Siemens."

Axelson

The Axelson air-cooled radial was built by the Axelson Machine Company in Los Angeles, and this seven-cylinder engine of 115 to 150 hp was the former "Floco" manufactured by the Frank L. Odenbreidt Company. Relatively few saw service.

By 1930 there were 215 aircraft manufacturers in the U.S. either in production or hoping to be, and many—particularly those offering civil biplanes—gave their customers a choice of engines in order to attract the widest market since there actually wasn't too much difference between many of the airplanes. A choice of engines provided some pricing flexibility and allowed customers to buy the performance best suited to the needs of each. For example, the Travel Air biplane was available with engines ranging from 90 to 420 hp during its five years in production.

Most of these engines did not survive the Great Depression, the worst period of which was 1931 through 1934 inclusive. (Actually, the Great Depression of the '30s did not completely end until the demands of WWII brought jobs with the increased demands placed on American industry.)

Other air-cooled aircraft engines that briefly appeared in civil airplanes during the late '20s and early '30s included the Comet, a seven-cylinder radial of 115 to 165 hp built in Oakland, California; the Walter, a Czechoslovakian import of five, seven, and nine cylinders, 60 to 135 hp, the 120 hp version of which first appeared on the Spartan C-3 in 1928; and the French Anzanis, which were produced in three, six, and ten cylinders (the latter two being twin row engines), ranging from 25 to 120 hp. The early Cessna AW and Bellanca CF were fitted with Anzanis.

Cirrus

The little four-cylinder in-line Cirrus engines, which enjoyed their greatest exposure (if questionable success) in the popular Great Lakes 2T-1 Sport-Trainer, were British designs produced by Cirrus Aero Engines, Ltd., a subsidiary of A.D.C. Aircraft, Ltd. A.D.C. was formed in 1920 and its Cirrus spin-off in 1927. The American Cirrus Engine Company was licensed to build the 85-hp Mk III, as well as the inverted version of 90 hp known as the Cirrus Hi-Drive.

The deHavilland Gipsy was another British design and was very similar to the Cirrus. It was fitted to the American Moth, which was the deHavilland Gipsy Moth biplane briefly built in the U.S. by a subsidiary of the Curtiss-Wright Aircraft Corporation.

Ranger

The most successful of the in-line air-cooled aircraft engines was the Ranger. Originally the Fairchild 6-390, it was a six-cylinder inverted engine of 125 hp. In 1931, the American Airplane and Engine Corporation was formed as a Fairchild subsidiary to produce this engine and it was then named the Ranger. The Ranger was initially tried in the Consolidated XPT-4 primary trainer in 1929. Its best known civil application prior to WWII was in the Fairchild Model 24 cabin monoplane. In 1940 it appeared in the Fairchild PT-19 primary trainer and was rated at 175 hp as the Ranger L-440-1. The follow-on version was the L-440-3 of 200 hp.

Meanwhile, a couple of the most enduring names in the civil aircraft piston engine field had appeared with air-cooled radials—Continental, in 1927, and Lycoming, in 1928.

Continental

Continental Motors Corporation, one of the largest automotive manufacturers in the world in the late '20s, produced its first aircraft engine, a 220-hp nine-cylinder air-cooled radial, in 1927. It was based on a design purchased from the Argyll Company, and served only to push Continental into the aircraft engine business. Founded in Detroit in 1900, the company's "Red Seal" engines powered cars, trucks, and boats worldwide by the time it built the Argyll engine. That project was scrapped, however and in late 1929 Continental engineers produced the Model A-70 air-cooled radial of seven cylinders and 165 hp (R-545 military designation). This was soon fitted to three WACO models, the Swallow Sport, and the Kellett Autogiro, among others. The Continental R-670 of 210 hp followed in 1931, and that engine, soon rated at 220 hp, would power the famed Stearman PT-17 primary trainer of WWII. The first of the "modern" Continental opposed engines was the A-40 of 1931, rated at 37 hp and fitted to the Taylor Cub (immediate forerunner of the J-3 Piper Cub).

Lycoming

Lycoming began building airplane engines in 1928 when engineer Val Cronstedt designed a seven-cylinder air-cooled radial of 645 cubic inches. It was less than a rousing success. Cronstedt reworked that engine to obtain a really great one—the R-680. This was initally rated at 200 hp, but was eventually produced in 240, 260, 290, and 300-hp versions.

The first opposed Lycoming was the 50-hp 0-145A rated at 2300 rpm. The appearance of this engine was concurrent with that of the 50-hp Franklin and 50-hp Continental in 1938. The 65-hp versions followed in 1939.

Lycoming was formed in 1908 at Williamsport, Pennsylvania, to produce engines for early-day automobiles, and by the mid-'30s Lycomings powered such cars as the Duesenberg, Auburn, and Cord. In 1939 the company officially became the Lycoming Division of AVCO, although it had by then been in the AVCO orbit for ten years. AVCO (Aviation Corporation) was organized in 1929 by G.B. Grosvenor of Fairchild, a covey of investment bankers led by W.A. Harriman and Robert Lehman, Robert Dollar of the steamship line, former Postmaster General Harry S. New, and Gen. Mason Patrick, U.S. Air Service, retired. AVCO's initial two million shares of stock were underwritten by the banks at $37.50 per share, giving AVCO $35 million in working capital—an impressive sum in 1929 dollars. AVCO was strictly a holding company, and it immediately began buying control of dozens of aviation properties, including American Airways (later, American Air Lines). Lycoming fell under AVCO's spell by way of a stock-swap deal with E.L. Cord, who headed Auburn Automobile Company and a couple of small airlines that American Airways needed to fill out its routes. Although AVCO was forced by law (1934 Air Mail Act) to dispose of its interest in American Air Lines, it held onto its interest in Lycoming and, in 1939, on the eve of WWII, Lycoming became an AVCO "Division."

Morehouse

The stories of the Continental and Lycoming opposed engines must include mention of Harold E. Morehouse, who markedly contributed to both.

Morehouse was 18 years old when in 1912 he became a frequent visitor at Chicago's Cicero Field where Laird, the Stinsons, and other air pioneers were venturing into the sky. He had left the family farm in southwest Michigan after completing a correspondence course in mechanical engineering and, after gaining experience as a draftsman for a company manufacturing marine engines, obtained a job with the WWI Dayton-Wright company. At Dayton-Wright, Morehouse designed a small air-cooled V-type engine for an experimental unmanned "flying bomb" that was not produced. In 1920 he joined the Air Service's research facility at McCook Field, where he worked under Sam D. Heron and Lt. Ed T. Jones on air-cooled cylinder development, and followed Heron and Jones to Wright Aeronautical in 1925 to work on the first Whirlwinds.

In 1929, Morehouse went with the newly-formed Michigan Aero Engine Company to design a four-cylinder inverted, air-cooled in-line engine of 75 hp known as the Rover. The Rover was the Fairchild 22's original powerplant, and was also fitted to the Driggs Dart.

When Michigan Aero Engine disappeared into the maw of the Great Depression in 1932, Morehouse moved on to Continental Motors Corporation in Detroit, where he worked on the A-40 and A-50, before ending up at Lycoming in 1939 to design the prototype of the ageless 0-235 (an engine still in production as this is written), and worked on subsequent opposed-type "Lycs" of four and six-cylinder configuration from which today's light-plane recips were developed. Morehouse retired from Lycoming in 1965.

We have included this brief outline of Morehouse's record not because he was unusual, but rather because he was *typical* of a host of others unnamed here. All of us who fly, or who take our sustenance from this industry, are in their debt. Most of us give little thought to the technology that produced the engines that power our flying machines, but that technology has always set the pace in aircraft development. Airframe design advances on the backs of the enginemen.

While Wright (with its Whirlwinds and Cyclones) and Pratt & Whitney (with its Wasps and Hornets) dominated the air-cooled radial aircraft engine market from the mid '20s onward, Curtiss tried such designs of its own. Curtiss aimed its 180-hp Challenger at the Whirlwind market, and its 650-hp Chieftain at the Cyclone and Hornet market. Both the six-cylinder Challenger and the 12-cylinder Chieftain were twin-row air-cooled radials, and both were in production by 1928.

The Challenger could be described as two three-cylinder engines in tandem, with the three rear cylinders spaced between the front three for maximum cooling. The Chieftain, however, had its cylinders mounted in pairs, with the rear six directly behind the front six cylinders. That resulted in built-in cooling problems with the rear jugs. Although that unusual configuration was intended to give the Chieftain (H-1640) maximum power with a minimum of frontal area.

The Chieftain was not successful, while the Challenger enjoyed modest success, mostly in Curtiss airplanes. After the Curtiss and Wright merger in 1929, all engine development was soon in the hands of the engineers at Wright Aeronautical Division of Curtiss-Wright, and the Challenger, too, was largely forgotten.

Allison

At almost the same time that Curtiss and Wright merged, General Motors purchased the tiny Allison Engineering Company from Fisher Brothers Investment Corporation. Allison had been formed in Indianapolis in 1915 to machine parts for race cars, and throughout the '20s had been little more than a large and well-equipped precision job shop. It got into the airplane engine field by converting WWI surplus Liberty engines into air-cooled configuration for Army and Navy Loening amphibians. These engines were designated V-1410s, and the modifications were engineered by the Air Corps.

Allison also built some experimental marine engines, and had tried a six-cylinder, 765-hp two-stroke diesel for the Navy's airship program. The latter ran well, but was not produced because it did not provide enough water from its exhaust to maintain lifting balance in a dirigible as its fuel burned off. Then, when founder James Allison died early in 1929, the company was sold to Fisher Brothers, who in turn sold it to GM later that year.

After GM took control, Allison manager N.H. Gilman was allowed to design a liquid-cooled V-12 aircraft engine of 750 hp when it was learned that the Air Corps was interested in such an engine.

The Air Corps had come to believe that a liquid-cooled high-horsepower engine would offer important advantages over the big radials in the fighter airplanes the Army hoped to have, say, five years hence. In addition to smaller frontal area and better fuel efficiency, the use of ethylene glycol as a coolant would permit higher operating temperatures, allowing in turn higher compression ratios and more power per cubic inch of displacement. Also, the big radials were encountering cooling problems above 15,000 feet. Air Corps thinking at that time was that the new engine should not be of *en bloc* construction, but of individual cylinder arrangement.

The Air Corps had asked the Wright Aeronautical Division at C-W to develop the engine it envisioned as a replacement of the Conqueror—and Wright refused, saying that henceforth they would build only air-cooled engines. Packard also declined the Air Corps' proposition, despite the fact that Packard had built a series of more or less successful high-horsepower liquid-cooled aircraft engines for the military during the early and mid '20s.

Wright and Packard could hardly be blamed for their positions. The Air Corps had almost no money to spend, and therefore could offer a manufacturer little hope of a production run sufficient to cover development costs.

At Allison, Norman Gilman saw opportunity

where Wright and Packard did not. He took the drawings of his proposed V-12 to Wright Field (which had replaced McCook Field as the Air Corps research center in 1927) where, to his dismay, the Air Corps refused support, probably because Gilman's engine would be of *en bloc* construction, and also because the Air Corps was reluctant to believe that Wright would not, in the end, develop the desired engine.

Undaunted, Gilman went to the Navy with the plans for his engine, which he called the V-1710. Although the Navy had no interest in liquid-cooled engines for its airplanes, it was receptive to the possibility of replacing the German Maybachs that powered its dirigibles with American-built engines. The big radials could not be cooled at dirigible airspeeds.

On 28 June 1930 the Navy gave Allison a contract for one engine to be rated at 650 hp, and that engine—the V-1710-A—would make possible a test of the basic design before development of a specialized version for airship use. It was first run in August 1931, after various changes had been made and its blower ratio raised from 7:3 to 8:1 (the integral supercharger or blower in these engines supplied no more manifold pressure than could be used at sea level). The engine passed a 50-hour test in 1932, with a club propeller on a rigid stand, producing 750 hp at 2400 rpm.

On 24 January 1933 the Navy gave Allison a contract for three reversible Model B engines, with no blower at all, for airship use. The Navy promised to buy an additional 20 engines as soon as these experimental powerplants were proven, and it was only because Allison had that promise that GM was willing to allow Allison to accept the three-engine order.

The first V-1710-B was built in 1933, but various problems delayed its successful completion of the necessary tests until September 1934. Then, on the day the two B engines were ready to ship—12 February 1935—the Navy's dirigible *Macon* was lost and with its loss the Navy's airship program ended. The promised production order was forgotten.

Meanwhile, however, the Air Corps had become interested in the V-1710 because it was apparent that the engine it thought it wanted appeared as far away as ever. That engine was the 12-cylinder 0-1430, designed at Wright Field and contracted to Continental Motors after the Air Corps had at last become convinced that Wright would not develop it. Continental was moving very slowly with the 0-1430 (it would eventually appear in the experimental McDonnell XP-67 Tornado twin-engine fighter, only one of which was built in 1942; it was also tried in the XP-49, a P-38 variant, and planned for the XP-52 and XP-53, neither of which was built), and so Wright Field decided that prudence dictated that the Army hedge its bet on a high-horsepower fighter airplane engine. Therefore, in December 1932, the Air Corps had gone to Allison and ordered a slightly modified version of the V-1710-A, requesting a larger impeller and other changes aimed at ultimately making it into a 1000-hp engine.

The Army's first Allison of this series was the V-1710-C1, which was delivered to Wright Field in June 1933, and the following year passed a 50-hour development test at a rating of 800 hp at 2400 rpm. The Army then ordered eleven additional engines.

In the spring of 1935, the Army was successful in operating a C1 engine at 1000 hp during a 50-hour developmental test, and that encouraged a try for a 1000 hp Type Test, but the engine broke down running with a full flight propeller, and the rest of that year was spent in other futile attempts before it was at last decided that major design changes were required.

The most serious problems with the C1 engine were breakage of the crankshaft and reduction gear (which used an overhung pinon), and with cylinder block cracks. Those parts were completely redesigned early in 1936, but the resulting engines broke down after 20 to 25 hours.

At that point a new engineer, R. M. Hazen, was placed in charge of the V-1710 project, and the engine was again redesigned. The most significant changes were a new intake manifold system, alteration of the coolant passages, new combustion chamber, and new pistons. The resulting V-1710-C6 operated for 141 hours of a Type Test at 1000 hp in the latter half of 1936, and in December extensive flight testing of the engine was begun in a Consolidated XA-11A two-seat fighter (the PB-2A without supercharger).

Detailed development of the C-6 produced the V-1710-C8, which passed its Type Test in March 1937. Normal rating on 87 octane fuel was 1000 hp at 2600 rpm at sea level; the engine weighed 1230 pounds. That was only a few months after the Rolls Royce Merlin passed a Type Test on the same fuel at the corresponding international rating of 990 hp at 2600 rpm. The Merlin's weight was given as 1335 pounds.

It should be remarked that the Allison engineering staff contained less than 25 people assigned to the V-1710 project, along with an experimental crew of 100, in 1937. Total cost of V-1710 development through the 1000-hp Type Test was approximately $2 million, with roughly half of that furnished by the military.

It may be interesting to note that Arthur Nutt and the enginemen at Curtiss, as a result of their experience with the D-12 and Conqueror, could reasonably have been expected to produce an engine at least as good as the V-1710, and in less time, if given the opportunity. It didn't happen because no new engine projects were begun by the Curtiss engine division after the Curtiss and Wright merger in 1929; the Curtiss engine people were eventually sent to the Wright Aeronautical Division of C-W in 1931 where they had little influence.

Merlin

Such speculation deserves a passing thought, perhaps, because a look at the Rolls-Royce Merlin's birth points to the Curtiss D-12 as the Merlin's true parent.

The Curtiss D-12 engine powered the military racers of the early '20s that dominated the international racing scene, and that led C.R. Fairey of Britain's Fairey Aviation Company to negotiate a license to manufacture the D-12 in England. The D-12, which Fairey called the "Felix," performed so well in the Fairey Fox, a two-place light bomber, that the British Air Ministry decided to finance development of a similar engine by a major engine builder in England.

Fairey did not get the assignment. It was first offered to Napier, but Napier refused and the proposal went to Rolls-Royce along with ample development funds and a D-12 engine for study.

Rolls-Royce, Ltd., was formed in 1906 to build motorcars. In 1910, Charles Rolls established Britain's first military flying school, and was killed flying a Wright pusher in July of that year. The company built its first V-12 water-cooled aircraft engine, the 275-hp Falcon, in 1916, which was based on a Daimler engine designed for a 1914 Mercedes racing car. During WWI, Rolls-Royce produced the 360-hp Eagle and 525-hp Condor, both V-12s of individual cylinder construction. Both remained in production into the late '20s.

Henry Royce and his staff began design work on a 1295 cubic inch V-12 early in 1925, and they had a 480-hp, 900-pound engine ready for production three years later. At first called the Falcon X, this engine was the Rolls Royce Model F when it entered service. By 1931, the Model F had been uprated to 600 hp and was renamed the Kestrel. The Kestrel XVI was up to 745 hp as the powerplant for the Hawker Fury biplane fighter in 1937. But in the meantime, development of the Merlin as a scaled-up Kestrel (from 1295 to 1649 cubic inches) was begun in 1933, and the Merlin's progress paralleled that of the Allison V-1710 very closely, including a major redesign of each in 1936. A.G. Elliot was the chief engineer in charge of the Merlin project.

A significant difference between the Merlin and Allison was due to the fact that the British favored gear-driven superchargers as *integral* parts of their aircraft engines, while in America the "bolt-on" exhaust-driven supercharger received whatever development funds the Air Corps could scrape together for the purpose. In service, the supercharger systems of the Allison and Merlin marked the only practical difference between them. We will have some interest in that because two versions of the Merlin powered some important U.S. fighter airplanes, while Allisons powered others. In each case, it was the supercharger system that determined the role of those machines in combat.

During WWII, Packard built the Merlins in the U.S.—originally the Merlin 28, V-1650-1 USAAF designation, which went into the Curtiss P-40F and P-40L. It had a single-stage supercharger which made it a medium altitude engine, rated at 1010 hp at 16,000 feet, and it was only marginally better than the Allison V-1710s that powered most P-40s and all P-39s. The Merlin 60 series (V-1650-3, -7, and -9 USAAF designations) were rated at 1300 to 1490 hp and had a two-stage, two-speed supercharger. This was the high-altitude version that went into the P-51 Mustang (B models and after).

The V-1710s that were fitted to the P-38 Lightnings ranged from the V-1710-27/29 (one left-hand and one right-hand rotation) in the first 30 production P-38s, to the V-1710-111/113 in the P-38L. Horsepower went from 1150 to 1475. Almost 70,000 V-1710s were built by war's end, and production finally stopped in December 1947, following an order to equip the F-82 Twin Mustang. By that time, Allison had been delivering jet engines for almost two years.

The Rotaries

The early air-cooled rotary radials had some influence on the development of American aircraft engines and therefore are entitled to brief mention. Although the rotaries all had foreign names, the first one was conceived, designed, and built in the U.S. It was the creation of F. O. Farwell, who planned his engine as a truck and bus powerplant before the Wrights ever flew. The Adams & Farwell Company found little market for its product in America, so Farwell went to France and the *Societe des Moteurs Gnome* where Laurent Seguin and his brother improved and marketed the engine as the "Gnome Monosoupape," so called because of its single-valve action.

Actually, the original seven-cylinder 50-hp Gnome received its fuel through a hollow crankshaft and automatic valve in the piston head; exhaust was through a mechanical valve in the top of the cylinder. The true Monosoupape drew fuel from holes in the cylinder walls when the piston reached bottom dead center, while air was drawn in through a single valve at the top. The Gnome, as other rotaries, always operated at full power, unless the pilot cut the ignition with the "coupe" button, which was usually mounted on the control stick. It was possible to cut out three, five, or all cylinders to control thrust.

All other rotaries that followed introduction of the Gnome in 1909 were copies, although some differed in their fuel intake systems. The LeRhone, for example, had two valves and a carburetor. But the central fact about the rotaries—and the reason they were called rotaries—was that the crankshaft was solidly bolted to the airplane's firewall, and the rest of the engine, to which the propeller was permanently affixed, rotated about the crankshaft. That ensured good cooling; it also meant that any unburned fuel, along with the castor oil used for lubrication, was flung out the valve (exhaust) in the top of each cylinder when the engine was in operation. There was no means of returning the oil to the oil reservoir. That is why all rotaries had cowlings—and why their pilots were allegedly advised to eat a lot of cheese.

Other rotaries included the French LeRhone and Clerget, the British Bently, and the German Siemens-Halske and Oberusel. Most were of single-row nine-cylinder configuration, and in the 80 to 100-hp range. Twin-row versions produced up to 160 hp. These were all four-cycle, five-impulse engines: intake, compression, ignition, power, and exhaust. Fuel consumption was about six gallons per hour on a typical 80-hp rotary, and oil consumption was two gph. A single magneto was employed, and weight-to-horsepower ratios were between 3.5 and 4.0 lbs/hp. Their recommended time between overhaul was 16 to 30 hours. Rotary engines powered many of the famous WWI airplanes, including the Sopwith Camel and Fokker Dr I (triplane). The 80-hp LeRhone was produced under license in the U.S. by the Union Switch & Signal Company of Pennsylvania, which delivered 450 LeRhones in mid-1918 for the Thomas Morse S-4C. In 1927,

Tips and Smith, Incorporated, of Houston, Texas, announced the "Super-Rhone," which was the 80-hp LeRhone obtained on the war surplus market and modified into a static radial. The Super-Rhone was rated at 120 hp at 1400 rpm and was priced at $750. It was not successful, possibly because the surplus engines had been rejects at the time of manufacture.

The Wright brothers' first engine weighed 179 pounds and produced 12 hp, or almost 15 pounds per horsepower. (NASM)

The Manley engine, a water-cooled radial, was far superior to the Wright engine, but was wasted on the Langley machine, which did not fly. (NASM).

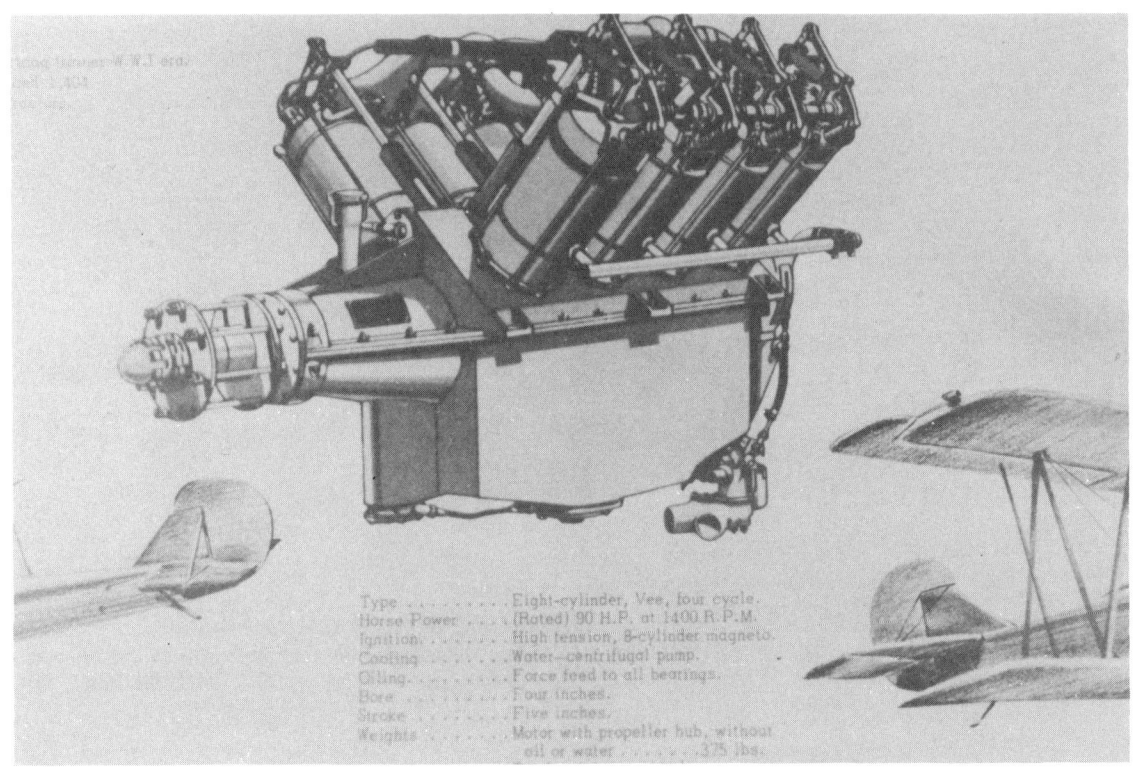

```
Type............Eight-cylinder, Vee, four cycle.
Horse Power.....(Rated) 90 H.P. at 1400 R.P.M.
Ignition........High tension, 8-cylinder magneto.
Cooling.........Water—centrifugal pump.
Oiling..........Force feed to all bearings.
Bore............Four inches.
Stroke..........Five inches.
Weight..........Motor with propeller hub, without
                oil or water........375 lbs.
```

The Curtiss OX-5, rated at 90 hp at 1400 rpm, was a water-cooled V-8 that entered production in 1916. Surplus OX-5s powered civil biplanes well into the '30s.

The Curtiss D-2 (V-1150) of 400/435 hp evolved from Charles Kirkham's K-12, built while Kirkham was Curtiss' chief motor engineer during WWI. The D-12 featured *en bloc* construction, a concept pioneered by the Hispano-Suiza V-8s. (Curtiss-Wright)

Turbosupercharger installation on the Curtiss PW-8 with intercooler mounted ahead of the turbo unit and extra radiator beneath the nose to supplement cooling provided by the wing-skin radiators. Service ceiling was 30,350 feet with 160 mph and 2160 rpm. Supercharger development began at the Air Service's McCook Field in 1918. (Air Force Museum)

The Curtiss Conqueror V-1550 of 600 hp. The Conqueror weighed 750 pounds dry and consumed 53 lb. fuel per hp/hr. The last ones were V-1570s rated at 650 hp. This engine appeared in 1925 and was designed for glycol cooling. (Curtiss-Wright)

The 1927 Wright J-5 Whirlwind of 200/225 hp followed the J-4 and was the engine that powered the *Spirit of St. Louis.* (Curtiss-Wright)

The Wright J-6 series Whirlwinds were offered in five, seven, and nine cylinder versions, rated at 165, 225, and 300 hp. Pictured is the J-6-9. The J-6 series entered the market late in 1928, and the J-6-7 replaced the J-5. (Curtiss-Wright)

The J-6-7 Whirlwind. Magnetos were moved to the rear of the case on the J-6 series. (Curtiss-Wright)

The Pratt & Whitney Wasp R-1340. At first called the Wasp A, it was rated at 400 to 425 hp at 1900 rpm depending upon carburetion and minor valve differences; the B was rated at 450 hp at 2100 rpm.

An R-1340 Wasp powered this Boeing P-12 Air Corps pursuit, producing 425 hp. This engine was rated at 600 hp by the time it was fitted to the North American AT-6 twelve years later.

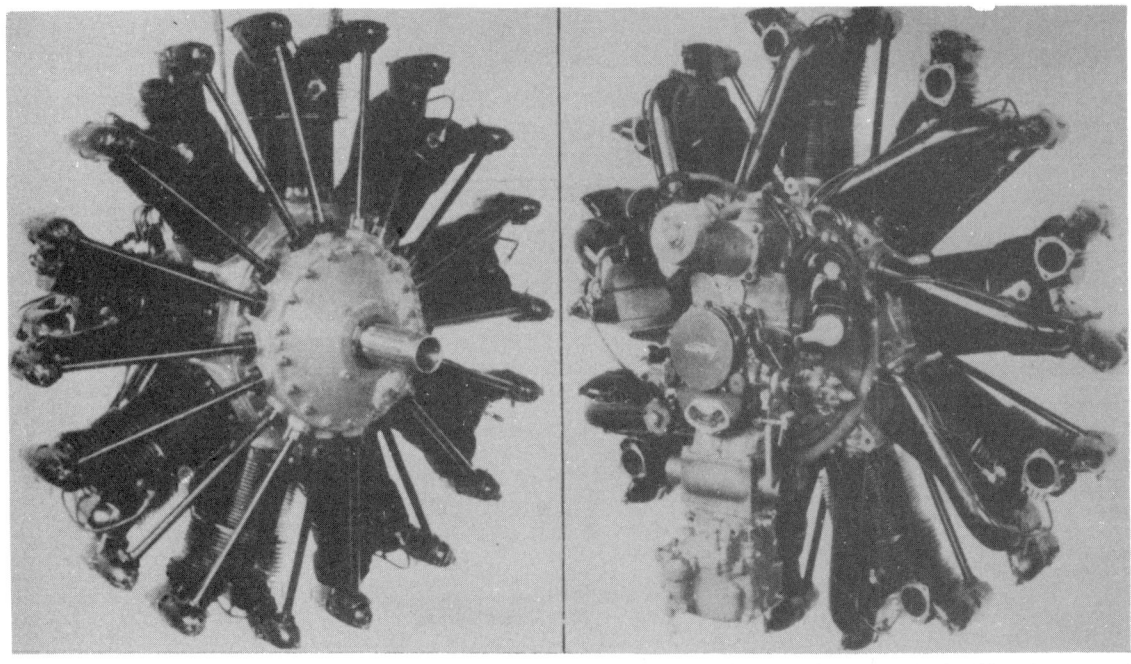

The Wright R-1750 Cyclone was developed concurrently with the J-6 series, and began life rated at 525 hp. The phenomenal success of both Wright and Pratt & Whitney radials was primarily owed to the work of Sam Heron and Ed Jones at McCook Field, who developed the sodium-cooled exhaust valve and the first truly reliable air-cooled cylinders. (Curtiss-Wright)

The P&W R-1830 Twin Wasp of 1050 hp entered service with the P-35 Seversky and P-36 Hawks in 1937. P&W engineers worked out the formula for heavy-duty engine bearings with this engine. (Pratt & Whitney)

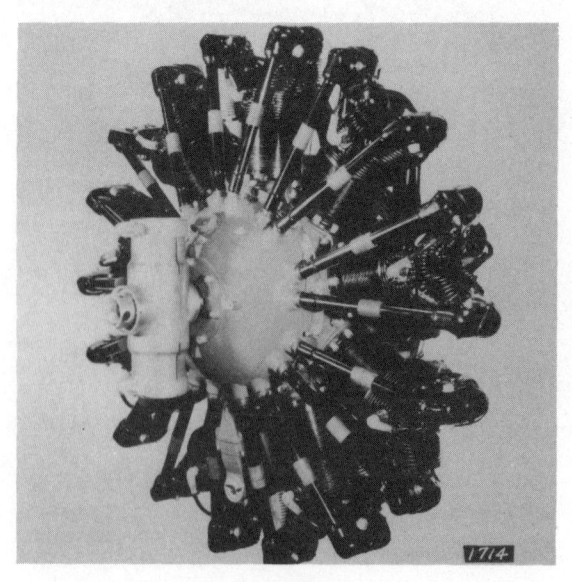

The P&W Wasp Junior, R-985, appeared in 1930; originally rated at 300 hp, it was almost immediately called upon for as much as 525 hp in the Thompson racers and was soon marketed at 450 hp. (Pratt & Whitney)

The Siemens-Halske as it appeared in a WACO 10, 1927. These engines, in five, seven, and nine-cylinder configurations, 65 through 125 hp, were built in Germany and distributed in the U.S. by Ryan.

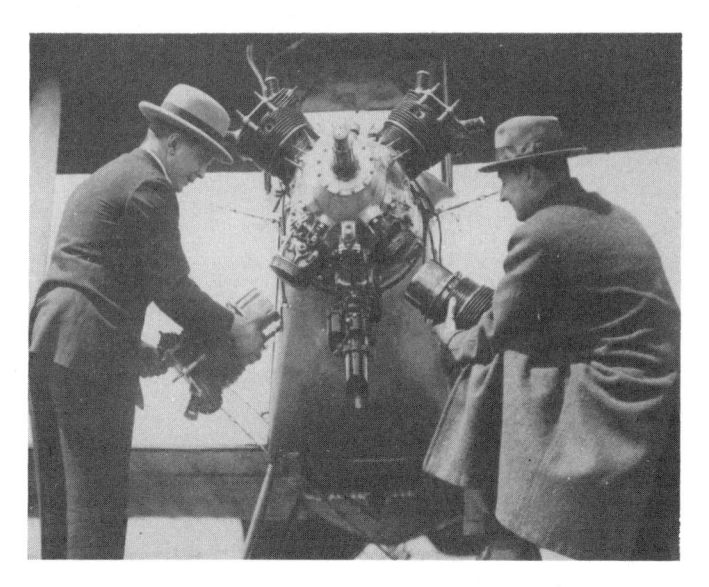

The Fairchild-Caminez of 135 hp was introduced in 1928 had no crankshaft; power was converted through a huge two-lobe cam. Inventor Harold Caminez (L) is seen with Sherman Fairchild in this publicity photo. (Fairchild)

The air-cooled Anzani radials, which first appeared in 1909, were offered in three, six, and ten-cylinders as twin-row engines. Pictured is a 1919 Farman fitted with an early single-row Anzani. Clyde Cessna used Anzanis in his first monoplanes in the late '20s. (Charles Meyers)

The A-65 Continental, rear view. If this engine were available today at a realistic price, it could do for this generation what it did for entry-level pilots of the late '30s. The A-65 powered the Cubs, T-Crafts, Luscombes, and Champs—plus a lot of homebuilts until the supply ran out. (Teledyne Continental)

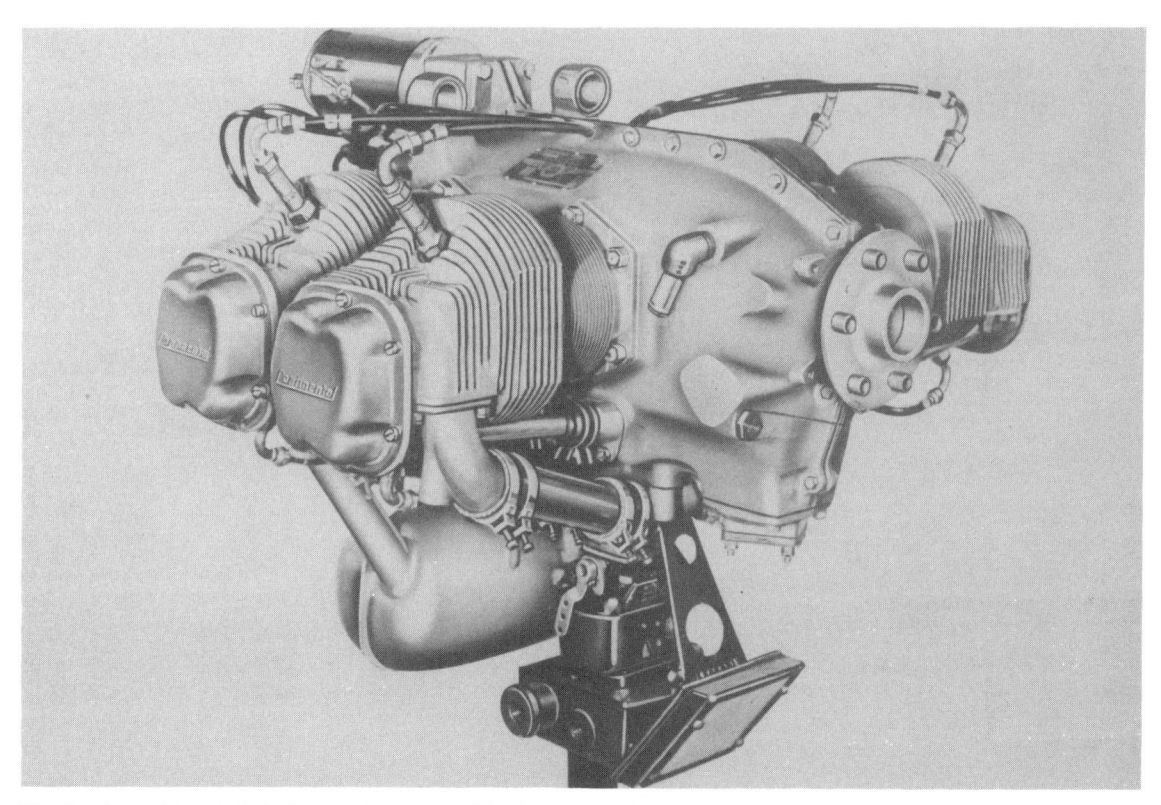

The Continental 0-200 of 100 hp was the engine of the Cessna 150 for many years; both it and the A-65 were designed for 80 octane fuel. Today's Lycoming 0-235, modified for low-lead 100, is the only low-horsepower production engine—108-117 hp. (Teledyne Continental)

The Curtiss H-1640 Chieftain is noteworthy because it was a twin-row with in-line cylinders. It was an unsucessful 1927 design rated at 600 hp. (Curtiss-Wright)

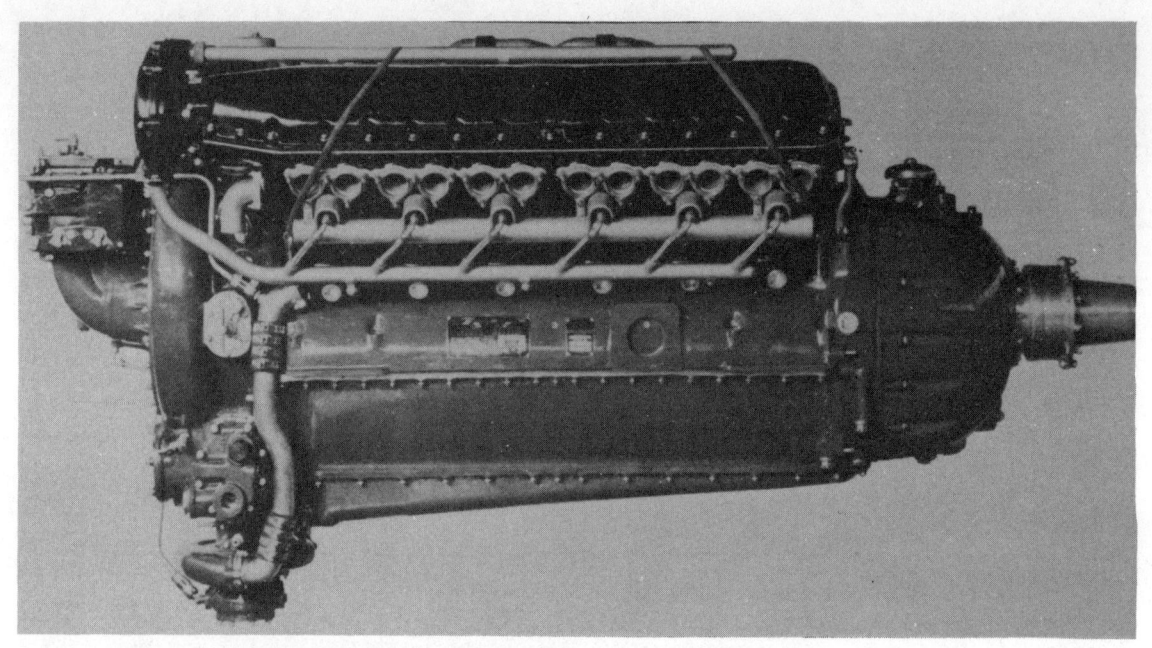

Originally designed in 1930, and at first developed on faith with company funds, the Allison V-1710 was an excellent engine, initially penalized in service by lagging supercharger development and imperfect intercooler and oil cooler systems. (Allison)

The F series V-1710s had raised thrust line, the −115 versions rated at 1475 hp. The last ones powered the F-82 Twin Mustang. (Allison)

The Packard-built Rolls-Royce Merlin V-1650-1 (Merlin 28), which powered the P-40F and P-40L, had a single-stage gear-driven supercharger and was therefore a medium-altitude engine. The two-stage, two-speed supercharger on the Merlin V-1650-3 and -7 (Merlin 60), which powered the P-51 Mustang, was a high-altitude engine.

A 130-hp Clerget rotary engine. On rotaries, the cylinders and propellor rotated around the stationary crankshaft. Average time between major overhaul was 15-20 hours. (Charles Meyers)

The 28-cylinder P&W R-4360 Wasp Major running in test. It entered production early in 1945 rated at 3500 hp. An improved version produced 4000 hp by 1950, when it went into the Douglas C-124 and Boeing C-97A. (Pratt & Whitney)

Chapter 4

The Airplane Goes to Work

Various aviation writers have characterized the 1926 Air Commerce Act as "U.S. Civil Aviation's Magna Carta." Others have pointed to the 1925 Kelly Bill as the true impetus behind the great boom in aviation that was touched off by Charles Lindbergh's solo nonstop New York-Paris flight in May 1927. A few have even recognized that it was Lindbergh's Wright J-5 engine—the first truly reliable aircraft engine—that convinced massive risk capital that the airplane had significant profit potential.

Lindbergh Boom

Actually, all of those factors contributed to the so-called "Lindbergh Boom," but the truth is that the airplane's role in U.S. commerce was inevitable, and had not these accelerating events occurred as they did, the airplane would nevertheless have established its value as a profit-making machine in another, perhaps more orderly (and less traumatic for many investors) fashion.

Once the airplane had evolved into a reasonably reliable, efficient, and predictably safe method of transport, it would establish its own unique place in the affairs of the nation. That's the way our capitalistic system works—unless our lawmakers erect enervating legal barriers or legislate artificial incentives that contravene the natural market forces. Three years before the Lindbergh Boom burst upon the struggling civil aircraft industry, the trend was discernable. In 1924, 60 new civilian airplanes found buyers; in 1925, 300, and in 1926 a total of 650 new planes were delivered.

Lindbergh flew to Paris in May 1927, and 1000 new civil airplanes were sold that year. In 1929, at the peak of the boom, more than 6000 civilian airplanes left the factories. Student pilot permits jumped from 575 in 1927 to 20,400 in 1929.

The Lindbergh flight was the first of a long series of ocean flights that continued until the United States and France placed restrictions on such activity. From the time of Lindbergh's hop until curbs were finally set up in 1933, no less than 56 Atlantic crossings were attempted by heavier-than-air craft. Of those, 22 were more or less successful; 21 went down short of their intended destinations without loss of life, and 13 airplanes, carrying a total of 27 people, disappeared. Ruth Elder and George Haldeman were among those who fell short but lived. They were fished from the Atlantic Ocean near the Azores.

These ocean flights, as well as a spate of endurance and long-distance flights, reflected the atmosphere of the times and constituted a phenomenon that is more easily explained by sociologists, perhaps, than by aviation writers. The country's mood during the heyday of the flapper (the hep chick of the late '20s) and the speakeasy seems unreal today, but it may have been a natural result of the sudden realization that the airplane had become something more than a daredevil's vehicle.

Within the industry were many who had known that all along— who believed that the airplane's great promise would be realized as airframes and engines became ever more efficient. It was only a matter of time. During the early '20s, they built for whatever market there was and looked to the future with confidence. Others followed as the market's upward trend became unmistakable in 1924-1925.

Eaglerock

Eaglerock Aircraft Company of Denver, Colorado, was formed by J.D. Alexander of the Alexander Film Company, makers of short motion picture commercials. The Eaglerock, a three-place open biplane originally powered with the OX-5, was designed by Dan Noonan and Al Mooney and priced at $2475. It sold well from the start, the first one being delivered in the fall of 1925.

Pitcairn

At about that same time, Harold Pitcairn of the Pittsburgh Plate Glass Company entered the field with a biplane designed by Agnew Larson. Called the Fleetwing, it resembled a five-place Curtiss Hawk and was fitted with the Curtiss C-6 engine of 160 hp, a 1919 water-cooled six-cylinder powerplant. The Arrow, Orowing, and Fleetwing II followed with a minimum of success until the Pitcairn PA-5 Mailwing, powered by the Wright J-5, appeared in 1927. The Mailwing series was operated on several of the early civilian-contracted airmail routes.

Pitcairn would later produce autogiros.

Buhl-Verville

The Buhl-Verville Airster also came into the market late in 1925, and it was still another three-place biplane fitted with an OX-5 engine, although it was unusual in that it featured folding wings. The later Airster, Whirlwind-powered, abandoned that innovation. Fred Verville and Lawrence Buhl went their separate ways after a short time, and Buhl Aircraft Company went on to make the sesqui-winged Airsedan and Bull Pup, while Fred Verville, who had built his first airplane in 1915, later emerged to produce the Verville Air Coach, which resembled a trim Travel Air 6000.

Fokker

Tony Fokker also began new airplane manufacturing in the U.S. during 1925. Actually, Fokker had put his foot in the door early in 1921 with establishment of the Netherlands Aircraft Manufacturing Company at Amsterdam, N.Y., but that was merely a sales office and information bureau run by Bob Noorduyn in a search for U.S. sales of Fokker airplanes being built in the Netherlands. In 1924, as the Atlantic Aircraft Manufacturing Company, a contract was obtained to convert the fuselages of 100 Army Air Service DH-4s from wood to steel tubing, and that led to establishment of the Fokker factory in America at Hasbrouck Heights, N.J. Anthony Fokker returned to the U.S. (after three previous visits) early in 1925 to make a personal assessment of the American market. In 1922, Noorduyn had sold eleven Dutch built Fokker F.IV and D.IXs (PW-5, -6, and -7) to the U.S. Army Air Service, along with a pair of Fokker T-2s, and although no Fokkers had been sold in the U.S. civilian market, that possibility appeared imminent.

Tony Fokker settled on a tri-motor design that would employ the J-4 Whirlwind engines (the J-5 was not yet in production), and sent appropriate instructions back to his chief engineer,

Reinhold Pfalz. The resulting airplane—the Fokker F.VIIa-3m, based on the single-engine F.VIIa— would be the prototype of a series of famous Fokker Trimotors. It was first flown in the U.S. late in 1925, and the company name was changed to Fokker Aircraft Corporation as the new trimotor and other Fokker designs were readied for production at Hasbrouck Heights. Some of Fokker's key craftsmen were brought to America to build the Fokker's thick, all-wood cantilever wing.

Anthony Herman Gerald Fokker was born in 1890 in Java, Dutch East Indies, where his father was a prosperous planter. He grew up in the Netherlands (he was never a German citizen), learned to fly, and designed his first airplane in Gonsenheim, Germany, in 1911. In 1913, the German Government provided a factory for Fokker at Schwerin in Mecklenburg, and his fighter airplanes were the best possessed by Germany during WWI. At the end of the war, Fokker escaped to his native Holland with some difficulty, but soon had a factory there.

By 1927 the American Fokker company had factories at Passaic, N.J., and Glendale, West Virginia, in addition to the Hasbrouck Heights facility. At the height of the Lindbergh Boom—May 1929—General Motors acquired 40 percent of the Fokker Aircraft stock and a year later the company was renamed the General Aircraft Corporation with Charles F. Kettering and friends on its board of directors when it became a subsidiary of North American. A year after that, Tony Fokker (who held 20 percent of the stock), left the company and, in 1934, it became the manufacturing division of North American Aviation under the direction of James Howard "Dutch" Kindelberger. A number of the original Fokker craftsmen, brought to America from Holland, went to California and remained with North American for years. Tony Fokker died in 1939.

Ford-Stout

Fokker's chief competitor for the transport airplane market was Henry Ford. Ford, and son Edsel, had an early interest in aviation, and became directly involved as the result of an unlikely proposal from William B. Stout. Stout was an extrovert who started his aviation career in Chicago and was one of the charter members of the Illinois Aero Club in 1910. His Stout Metal Plane Company was organized in 1922 and financed by the straightforward expedient of writing letters to some Detroit industrialists, asking each for $1000—and warning that they should not expect to see the money again. Twenty-five sent checks, including the Fords, and Stout produced the Air Sedan, an OX-5 powered all-metal craft developed from his earlier Batwing. With engineering help from George Prudden, the Air Sedan evolved into the Air Pullman, later called the Model 2-AT, and the 2-AT in turn fathered the first Ford Trimotor, 4-AT, following an embarrassment known as the 3-AT.

Henry Ford purchased the Stout Metal Plane Company in July 1925, almost a year before the 4-AT Trimotor made its maiden flight, and meanwhile began a private airline flying the single-engine (Liberty powered) 2-ATs between Detroit and Cleveland, and Detroit and Chicago, carrying automobile parts and company executives. "Shorty" (he was six foot four) Schroeder was Ford's chief pilot.

The Ford-Stouts were all-metal with full cantilever wings, a concept pioneered by the German Junkers company during WWI, but one that American designers had largely ignored (the Thomas Morse company was building experimental single-engine military aircraft of all-metal construction in 1922), and which was considered a daring innovation in America in the mid-'20s.

A total of 199 Ford Trimotors was built, following the eleven single-engine 2-ATs. The 12-passenger 4-ATs were normally fitted with Wright J-5 engines, while the 14-passenger 5-ATs were powered with the R-1340 Wasp. Inevitably, other engines and combinations of engines ended up in many Ford Trimotors, the 300-hp Wright being a popular option. The 4-AT sold new for $42,000, and the 5-AT for $65,000. The "Tin Goose," as all Ford Trimotors came to be called, was produced from 1926 to 1932.

During this period, Ford aircraft engineers (including George Prudden, Otto Koppen, John

Lee, and Tom Towle) produced six or eight experimental aircraft, including a 70-passenger airplane that never flew, and a pair of single-place 36-hp "Flivver" planes that flew well but were not produced.

Henry Ford was said to have lost as much as $3 million manufacturing the Tin Goose, but it was a sturdy, honest airplane, its only fault being that it was not economically efficient—its seat-per-mile operating costs were too high, a fault shared by all other airliners until the Douglas DC series appeared in 1934.

Stinson

Stinson would produce some popular low-cost trimotor airliners, beginning in 1930 with the 10-place SM-6000, priced at $18,000 and fitted with three 215-hp Lycoming R-680 engines.

Eddie Stinson had started the company that bore his name late in 1925 when he built the first of his famed "Detroiters" in a Congress Street loft in the Motor City. It was a four-place cabin biplane with wheel brakes, cabin heat, and an electric starter. It was fitted with a J-4 Whirlwind engine and made its first flight at Selfridge Field, Michigan, on 25 January 1926. Designated the SB-1, this craft's performance sufficiently impressed Eddie's potential backers, a small group of Detroit industrialists, that an empty factory was obtained in nearby Northville, and the Stinson Aircraft Corporation organized. Production of the biplane Detroiter began in August 1926, and the first Stinson monoplane, the SM-1 Detroiter, followed eleven months later. The SM-1 was six-place, powered by the Wright J-5 Whirlwind, and was priced at $12,000. These craft and follow-on Stinsons sold well throughout the '30s.

Many have assumed that Eddie's sisters taught him to fly, but Marjorie Stinson has stated that Eddie's mother (Emma) paid for his instruction at the Wright school in 1915. Eddie served as an instructor at Kelly Field, Texas, during WWI, and barnstormed in the early '20s. He also did some charter flying with a WWI all-metal Junkers that he bought at a bargain from the Post Office Department. Neither Katherine nor Marjorie Stinson resumed flying after WWI. Katherine married and moved to Santa Fe, New Mexico, where she lived quietly, almost anonymously. Marjorie did drafting work for the Navy until her retirement in Washington, D.C.

Stinson Aircraft survived, and even grew, during the Great Depression, perhaps partly because automobile magnate E.L. Cord bought out Eddie's original backers early in 1929. Cord controlled Lycoming and, by 1932, American Airways by way of his holdings in AVCO. Thus Stinson had at least one captive market: American Airways flew Stinson airliners, fitted with Lycoming engines.

Travel Air Spawns the Cessna and Stearman Companies

Meanwhile, down in Wichita, the Travel Air company flourished. By late summer of 1926, Travel Air was well into production of the 46 airplanes it would sell that year, and Walter Beech had won air races at Wichita and Flint, Michigan, then capped the season by taking first place in the second annual Ford Reliability Tour with a Model 4000 biplane. Travel Air dealer Fred Hoyt won the "On to Sesqui" race, a feature of the National Air Races, by flying a Travel Air between Eureka, California, and Philadelphia in 31 hours flying time. The company acquired a plant site and airfield on East Central Avenue (site of today's Beechcraft facilities in Wichita), and during 1929 would produce 1000 biplanes and cabin monoplanes with a work force of 650, although both Cessna and Stearman had long since departed Travel Air by then.

Lloyd Stearman pulled out a few days after the National Air Races in 1926 to join with Travel Air dealer Fred Hoyt and engineer Mac Short for the purpose of establishing the Stearman Aircraft Company in Venice, California. The first Stearman, the C-1 (another entry into the OX-5, three-place open biplane category), first flew from Santa Monica's Clover Field early in 1927. Several of Lloyd's first production aircraft went to Walter Varney, who operated one of the first civilian-contracted airmail routes.

Stearman's original company was underfinanced and suffering cash flow problems, despite a good product, so Walter Innes, Jr. raised fresh capital in Wichita and returned Lloyd to the "Air Capital" on the Plains. The new Stearman company began operations in November 1927 in an empty factory building formerly occupied by the Bridgeport Machine Works—the same facility, by the way, used by the Jones Motor Company years earlier and in which Clyde Cessna constructed the first airplane ever built in Wichita. Stearman's first production craft at Wichita was his C-3, a very successful design. Others would follow, including the famed PT-17 series of WWII.

Stearman, however, would sell out to United Aircraft in 1929 (he would soon participate in the reorganization of Lockheed and remain with Lockheed until his retirement). United would be Boeing's parent corporation by that time, and the Stearman company in Wichita would eventually end up as Boeing's Wichita Division. So the PT-17s, so fondly remembered as "Stearmans," were actually Boeings. Lloyd Stearman had been gone for years by the time the PT-17 was designed at Boeing's Stearman Division.

While Stearman was still in California building the C-1—in April 1927, to be exact—Clyde Cessna and Walter Beech split up at Travel Air, apparently as the result of a dispute over the kind of monoplanes Travel Air should build. (At least, that is the accepted, time-worn story around Wichita, and we could find no evidence to the contrary.) The Model 5000 Travel Air, then in the market, and the Model 6000, on the drawing boards, were sturdy, deep-bodied machines with external wing bracing. But Clyde's ideas, soon to show up in his clean, cantilever-winged monoplanes, were basically opposed to the design concepts reflected in the Travel Airs.

Some accounts hold that the Travel Air 5000, the first civilian airplane to fly to Hawaii, was in truth an airplane designed and built by Clyde Cessna outside the Travel Air factory and on his own time, and which was modified at Beech's order in an effort to sell it to an air mail contractor. But that seems highly unlikely because that Cessna airplane had delicate lines very similar to the 1924 Bellanca CF, and in no way resembled the robust 5000.

Clyde appears to have financed the first Cessna commercial airplane himself. It was built during the summer of 1927 and again, this machine, the Cessna Model A, very much resembled the earlier Bellanca CF. On 22 August, the company became known as the Cessna-Roos Aircraft Company when Victor Roos joined Clyde. Roos had been associated with Guiseppe Bellanca in the short-lived Bellanca-Roos company of Omaha. But Roos' tenure with Cessna was brief; he sold out his interest to Clyde, George Siedhoff, and M.T. Hargiss at the end of the year to become general manager at Swallow. Cessna, Siedhoff, and Hargiss reorganized as the Cessna Aircraft Company, working from a modest plant at 200 East Douglas in Wichita.

The Model A Cessna, originally called "Type 3-120," was at first powered with a 120-hp Anzani engine. It was four-place, had a top speed of 120 mph, and sold for $3575 less powerplant. The French Anzanis were generally considered cantankerous beasts with excessive vibration by American pilots, but Clyde had flown them for years and had worked out modifications that made the Anzani Cessnas reasonably reliable.

Nevertheless, aware of the Anzani's reputation, Clyde also offered his new ship with a choice of engines. The Warner-powered Cessna AW, and the Whirlwind-Cessna, the latter certificated in 1928 as the Model BW, became the most popular. The BW, with 225 hp, had a top speed of 150 mph, and landed at 45 mph; its initial climb rate was 1400 fpm.

In 1929, Cessna purchased 80 acres at Cessna's present main plant site and constructed five buildings. Clyde, assisted by his son Eldon, designed the Cessna DC-6; it entered production that year. The DC-6 series were larger and heavier than the A Models, and powered with 250-hp and 330-hp Whirlwinds. By mid-summer, Cessna had 220 employees and was producing one airplane per day, and it was at that point that the company was refinanced by a group of Boston Bankers, and Clyde

79

announced that he would build 545 airplanes during the last half of 1929.

That program was never realized. The props were swept from beneath the nation's economy in October and the booming civil aviation market disappeared almost overnight. In an attempt to keep the plant open, Clyde made and sold about 300 primary gliders priced at $395 each. Eldon experimented with an ultralight airplane, and the company received an order for a small racing plane from a group of Blackwell, Oklahoma, businessmen who had engaged Stan Stanton to fly it in the 1930 cross-country Cirrus Derby. This craft, the mid-wing GC-1, was the first of several pure racers built by Cessna during the 1930-1933 period. Part of that time, the factory doors were closed and the plant idle.

But for an event that had taken place nine years earlier, one can only guess at what would have happened to the company. Clyde was nearing retirement age, and other airframe makers were being liquidated almost daily as the economic disaster deepened. But in 1923 Clyde had given an airplane ride to his 12-year-old nephew, Dwane Wallace, and that had determined young Wallace to pursue a career in aviation.

Wallace graduated from Wichita State University in 1933 with a degree in aeronautical engineering. He worked briefly for Walter Beech, then pursuaded Cessna shareholders to reopen the plant. Dwane had plans for a new five-place Cessna that he called the "Airmaster," and the C-34 Airmaster would keep the company afloat, with average sales of 30 units per year, until the demands of an approaching war in Europe signaled an abrupt change in the Cessna company's fortunes.

Clyde would retire to his farm in 1936 (he died in 1954), and Dwane Wallace would remain at the company throttle until forced into retirement in 1983.

Beechcraft Formed
T.A. Mystery

Event No. 26 at the 1929 National Air Races received an unusual amount of attention because of the airplane that won it, the Travel Air Model R. Had it not been for that entry, Event 26 would have been little different from the many free-for-all air races that preceded it, and the Thompson Trophy classic that it spawned may never have resulted.

The Travel Air entry focused wide attention on the race well in advance. Walter Beech, a born public relations man, built the Model R in secret, and when the craft was flown to Cleveland prior to the race, he ordered it whisked into a hangar and covered with a canvas — although not until he was sure that the newspaper photographers had obtained some good pictures. Inevitably, Beech's actions gained the name "Mystery Ship" for his trim, low-wing monoplane, and that undoubtedly made Walter very happy.

He fostered the idea that his little plane — less expensive and less powerful than most of the other entries — was sort of a sheep among wolves, and that gave the crowds an "underdog" to favor. Some underdog — it was at least 30 mph faster than its closest racing rival. All of which takes nothing from Walter or his airplane (or its designers: Herb Rawdon, Ted Wells, and Walt Burnham). Walter knew he had a winner and meant to get as much return as possible from it. But the golden age of air racing is viewed in a subsequent section.

A month earlier, Beech also got a pretty good return from his investment in the Travel Air company. In mid-August, the Curtiss-Wright Corporation offered to trade 175,000 shares of C-W common stock for the 100,000 shares of Travel Air stock then outstanding. Walter, with Travel Air shareholders concurring, accepted. In the merger, Beech became president of the C-W Travel Air Division, the twelfth unit of the growing C-W complex, as well as president of the C-W Sales Corporation.

However, by 1932 few civil airplanes of any kind were finding buyers, and that included the once popular Travel Airs. C-W had closed the Travel Air plant in Wichita and the relative handful of Travel Airs that could be sold were made in C-W's Curtiss Robertson plant at St. Louis where the Curtiss Robin monoplane was built.

Predictably, Walter Beech was unhappy selling someone else's machines. At C-W he spent much time promoting such planes as the Curtiss Kingbird, a twin-engine cabin monoplane designed for the fledgling airlines, and the Curtiss Teal, a small amphibian. At best, some of the C-W offerings under the Travel Air name were poor relatives of the amply-proportioned airplanes that had established the Travel Air reputation. The C-W Travel Air 12Q, for example, was an anemic, spindley-legged biplane that bore no resemblance whatever to its Wichita predecessors.

Walter stuck it out until early in 1932. Then he resigned, went back to Wichita and, in concert with half a dozen local businessmen, organized Beech Aircraft Corporation in April of that year. For a while, Beech worked in rented space in the idle Cessna factory, then leased the Travel Air facility from C-W (Beech bought the property in 1937 for $150,000) and, near the low point of the worst financial depression this country has experienced, boldly sent Ted Wells and Herb Rawdon to their drawing boards with instructions as to the kind of airplane he wanted—reportedly, " . . . a reverse stagger cabin biplane that will carry five people and go like Hell."

He got it—the classic Beechcraft Staggerwing, Model 17. And Beech Aircraft has been going like Hell ever since.

Great Lakes

Ten months before Beech sold Travel Air—in October 1928, to be exact—WACO's sales manager, Charles van Sicklen, resigned from the Advance Airplane Company to help form Great Lakes Aircraft Corporation in Cleveland. He invited Charlie Meyers to go along as test pilot and design engineer.

The founders of Great Lakes, learning that Glenn Martin wanted to move to Baltimore, purchased the Martin plant in Cleveland, including the balance of a Martin contract with the Navy for a number of torpedo planes. Col. Ben F. Castle was president of the new company, van Sicklen vice president in charge of sales; former Navy Capt. H.C. Richardson was another vice president, while P.B. Rogers, Earl Stewart, and Cliff Leisy were project engineers.

While much of the GLAC work force concentrated on the torpedo bomber production (Martin T4M-1—the TG-1 as built by GLAC) and the conversion of two TG-1s into eight-place civil transports, Meyers and Leisy designed a two-place biplane fitted with the 95-hp Cirrus engine. This was designated the 2T-1 Sport-Trainer.

The 2T-1 prototype was ready for test in March 1929, and Meyers discovered that it was markedly tail-heavy. The easiest and cheapest way to solve that was to give the upper wings nine degrees of sweepback, and that expedient resulted in another classic airplane. About 500 were built, initially priced at $4990, later lowered to $3985. Production ended in 1933 following the failure of the Great Lakes amphibian, a twin-engine design.

Early Piper and Taylorcraft

Meanwhile, almost unnoticed, William T. Piper, Sr. entered the airplane manufacturing business. He sort of slipped in with the Great Depression, and with an investment that wouldn't have started a decent neighborhood grocery. Piper, a 48-year-old Bradford, Pennsylvania, oil man, and his partner, Ralph Lloyd, each invested $800 in the Taylor Brothers Aircraft Corporation when that company moved from Rochester, New York, to Bradford in search of fresh capital in 1929.

The Taylor company was headed by C.G. Taylor, a self-taught aeronautical engineer who, along with his late brother, Gordon, had developed the Taylor Chummy two years before. The Chummy was a two-place, side-by-side parasol monoplane of 90 hp, and proved as unattractive to buyers at Bradford as it had in Rochester. At Piper's suggestion, Taylor abandoned the Chummy in favor of a lighter, simpler design that he called the E-2. The E-2, with minor modifications, would eventually become the J-3 Cub, but in 1930 it was a failure because it lacked a suitable powerplant. The E-2's "Kitten" engine, made by the Brownbach company of Pottstown, Pennsylvania, contributed only the inspiration for calling the E-2 the "Taylor Cub." Its 20 hp was insufficient to even prove the Cub's concept.

The company was declared bankrupt early in

1931, whereupon Piper bought its assets—including the prototype Cub—for less than $1000. He retained Taylor as engineer of the new Taylor Aircraft Company. Then, Continental Motors of Detroit brought out a 37-hp opposed air-cooled engine called the A-40, and when the A-40 was fitted to the Cub airframe, a reasonably dependable, simple, inexpensive, and fun-to-fly lightplane resulted. The 40-hp E-2 Cub was certified by the Bureau of Air Commerce in November 1931 and 24 E-2s were sold in 1932 priced at $1325. The Cub was truly a child of the Great Depression.

Sales slowly improved through 1935, by which time a total of 351 Cubs had been sold—some, beginning in 1935, with optional 40-hp Aeromarine or Szekely engines.

Gilbert Taylor, obviously aware that the Cub concept could encompass more than one design, pulled out of the company in 1935, hied himself to Alliance, Ohio, and founded the Taylor-Young Airplane Company. His Taylorcrafts reverted to Taylor's original two-place side-by-side seating arrangement, and were quite successful.

The first T-Craft was the 40-hp B-12 of 1937. It received the 65-hp Continental engine in 1940—as did the Cub—and 1949 were purchased during WWII by the USAAF as the 0-57 and L-2. Production for the civil market resumed in 1946 as the BC-12D, and a total of 3000 were built before the company folded in 1947.

With the same horsepower, the T-Crafts were faster than the gentle, pokey J-3s due to the use of a more efficient airfoil shape chosen for its wings.

In 1974, a new company in Alliance announced production of the F-19 Taylorcraft Sportsman, a re-engineered version of the BC-12D fitted with the 100-hp Continental 0-200 engine.

In 1936, after Taylor's departure, Piper slicked up the E-2 with rounded wingtips and wider landing gear tread to get the Piper J-2, priced at $1470. Then, following the loss of the Bradford factory to fire in 1937, Piper moved to Lock Haven, Pennsylvania, and changed the company name to Piper Aircraft Corporation.

The J-3 Cub appeared in 1938 when the vertical tail was reshaped to its now-familiar form and wheel brakes were added, but only a few 40-hp J-3s were sold before Continental, Lycoming, and Franklin introduced 50-hp engines, all of which were available on the J-3. That upped the price to $1617, and 736 were sold that year.

Actually, about 100 of those were J-4s, a plushed-up side-by-side version known as the Cub Coupe. The Coupe was priced as $1995 and 1250 were sold 1938-1942 inclusive.

The 50-hp 1938 J-3 Cubs were the first to standardize on the now famous "Cub Yellow" paint scheme, and this finish, like Henry Ford's "Model T black," remained with the Cub until the last J-3 was produced in 1947. Altogether, 14,125 were built.

In 1940, the Cub was certificated with 65-hp Continentals and Lycomings, and production jumped to more than 3000 units per year as the Civilian Pilot Training Program (CPTP)—later, the War Training Service—was begun in colleges across the country to give flight training to thousands of young Americans on the eve of WWII. When that program ended in mid-1944, 1132 educational institutions had trained more than 430,000 primary flight students.

During WWII the Army procured 5671 Cubs, which were designated 0-59s and L-4s. These machines are included in the 14,125 listed above.

Monocoupe—Luscombe

The Monocoupe and the Taylor Chummy originally appeared at approximately the same time in 1927. Both were two-place (with side-by-side seating) 60-hp, high-wing monoplanes, and both were intended for the low-cost personal flying market. They shared a design concept with the 55-hp Star Cavalier and St. Louis Cardinal, and the Monocoupe was no more successful than the others until it received a proper engine—and until the market for such craft developed.

Designed by Don Luscombe and Clayton Folkerts, acting as the Central States Aero Company in Davenport, Iowa, the Monocoupe was first tried with a six-cylinder Anzani engine, and then the five-cylinder Air Cat of 65 hp built in Detroit by Glenn Angle. In mid-1928, Central States Aero was reorganized as Central States Aircraft Company and then, with new capital and a move to Moline, Illinois, later in 1928, it became the Mono Aircraft Company. The Monocoupe received a five-cylinder Seimens-Halske engine and an 85-hp Cirrus in turn. Not until the 60-hp Velie was fitted late in 1928 was the proper combination of engine and airframe achieved, and from that time the Monocoupe began to sell.

Also at that time, former car manufacturer W.L. Velie took over the company as president (presumably with an appropriate cash investment), and Mono Aircraft became a subsidiary of Allied Aviation Industries. Luscombe remained as a vice president, and Folkerts became chief of design. Known as the Model 113, this first successful Monocoupe was produced in 1930 and sold for $2675.

The Monoprep, introduced in 1929, was a popular open-cockpit version of the Monocoupe sold primarily as a trainer. A less successful offering was the Folkerts-designed Monocoach, a four-placer powered by the J-5 Whirlwind. The Monosport was essentially a Model 113 cleaned up aerodynamically and given a 110-hp Warner Scarab engine. The Monocoupe 90 was announced in 1930 with a 90-hp Lambert engine.

Apparently, W.L. Velie lost interest in the company when sales of the Velie-powered Monocoupe ended. In any case, Velie left, Monocoupe moved to Lambert Field, St. Louis, late in 1931, and Luscombe effected another reorganization, emerging as president of the new Monocoupe Corporation. In 1935, Luscombe resigned and went to Kansas City to begin work on the Luscombe airplanes, and shortly thereafter Al Mooney became chief engineer at Monocoupe. Mooney designed the low-wing Monocoupes; these were not produced but sold to the Culver Aircraft Company of Columbus, Ohio, which produced the Mooney design as the Dart. Mooney went with Culver in 1938, where he designed the Cadet. Monocoupe was dissolved in 1940 and its assets purchased by the Universal Moulded Products Corporation of Orlando, Florida. Al Mooney, of course, would be heard from again after WWII.

Aeronca

Aeronca is one of the truly historic names in civil aviation. As this is written, the company is not building an Aeronca airplane, and has not since the Aeronca Champion and Sedan of the late '40s, but produces subassemblies for others. The original Aeronca company was formed in 1928 in Cincinnati by Taylor Stanley and Conrad Dietz. Production of the Aeronca C-2 began in mid-1930, but the prototype airplane had been designed in 1925 as the Roche-Dohse "Flivver" plane, a minimum airplane that would provide the cheapest possible fun flying.

Jean A. Roche was a gifted engineer and, according to Donovan Berlin (the P-36/P-40 designer who had worked with Roche at McCook Field during the early '20s), Roche was a "character," given to nude swimming with a circle of equally daring friends, along with other interesting hobbies. Roche's original design was fitted with a two-cylinder Morehouse engine, but this very light airplane was well ahead of its market. Roche slicked it up a bit in 1927 as the Lindbergh Boom gathered steam, but the concept was not attuned to the heady financial climate of that time. The WACOs, Travel Airs, and Stinsons—the conventional-size private airplanes—were the machines then in demand.

Nevertheless, the above-mentioned principals founded the Aeronautical Corporation of America (hence, "Aeronca") at Cincinnati's Lunken Airport the following year and maintained their faith in the homely little aircraft until the Great Depression created a market for it. As initially produced, the Aeronca C-2 was a single-placer of 400 pounds (empty) powered by a 30-hp Aeronca engine. Its price at the factory was $1555, later reduced to $1245. It cruised at 65 mph on two gallons of fuel per hour, and landed at 35 mph.

In March 1931, the two-place, side-by-side Aeronca C-3 was introduced and an improved version of its engine produced first 36 hp, and later 40 hp. The C-3 trainer version sold for $1695, and a number of flight school operators bought the C-3 and offered solo student time as low as $4 per hour.

Subsequent Aeroncas included the Model K and the Chief, side-by-side two-place cabin monoplanes, and eventually the two-place tandem-seat Champion, a version of which served the USAAF during WWII as the O-58 and L-3, powered with the 65-hp Continental engine.

The Aeroncas were trainers and personal, fun airplanes, as were most of the others listed above with less than 200 horsepower, but many of the three-place biplanes were put to work by the civilian contractors for the air mail during the late '20s. Others were used in aerial advertising (Baby Ruth candy bars, dangling from tiny paper parachutes, were rained on American cities); skywriting—and even loudspeaker systems mounted in airplanes—promoted various products. Many airport operators offered charter and air taxi service, while several of the major oil companies acquired small fleets of airplanes in the early '30s. The U.S. Forest Service had begun employing Army DH-4s for fire patrol in the Pacific Northwest as early as 1921.

The First Dusters

The first aerial crop dusting experiments in the U.S. also took place in 1921 when an Army Air Service Hisso-powered Jenny was used to dust a catalpa grove near Dayton, Ohio, with lead arsenate to combat an infestation of caterpillars. Under the direction of C.R. Neillie and J.S. Houser of the Agriculture Department, a hopper was rigged outside the rear cockpit on the right-hand side of the airplane. The airplane was flown from the front cockpit, and the hopper operator in the rear turned a crank by hand to distribute the pesticide. The Agriculture Department described the experiment as "moderately successful."

Concurrent with the Dayton experiment, Dr. B.R. Coad of the Agriculture Department, who had been attacking the cotton boll weevil in the Southern States with calcium arsenate using ground-mounted equipment, followed up on the Dayton operation by obtaining a pair of Air Service Jennies and pilots to try aerial dusting.

Dr. Coad's hoppers were installed inside the Jennies' rear cockpits and operators were required to crank the mechanisms, which discharged the dust through the bottoms of the fuselages. At first, the airstream interrupted the flow of the dust, but a venturi solved that, and when an agitator was added inside the hopper on each plane, some basic principles were established for dispensing the chemicals that are still in use today.

One result of Dr. Coad's research was the formation of a privately owned aerial dusting company in 1923. The Huff-Daland Company of Ogdensburg, New York, builders of bombers and ing airplanes for the Air Service, designed a biplane expressly for the purpose of dusting crops, and when it proved the concept, constructed a fleet of 14 such machines for a subsidiary company, Huff-Daland Dusters, Incorporated at Monroe, Louisiana, in 1925. The Huff-Daland Petrels (commonly called "Puffers" by their crews), were Hisso-powered, had hopper capacities of 800 pounds, and cruised at 80-90 mph.

C.E. Woolman's Delta Air Service grew from the Huff-Daland Dusters after the parent Huff-Daland Company was absorbed by Keystone Aircraft Manufacturing Company of Bristol, Pennsylvania, in March 1927 (Keystone was in turn merged into the Curtiss-Wright Corporation two years later), and Delta Air Corporation (Delta Air Lines in 1934) began airline operations in June 1929 between Birmingham and Dallas.

Daland Dusters, meanwhile, had accomplished a great deal of valuable work. In 1922 about 2000 acres were effectively treated. In 1923, approximately 7000 acres were covered; in 1924, almost 20,000 acres. This was up to 50,000 acres in 1925, and in 1926 100,000 acres were dusted by air.

Aerial dusting was well established by 1930, with more than 25 companies operating almost 100 airplanes. Paris green (organic lime and sulphur) in dust compounds was the most commonly used chemical, along with arsenates, until DDT appeared in the early '40s.

After WWII, when PT-17s, N3Ns, and BT-13s

were available in the surplus market for as little as $200, aerial dusting companies proliferated, including some shoestring operators who were less than responsible. But the nature of the business, including a generally tough-minded clientele, eliminated the inefficient and the get-rich-quick types; the low-cost airplanes disappeared quickly—and in any case were not low-cost in maintenance—while the individual states increasingly regulated the industry. These factors, combined with improved chemical compounds and continuing research, made aerial application a highly specialized and very demanding business.

Piper led the field in offering new replacement equipment for the converted war surplus airplanes. Not counting the PA-18 Super Cub (which appeared in 1949 available with spray booms or dust hoppers), the PA-25 Piper Pawnee, specially designed for agricultural flying, entered the market in 1959, and its success encouraged other airframe builders to follow with ag-plane designs that continue to become ever more efficient.

From the pilot's point of view, agricultural flying in recent years pays well, but the hours are long and the precision flying required fatiguing.

Odd Jobs

During the early '20s the airplane was put to work by enterprising pilots in ways that were especially suited to its unique capabilities. Payrolls were flown to remote mining sites in Mexico to confound the *bandidos*. Whiskey was flown into legally-dry America across both northern and southern borders; the motion picture newsreel companies covered (and manufactured) newsworthy events filmed from airplanes. The first cloud-seeding experiments were carried out in 1924 under the direction of Dr. E. Francis Warren when an Air Service Pilot sprinkled electrified sand from a DH-4 into an area covered with fog. And, of course, the infant motion picture industry discovered the airplane very early and they matured together as Hollywood, in many hundreds of screenplays, depended upon the flying machine and the portrayal of flying people to provide the action and drama that spelled box office success. In 1915, Glenn Martin was paid $700 per day for his performance with his airplane in the Mary Pickford movie *The Girl From Yesterday*.

Technological Advances of the '20s

Significant technological advances in aviation were made during the '20s, and some, begun a decade before, came to fruition. The Army's aviation research and development center, McCook Field, on the outskirts of Dayton, Ohio, was easily the most important place in American aviation during the first half of the '20s. Almost all new aerial hardware was either developed or tested there. The first useful handbook on aerodynamics was written there by Lt. Charles N. Monteith (years later, Boeing's top engineer). Alfred Niles and Joe Newell wrote the first structural design handbook. In the Powerplant Section, the moody, brilliant Englishman, Samuel D. Heron, along with Lt. Edward T. Jones, designed the first truly practical cylinder for air-cooled aircraft engines. Heron and Jones worked under Maj. George E.A. Hallett, and Hallett, too, was something of an engine genius, having started as a mechanic for Glenn Curtiss in 1911.

Floyd Smith's manually-operated free-fall parachute was perfected at McCook under the direction of Maj. E.L. Hoffman.

Col. Carl Green, head of the Structures Section, was the first to describe the phenomenon of aileron flutter.

In 1921, Lts. Donald L. Bruner and Harold R. Harris at McCook, along with civilian engineer Carl F. Egge, designed and operated an experimental lighted airway between McCook Field and Columbus, Ohio, a distance of 80 miles; by mid-1924, most of the transcontinental airmail route between New York and San Francisco was lighted.

The first successful air-to-ground wireless test was made by the Army at Ft. Riley, Kansas, in 1912. The pilot was Henry H. "Hap" Arnold. The use of radio for air navigation originally began as an experiment by the Germans during WWI in an effort to provide a directional signal to guide their giant rigid dirigibles in bombing raids over London. Learning of that attempt, the U.S. Bureau of Standards was enlisted to work with Capt. W. H. Murphy at McCook Field to perfect a directional radio beam for aircraft guidance. By 1922, two wire triangles—closed-loop antennae—with bases 300 feet long and 72 feet high, positioned at right angles to one another, allowed the broadcast of a signal in any chosen direction. From that beginning, the first network of low-frequency, ground-based radio navigation aids for aircraft was put in place about the U.S. during the early '30s. Each station, employing four antennae, broadcast a "beam" in four directions, and tracking on a beam a pilot received a steady aural tone in his headset. To the right of the beam, this signal became the Morse Code letter "N," and to the left, a letter "A." The omnidirectional VOR transmitters in use today were perfected in principle by 1940, but did not replace the old four-course transmitters until the decade of the '50s.

The all-metal airplane gained grudging acceptance in the United States by the late '20s. In addition to the Ford Trimotors and Thomas-Morse military designs, the Hamilton Metalplane Company entered the market in 1927 with four single-engine all-metal cabin monoplanes, the prototype of which placed second in the 1927 Ford Reliability Tour.

The metal that made these aircraft possible was duraluminum, which was first formulated in Germany and developed there by 1914 (later manufactured in the U.S. under the trade name Duralumin). It is an alloy of aluminum (92 percent) and other metals to achieve maximum strength with minimum weight.

In Germany, Dr. Claude Dornier built the first airplane to be constructed with the new material. In the U.S., former Chicago architect John B. Moisant designed and built two all-metal airplanes in 1910. (Moisant also organized a popular aerial exhibition team, which included his sister, Mathilde.) Duralumin would be the primary building material for airplanes for more than half a century after the Boeing, Douglas, and Lockheed airliners—and most military aircraft—appeared so constructed in the early '30s.

Another significant advance took place on 24 September 1929 when Lt. James H. Doolittle, using the Sperry Gyro Horizon and Directional Gyro, accomplished the first "blind" flight. Flying a Consolidated NY-2 biplane while under a hood that completely covered the rear cockpit, Doolittle took off from Mitchel Field, Long Island, flew a 15-mile course, and landed. Lt. Ben Kelsey rode in the plane's front cockpit as a check pilot.

As early as 1916, Lawrence Sperry had experimented with gyro-controlled pilotless airplanes—small biplanes fitted with OX-5 engines and envisioned as flying bombs. We can trace the ancestry of both the automatic pilot and the cruise missile back to Sperry's WWI concept without stretching things too much.

NACA

The National Advisory Committee for Aeronautics (NACA) was established by Act of Congress on 3 March 1915, and although NACA was directed by some of the nation's most learned educators, industry leaders, and military aviation chiefs, and given a broad mandate to research, develop, and advise in all technical areas of aviation, it is unlikely that the Congress envisioned the contributions NACA would make, even during its early years when it existed on meager budgets. Located at Langley Field, Virginia, its engineers initially concentrated on aerodynamics, particularly investigating and describing the properties of hundreds of subsonic airfoils. That proved of immense benefit to airplane designers both civilian and military. The "NACA Cowling" for air-cooled radial aircraft engines was developed in the NACA wind tunnel and ready for industry-wide use soon after the Whirlwinds and Wasps were in mass production. The appearance of the NACA Cowling, and its system of baffles for effective cylinder cooling, largely overcame the drag penalty originally paid by the radials compared to the in-line engines.

During the late '30s, the high-speed wind tunnel at NACA's Ames Laboratory provided engineers with visual evidence of the effects of compressibility as airplanes neared the speed of sound. Later, an aircraft test facility was established at Muroc Dry Lake, now Edwards AFB, California.

NACA became the National Aeronautics and Space Administration (NASA) in 1958, and has been preoccupied with America's space projects since that time.

Clearly, except for the composite materials used in the construction of some airplanes today, the microprocessor which effected a quantum advance in air navigation and communication systems, and, perhaps, supersonic airfoils, the basic principles that constitute the foundation of modern aviation and space technology were well understood 50 years ago.

What about space flight? Dr. Robert Goddard had that all figured out, except for a proper propellant, by 1935. (The Germans built on Goddard's discoveries.)

The jet aircraft engine? Well, Henri Coanda designed, built, and briefly flew a jet airplane (with full cantilever wings!) in France in 1910, and the operating principle upon which the jet engine is based was defined by Sir Issac Newton *300 years ago.*

General Aviation 1927-1931*

	1927	1928	1929	1930	1931
Airplanes (licensed & unlicensed)	2,612	4,779	9,315	9,218	10,090
Total accidents	253	1,036	1,586	2,033	2,205
Miles flown per accident	NA	57,915	69,357	53,256	42,786
Number of fatal accidents	95	215	287	301	253
Miles flown per fatal accident	315,789	279,070	383,275	359,700	372,989
Airports	1,036	1,364	1,550	1,782	2,093
Repair stations	0	0	0	29	85
Flight schools	0	0	35	38	28
Airplanes certified	1,908	3,165	6,803	7,354	7,553
Mechanics certified	0	4,383	7,701	8,993	9,016
Pilots certificated	1,572	4,887	10,287	15,280	17,739
Student pilots certificated	545	9,717	20,400	18,395	16,061
Aircraft produced (civil only)	1,374	3,127	5,516	2,690	1,988
Miles flown (thousands)	30,000	60,000	110,000	108,269	94,343

*Source: U.S. Department of Commerce

Charles A. Lindbergh's solo flight from New York to Paris 20-21 May 1927 covered a distance of 3600 miles in 33 hours, 30 minutes, 29.8 seconds. His Ryan monoplane was powered by a Wright J-5 Whirlwind of 200 hp; *Time* magazine reported in its 30 May edition that Wright Aeronautical stock had risen from 29¾ to 34⅜ within a few hours of Lindbergh's landing in Paris. The "smart money" recognized the key to Lindbergh's success. (Joe Reed)

The specially-built Ryan NYP cost Lindbergh and his St. Louis backers $10,580. It had a span of 46 feet and a fuel capacity of 450 gallons, which gave it a no-wind theoretical range of 4100 miles. Its average cruising speed was about 110 mph. (NASM)

The Alexander Eaglerock appeared in the fall of 1925, powered with the plentiful—and cheap—WWI surplus OX-5 engine. Later models were offered with a choice of engines, including the J-6 Whirlwind. The OX-5 Eaglerock was priced at $2475 in 1926.

A Western Air Express Fokker F-10 TriMotor. The F-10 was certificated in 1928. Its engines were 400-hp Wasps. It seated 14 passengers and cruised at 110 mph.

William Bushnell Stout's first airplane design was the 1919 Batwing, powered by an OX-5 engine of 90 hp. (Earl Reed)

The Stout Air Sedan of 1923 was the Batwing follow-on. It had corrugated aluminum skin and metal framing. Only one was built, but it convinced the Fords that Stout's approach to aircraft design was sound. (Charles Meyers)

The Ford-Stout 2-AT was Liberty-powered, and was America's first all-metal airliner. Eleven were built and all had perfect safety records, but the 2-ATs were not licensable for carrying passengers after the 1926 Air Commerce Act established airworthiness standards for airliners. (Ford)

The Model 3-AT was the first Ford TriMotor. Only one was built; performance matched its appearance. It was destroyed in a mysterious hangar fire. (Ford)

The first "Tin Goose" was little more than a three engined 2-AT. Designated the 4-AT, it retained an open cockpit in deference to pilots' prejudices of that day. This one belonged to National Air Transport, which would later become part of United Air Lines. (Ford)

The Ford 5-AT of 1929 had 450-hp Wasps and a passenger capacity of 15-17. Cruising speed was 122 mph. Price at Detroit was $55,000. (Ford)

93

Eddie Stinson learned to fly at the Wright brothers' school in 1915; he built the first of his Stinson Detroiters in 1926. Eddie was killed 26 January 1932 while demonstrating a Stinson Model R over Chicago. (Charles Meyers)

The Travel Air 5000 "Woolaroc" was flown to first place in the 1927 San Francisco-Hawaii Dole race by Art Goebel and William Davis; 26 hours, 17 minutes, 33 seconds. Ten lives were lost as a result of the race; only two airplanes of eight starters reached Hawaii. The race bankrupted Jake Moellendick at Swallow; his entry was one that disappeared somewhere over the Pacific. (Beech)

In January 1929 the Travel Air 6000 became the 100th civil aircraft certificated by the Bureau of Aeronautics since that law was enacted in 1926. The 6000 seated six passengers, cruised at 95-100 mph with J-5 Whirlwinds of 220 hp and J-6s of 300 hp. Price at factory was $12,000. (Beech)

The Cessna AW of 1928 was four-place and cruised at 105 mph with a 110-hp Warner Scarab engine. It was priced at $6900. (Cessna)

During the Great Depression of the early '30s, Cessna attempted to keep his Wichita factory open by selling primary gliders for $395. (Cessna)

The Cessna FC-1 was another depression project designed by Clyde Cessna's son Eldon in 1931 as a minimum-cost private airplane fitted with an inverted Menasco engine. It was not produced. (Cessna)

The Cessna EC-3 of 1932 was still another Eldon Cessna design, a two-placer suggestive of the Aeronca C-3, though a much cleaner aircraft with a cantilever wing. (Cessna)

The C-165 Cessna Airmaster of 1935 was a follow-on to the first Airmaster produced by Clyde's nephew, Dwane Wallace, in 1934. These craft, of which an average of 30 per year were built 1934-1938 inclusive, kept the Cessna Company afloat until the demands of WWII overwhelmed all U.S. aircraft manufacturers. (Cessna)

Clyde Cessna with nephew Dwane Wallace shortly before Clyde's death in 1954. Wallace remained Cessna's chief executive officer until 1983. (Cessna)

The Travel Air Model R "Mystery Ship," one of five built, 1929. Airline pilot Doug Davis flew this one to victory in the first Thompson-sponsored air race that year to mark the beginning of American air racing's ten most exciting years. (Beech)

Walter H. Beech with Charles Lindbergh and the Travel Air 6000 Lindbergh borrowed from Central Air Lines for a flight to Mexico City to visit his future wife, Anne Morrow, daughter of Dwight Morrow, Ambassador to Mexico (and chairman of the 1924 Morrow Board which attempted to justify the court martial of Billy Mitchell). Lindbergh flipped this airplane onto its back in Mexico City with Miss Morrow aboard after it lost a wheel on takeoff. (Beech)

The first production Model 17 Beechcraft had a fixed landing gear enclosed in a streamlined housing. Engine was a 420-hp Wright R-975 Whirlwind, a J6-9, and maximum speed was 201 mph. Known as the 17R, this craft was priced at $19,000 in January 1933. (Beech)

The Great Lakes 2T-1 Sport Trainer was designed by Charles W. Meyers and Cliff Leisy in 1929. Engine was the four-cylinder, air-cooled Cirrus of 85 hp, which gave the Lakes a cruising speed of 90-95 mph. About 500 were built, originally priced at $4990, and reduced to $2985 in 1931. (Charles Meyers)

The prewar Cubs, T-Crafts, and Aeroncas of 65 hp were purchased by the Army for liaison duty. This 1942 Taylorcraft L-2 was the military version of the 1941 DL-65 T-Craft. Restored, it was photographed at the 1966 Antique Airplane Association Fly-in at Ottumwa, Iowa.

Airline pilot Dave Jameson restored this 1930 Monocoupe 110, which is powered with a Warner Scarab of 110-125 hp. Maximum speed is 142 mph, and its price at the factory, Lambert Field, St. Louis, in 1930 was $4500. If that sounds like a bargain, we should note that haircuts were 25¢ and a hamburger 5¢ then, while aircraft production workers could expect 40¢ per hour.

The 45-hp Welch OW5-M could pass for an Aeronca C-3 at a glance. The depression years fostered a rash of minimum airplanes aimed at the private pilot. In the mid '80s, the concept has once again become attractive.

The Grumman Ag Cat, with piston engines up to 600 hp, was the only biplane agricultural aircraft in production in recent years. Since introduction of aircraft specifically designed for the aerial application of agricultural chemicals in the late '50s, the trend has been to airplanes with large load capacities operated by larger, multi-plane companies.

The Cessna Agwagon is an example of a modern agricultural aircraft, designed for dusting and spraying operations. The 300-hp Agwagon has a hopper capacity of 200 gallons. Ag-flying is demanding work, but it pays well. (Cessna)

The Laird Speedwings and Commercials of the late '20s and early '30s were advertised as the "Thoroughbreds of the Air" and were carefully-crafted machines. The LC-RW300 Speedwing (L), powered with the R-985 Wasp (then) rated at 300 hp, cruised at 150 mph. The LC-B300 Commercial (R) with a Wright J-6 engine of 300 hp cruised at 130 mph. In 1930 they were priced at $15,500 and $13,500 respectively. (E.M. Laird)

During the late '20s and into the '30s, Bill Lear developed some of the first aircraft radios, while Amelia Earhart was flying a Lockheed Vega to fame. (Gates Learjet)

A 1928 Whirlwind-powered WACO Taperwing; probably the best aerobatic civilian airplane produced in America at that time. Actually, the Curtiss Hawks were available in the civilian and export market, and Jimmy Doolittle performed the first outside loop in a P-6 Hawk in 1927. "Fearless" Freddy Lund was the first civilian to perform that maneuver, in a WACO Taperwing. (Charles Meyers)

The Lockheed XC-35 was developed in secrecy and delivered to the Air Corps in 1937. A modified Electra, it was the first airplane with a pressurized cabin to fly in the substratosphere, earning the Collier Trophy for the Army Air Corps. (Lockheed-California)

The Ryan Broughams, produced by the B.F. Mahoney Aircraft Company in San Diego, builder of Lindbergh's *Spirit of St. Louis,* were produced 1928-1930 inclusive. Most were fitted with 225-hp Wright Whirlwinds; the last ones were Wasp-powered. About 150 were built, priced at $12,500. (George Vaughn)

The WACO RNF, with the 110-hp Warner Scarab pictured, was also produced as the INF powered with the B-5 Kinner of 125 hp. Cruising speeds were 110 and 118 mph, respectively. At least 200 of this model WACO were produced 1930-1931. (Bill Selikoff)

Chapter 5

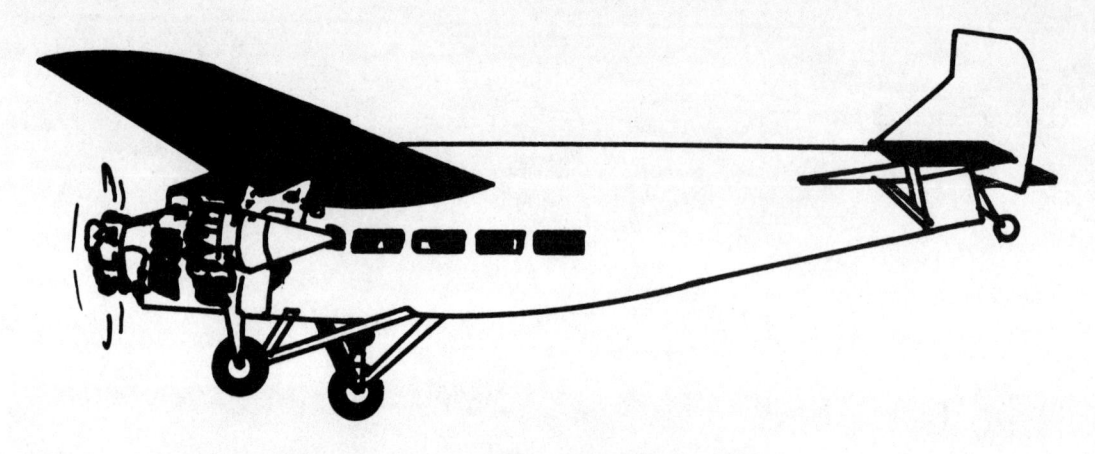

The First Airlines

If deregulation of the U.S. airlines visited a deep trauma upon the major air carriers during the early 1980s, that was a romp among the daisies compared to the vendetta against the industry mounted by President Franklin D. Roosevelt's "New Dealers" during the '30s. America's domestic airline system, carefully (if forcibly) pieced together following passage of the Kelly Bill in 1925, was far and away the best in the world by 1933. Then, overnight, the new administration in Washington tore it apart, in the name of "justice," but with an overkill that served neither justice nor the best interests of the nation.

The airlines evolved from the air mail routes, pioneered at great cost by the Post Office Department pilots (31 of the original 40 were killed flying the mail). The Post Office Department had flown the mail since 1918, and said all along that it would relinquish the routes to private contractors as soon as practicable—"practicable" meaning whenever the Congress decided that private enterprise could provide dependable service without too much government subsidy. America's overseas steamship lines and railway systems had been established with federal subsidies of one form or another. Indeed, the shipping lines continued to enjoy guaranteed profits by way of generous pay for carrying overseas mail.

By 1924, there was also pressure from the Railway Clerks Union to remove what they considered to be unfair competition from a department of the government, and that may be why U.S. Representative Clyde Kelly of Pennsylvania, known as the "Voice of the Railway Mail Clerks," introduced legislation that authorized the postmaster general to contract with civilian operators to fly the mail.

The Kelly Bill

The Kelly Bill, HR 7064, was passed 2 February 1925, and provided for payment to the operators of up to four-fifths of the postage revenue flown by each. Since the Post Office Department had never been able to break even with all the air mail revenue, that provision of the new law was not as generous as it may have sounded to some. It represented no direct subsidy at all. Such pay scale meant that the civilian contractors had to effect every possible economy in flying the mail, and that, in turn, meant that most would do it as the Post Office had done—in open-cockpit biplanes, which was no advance whatever. Both the Post Office Department and the Congress hoped to see the air mail routes develop into passenger-carrying operations, as had already happened in Europe.

Some of the civilian operators—perhaps most—envisioned the same. Therefore, most were willing to start under whatever terms they could get in order to establish an equity in their routes, confident that, once begun, the government would not allow an honest effort to fail or the routes to be abandoned. They were right.

Early Efforts

Actually, several attempts at establishing airline service had been made in America in the past, both with and without mail pay (a hundred or more "air lines" appeared briefly during the early '20s that were actually air taxi services; most failed). The first *scheduled* passenger line in the U.S. was the St. Petersburg-Tampa Airboat Line, which began operation 1 January 1914 with a pair of two-place Benoist flying boats. Pilot Tony Jannus flew the 46-mile round trip daily for three months, and then the tourists returned north in the spring and the St. Petersburg-Tampa Airboat Line went out of business.

In 1919, along with prohibition, Aeromarine Sightseeing & Navigation Company came into existence for the purpose of flying thirsty Americans between Key West and Havana in a couple of war surplus flying boats. A similar operation was started at about the same time by Aero Limited, flying between Miami and the Bahamas. These two services merged with a third. West Indies Airways, on 1 November 1919, to obtain a mail contract and began regular service to Havana as Aeromarine West Indies Airways, Inc. Despite the ready-made passenger market, Aeromarine was forced to suspend operations early in 1924 when its air mail contract was withdrawn. It could not show a profit with passenger fares alone. It had safely flown 30,000 passengers two million miles.

Both Edward Hubbard and Merrill Riddick obtained Post Office contracts to carry mail on short routes in 1919. Hubbard, who maintained his one-man operation for seven years, flew a small Boeing flying boat across Puget Sound from Seattle to Victoria, B.C., while Riddick flew from New Orleans to the mouth of the Mississippi River, each to meet steamships and to subtract one day from their overseas mail schedules both inbound and outbound.

The Seeds of Empire

The first Air Mail Act (Kelly Bill) was subtitled "An Act . . . to Encourage Commercial Aviation," and it did so. Within 60 days of its passage, the postmaster general received more than 5000 inquiries from interested parties. If there was money to be made in aviation, there would be no lack of promoters. But promoters and investors are not necessarily the same, and President Calvin Coolidge's Postmaster General, Harry S. New, had no intention of parceling out the air mail routes to speculators or irresponsible bidders. New made that clear when he advertised the routes, and gave prospective operators until September 1925 to submit their proposals. Even then, he withheld the major routes and offered only the shorter feeders in order to allow the successful bidders to gain experience before seeking longer runs.

Eight routes were advertised, but contracts were let on only five because bidders for the other three were not, in Postmaster General New's opinion, sufficiently responsible. The successful bidders were: Colonial Air Lines, which changed its name to Colonial Air Transport after its bid was accepted; Robertson Aircraft Corporation; National Air Transport; Western Air Express; and Varney Speed Lines.

Colonial was awarded Contract Air Mail Route No. 1 (CAM-1), New York-Boston. It was organized in Naugatuck, Connecticut, in 1923 as the Bee Line, Inc., an air taxi service. Bee Line expired of malnutrition but, in anticipation of favorable Congressional action on the upcoming Kelly Bill, was resurrected as Colonial by former Connecticut Governor John Trumbull, Harris Wittemore of

Connecticut Bond & Share, Sherman Fairchild, William A. Rockefeller, Cornelius Vanderbilt Whitney, and Juan T. Trippe, the 25-year-old scion of an investment banking family. That was the kind of "responsibility" the Postmaster General was seeking.

Colonial Air Transport began carrying the mail over CAM-1 18 June 1926, and would later provide an important part of the nuclei around which American Airlines was formed. Juan Terry Trippe would organize and build Pan American Airways.

The Chicago-St. Louis route, CAM-2, went to Robertson Aircraft Corporation, founded by Frank and William Robertson, WWI pilots. The brothers gathered financing in their home city of St. Louis for the purchase of 14 DH-4s and began operation on 15 April 1926. One of their pilots was Charles A. Lindbergh.

National Air Transport (NAT) won CAM-3, Chicago-Dallas. NAT was organized immediately after the Kelly Bill became law, and it acquired one of the potentially best routes because NAT's incorporators, too, possessed a lot of muscle. That power block included Howard E. Coffin of the Hudson Automobile Company (and the infamous WWI Aircraft Production Board); Col. Paul Henderson, who resigned his post as second assistant postmaster general to join this group; Clement M. Keys of Curtiss Aeroplane & Motor; Charles Lawrance of Wright Aeronautical; C.F. Kettering of GM (another of the WWI profiteers); William Rockefeller, and industrialists Philip Wrigley and Lester Armour, as well as Jeremiah Milbank (Allis-Chalmers). NAT began flying the mail between Dallas and Chicago 12 May 1926 with a fleet of ten Curtiss Carrier Pigeons, and would later form an important part of United Air Lines.

Western Air Express, later a parent of TWA, was awarded CAM-4, Los Angeles-Salt Lake City. This was actually the best of the feeder routes because it connected with the transcontinental air mail at Salt Lake City and the great number of Southern Californians with ties in the Northeast generated lots of mail. Therefore, Western Air Express (WAE) operated in the black from the outset.

WAE was the creature of Harris M. "Pop" Hanshue, a tough old former racing driver, and was backed by Harry Chandler, publisher of the *Los Angeles Times*, along with some other West Coast moneyed types. Hanshue began service 17 April 1926, flying new Douglas Cruisers (similar to the Army's Douglas 0-2H).

Varney Air Lines won the contract for CAM-5, Elko, Nevada, to Pasco, Washington. Walter T. Varney would soon extend his route in both directions to fly from Salt Lake City to Seattle and Spokane—after which Varney would be merged into United Airlines. Varney was a WWI pilot who operated a flying school and air taxi service out of San Mateo, California, and apparently assumed that, with only $30,000 in capital (NAT started with $10 million), he would not be considered for any route except that over the mountains in the Northwest, one that few others would want. Varney bought six new Swallows fitted with Curtiss C-6 engines of 150 hp and began flying the mail on 6 April 1926. The C-6 engines proved so unreliable, however, that all were replaced by J-4 Whirlwinds within a month. Later, Varney would fly Stearman biplanes. After selling out to United, Varney would reappear—several times—as an airline operator.

CAM-6 and CAM-7 were awarded to Henry Ford two weeks after the first five contractors were determined. These were the routes that Ford had been flying on schedule as his private airline, and therefore Ford was the first to actually begin flying the mail. Ford Air Transport (FAT!) was also the first to carry passengers, and the only one at the time with such equipment. Initially with the single-engine 2-ATs, and later in the Ford Trimotors, Ford flew between Detroit and Chicago, and Detroit and Cleveland.

In 1928, Stout Air Services took over the Ford air mail contracts, although it may have been a name change only. In any case, the Ford-Stout airline became part of United's fast-growing empire in 1929.

CAM-8 was the Los Angeles-Seattle route, won by Pacific Air Transport (PAT), a company founded by Vern Gorst, a successful West Coast bus line operator. PAT began service 15 September 1926, but early in 1928 was purchased by Boeing Air Transport (United Air Lines' main foundation), and Gorst ended up with Edward Hubbard's little mail route across Puget Sound to Victoria, Canada. PAT flew seven Ryan M-1s, for which it paid $7400 each. Interestingly, Gorst financed his airplanes through the Wells Fargo Bank in San Francisco, dealing with banker William A. Patterson. Later, Patterson would be president of United Air Lines.

CAM-9, Chicago to Minneapolis and St. Paul, had earlier been abandoned by the Post Office Department after losing four pilots to its violent weather, but Chicago seed magnate Charles Dickinson, who looked like Santa Claus and had been an ardent supporter of aviation since 1910, successfully bid the route—at a ridiculously low rate to "keep it out of the hands of Wall Streeters"—and began flying the mail 7 June 1926 with a pair of J-4 Lairds, an OX-5 Laird, and a homebuilt cabin biplane constructed by pilot Elmer Partridge. Partridge crashed and was killed on the first day, apparently the victim of carbon monoxide poisoning. He and Matty Laird were the only two to fly a scheduled air mail route in airplanes they had designed and built themselves (Matty flew several trips over CAM-9 during its first week of operation).

Dickinson maintained his schedules for three and a half months, losing money on every flight, then sold out to Northwest Airways, a company quickly put together by Col. L.H. Brittin of the St. Paul Chamber of Commerce, Frank Blair of Detroit Union Trust Company, William B. Stout (!), and Harold H. Emmons, the Detroit attorney of the WWI Aircraft Production Board and a crony of Edward Deeds, Howard Coffin, and Charles Kettering.

Northwest began service with three new Stinson Detroiters and soon added eight all-metal, eight-place Hamilton single-engine cabin monoplanes (designed by James S. McDonnell, who would later form his own aircraft manufacturing company which, merged with Douglas, would result in McDonnell-Douglas). Northwest's pilots included David L. Behncke, who would soon form the powerful Airline Pilots Association, and Charles "Speed" Holman. Northwest Airways would, of course, evolve into Northwest Airlines, and then Northwest Orient Airlines.

CAM-10, Atlanta-Jacksonville, was awarded to Florida Airways Corporation, organized by WWI combat flier Reed Chambers and John Harding. It began flying the mail in September 1926, equipped with two (or possibly three; the best sources differ) Ford-Stout 2-ATs. Florida Airways was absorbed by newly organized Pan American Airways in March 1927.

CAM-11, Cleveland-Pittsburgh, was the shortest feeder route contracted by the Post Office during that initial two-year break-in period, and a profitable one. It went to Clifford Ball, a McKeesport automobile dealer who had acquired (reluctantly) a half-dozen airplanes whose owners had defaulted on hangar/fuel/maintenance fees at Bettis Field, an airport south of Pittsburgh in which Ball had a significant investment. Since he had a base of operations and a fleet of airplanes—mostly WACO 9s—Ball bid for the route at the maximum rate, and seemed surprised when given the contract. Ball would sell out to Pennsylvania Airlines in 1930 (PA was a subsidiary of Pittsburgh Aviation Industries Corporation), which in turn merged with Central Airlines in 1936 to form Pennsylvania-Central Airlines; this was renamed Capital Airlines in April 1948, and absorbed by United 1 June 1961.

CAM-12, Pueblo Colorado, to Cheyenne, Wyoming, was given to Pop Hanshue's Western Air Express, which carried the first load of mail north through Denver to connect with the transcontinental route in Cheyenne at the end of December 1926.

These twelve routes made up the foundation of America's domestic airlines system (although the great transcontinental routes were yet to be contracted and those companies yet to be formed) that would follow—with a great deal of wheeling and dealing—during the next five years. But first the airways had to be regulated, then additional air mail legislation would encourage the contractors to obtain larger, passenger-carrying airplanes, while profits would be assured—if the operators filled out the route structure to please the postmaster general.

Air Commerce Act of 1926

As early as 1919 the NACA sent a bill to the Congress that would require the licensing of planes and pilots and establish aviation regulations, but it was not reported out of committee. Other proposed legislation of a similar nature was repeatedly offered during the first half of the '20s and none ever reached a vote. By mid-1925, however, pressure from several quarters—including investors with interests in the new contract air mail routes, and the unsettling public statements of Gen. Billy Mitchell—prompted President Coolidge to appoint a "President's Aircraft Board" that would recommend federal policy and legislation dealing with the "aviation question."

The "aviation question" at the time was mostly concerned with the role the airplane should play in the nation's defense (a very prominent one, as Mitchell advocated, or a relatively minor one, as the "saddlehorse generals and battleship admirals" believed), while regulation of civil aviation was more or less a side issue.

The President's Aircraft Board—popularly called the Morrow Board after its chairman, Dwight Morrow, an old school chum of the President's, a member of the J.P. Morgan banking house, and future father-in-law of Charles Lindbergh—included Howard Coffin, the perennial opportunist when money or influence was involved in aviation matters; Senator Hiram Bingham of Connecticut; the Navy's Adm. Frank Fletcher (ironically, best remembered in history for the defeat his carrier airplanes inflicted upon the Japanese in the Battle of Midway); and the Army's Gen. James Harbord, an old horse soldier, Pershing's chief of Staff in WWI, who had already spoken out against Mitchell.

The Morrow Board fulfilled its mission. After three weeks of hearings in the fall of 1925, it issued a report rejecting Mitchell's position, exactly what Coolidge and the military chiefs wanted, because that paved the way for Coolidge's order that Mitchell be court-martialed under the 96th Article of War—that is, publicly sassing his superiors.

All this had its effect on civil aviation because of the great amount of publicity it generated, and because the Morrow Board also made some sensible recommendations regarding the regulation of civil aviation.

Actually, those recommendations were taken from a bill originally introduced in the Senate in 1922 by Senator James W. Wadsworth of New York and sponsored in the House by Representative Samuel F. Winslow. It failed to come to a vote, and was reintroduced in 1924 after a NACA objection was honored (local control of airports, rather than federal control), and then structured by Secretary of Commerce Herbert Hoover. That proposed legislation was submitted to the Senate by Senator Hiram Bingham as the Air Commerce Act, where it was approved 16 December 1925 (one day before the guilty verdict in the Mitchell trial). It went on to the House, where it was sponsored by Representatives James Parker and Schuyler Merritt. Known as the Bingham-Parker-Merritt Bill, the Air Commerce Act was passed by the Congress 20 May 1926 and became law on 1 January 1927.

It was a good law, in the best interest of the public and civil aviation. It brought law and order—and responsibility—to the airways. It was civil aviation's constitution. But it should not be credited, as it usually is, to the Morrow Board. Bingham's sponsorship, following immediately on the heels of the hatchet job performed on Mitchell, could be viewed as an attempt to legitimatize the board's main purpose. In any event, none of the provisions of the Air Commerce Act originated with the Morrow Board.

Bureau of Aeronautics

The new Bureau of Aeronautics, within the Department of Commerce, was given an appropriation of $550,000, and President Coolidge appointed Washington attorney William P. MacCracken as chief of the new Bureau with the title of Assistant Secretary of Commerce for Aeronautics. MacCracken was secretary of the American Bar Association, and had taken an active interest in aviation matters since serving in the Army Air Service during WWI.

He selected Maj. Clarence Young of the Air Service as head of the aviation regulations division, and it was Young who wrote the first air regulations.

Since the Bureau of Aeronautics belonged to the Department of Commerce, the Airways Division was under the Bureau of Lighthouses, which maintained the 4000 miles of lighted airways and operated the 42 aviation weather reporting stations. Other important new divisions were Airway Maps and Aeronautical Research, the latter being responsible for the certification of new airplanes. The first new airplane to receive an Approved Type Certificate was the Buhl-Verville J-4 Airster, on 27 March 1927.

Payment to Operators and Postal Rates Changed

On 3 June 1926 Congress acceded to Postmaster General New's request for a change in the original Air Mail Act, and amended the law to allow payment to the contractors of up to $3 per pound for the first 1000 miles and 30¢ per pound for each additional 100 miles. The original rates, based on postal revenues, were cumbersome, requiring that each letter be tallied each time the mail was put aboard an airplane. Contributing to the problem was the postage zone system that divided the U.S. into three zones with air mail costing 8¢ per half-ounce to transit each zone, or 24¢ coast-to-coast.

Then, on 1 February 1927, the Post Office Department reduced the air mail postage to a flat 10¢ per half-ounce and abandoned the zone system. These measures combined to substantially improve the operator's profit potential. Few of the first dozen air mail contractors made a profit in their first year of service, but the 10¢ air mail stamp, which would deliver a letter anywhere in the country, immediately and dramatically increased the air mail loads.

This, of course, amounted to a direct subsidy for the operators, and the Post Office lost money on the service. That was expected and acceptable, because it was clear that a subsidy was necessary if the private contractors were to survive and, especially, take over the transcontinental route that had thus far been held back. Theoretically, if the operators could afford larger airplanes with provision for passengers, the subsidies should in time no longer be needed.

The Major Air Carriers Appear

Out in Seattle, William E. Boeing noted these changes and had, in fact, anticipated the increased incentives to the air mail contractors. His factory was supplying a few airplanes to the military, but he was also interested in a mail contract—not just *any* contract, but something substantial, like the transcontinental route. The Post Office, however, offered the transcontinental or "Columbia" route in two sections: New York to Chicago, and Chicago to San Francisco. These routes were advertised in mid-1926, and Bill Boeing bid for the western half. He apparently was so confident that his bid would be accepted that he directed his engineering staff—by then headed by C.N. Montieth—to produce a large, load-carrying biplane for service on the San Francisco-Chicago air mail route. The resulting Boeing Model 40A airframes, 25 of them, were ready soon after Boeing's bid for CAM-18 was accepted in mid-January, 1927. (Boeing later claimed that the 40As were designed, built, and certified by the Bureau of Aeronautics between the time his bid was accepted and the airplanes entered service in July 1927, but that was probably due to the fact that the 40A was designed for the P&W Wasp engine, and Fred Rentschler at P&W diverted 25 Wasps from the Navy's order for the first 200.)

Boeing bid the route at only $1.50 per pound for the first 1000 miles and 15¢ per pound for each additional 100 miles. This was just half the maximum allowed, and that occasioned howls of protest from other contractors, some of whom charged that Boeing's "reckless" bid made the rest of them look positively greedy. In retrospect, it appears that Boeing merely wanted to make sure that no one underbid him—that his objective was not immediate profits, but to pin down that important route.

Meanwhile, Boeing formed Boeing Air Transport with his old partner Edward Hubbard, who had been flying the mail across Puget Sound to Canada since 1919 as Hubbard Air Transport. Boeing Air Transport was a subsidiary (as was the Boeing Airplane Company) of Boeing's holding company, Boeing Airplane & Transport Company. Then Boeing Air Transport purchased Vern Gorst's Pacific Air Transport which had been flying the mail between Los Angeles and Seattle since September 1926, and Gorst ended up with the little Hubbard operation, evidently as part of the deal.

By early 1928, Boeing Air Transport was operating on the West Coast and across the country to Chicago, while National Air Transport and another bidder had fought through the courts for the Chicago-New York segment of the transcontinental route. The case was decided in favor of NAT, the postmaster general's choice, and NAT began flying the mail in Conqueror-powered Curtiss Carrier Pigeons between the Big Apple and the Windy City in September 1927.

With the transcontinental route in the hands of private companies, a second amendment to the Kelly Bill was passed 17 May 1928. This cut the air mail postage rate to 5¢ per ounce (from 10 per half-ounce), authorized the postmaster general to issue ten-year Route Certificates to contractors who performed satisfactorily for two years or more, and allowed the postmaster general to periodically negotiate a reduction of mail payment to an operator when, in the opinion of the postmaster general, such action was justified.

That was an important change, because it not only greatly increased mail loads and therefore profits to the operators, but gave the operators the security they needed for long-term planning. The fact that the postmaster general could arbitrarily demand a cut in those profits was the bitter that came with the sweet, but everyone knew—or should have known—that the government would not subsidize the air mail contractors forever.

Nevertheless, stability, with a steady and reasonable growth for the airlines, was clearly the intent of the Post Office Department and the Congress, and while the 1928 amendment to the Air Mail Act surely would have been all the stimulus the budding airlines needed in normal times, taken together with the Lindbergh Boom (which was gaining strength daily throughout 1928, with investors eagerly buying almost any stock offered by an aviation enterprise), it made the future of America's air transport industry seem promising indeed. The euphoria wouldn't last, but while it did it generated some wheeling and dealing of heroic proportions.

The air transport scene in 1928 was no place for the timorous.

United Formed

Whatever else may be said about Bill Boeing and Fred Rentschler, they were not timid. In the roseate financial climate of 1928 they recognized both the opportunity and the need to combine their respective strengths in order to achieve what neither could do alone.

First, the Boeing Airplane & Transport Company purchased the Chance Vought Corporation of Long Island City, New York (the Navy's first Wasps were going into Vought's new 02U-1 Corsair observation planes). Then, with the incorporation of their new holding company, United Aircraft & Transport Corporation, Rentschler brought Pratt & Whitney in with an exchange of stock. Boeing did the same, then United bought Hamilton Propeller, and Standard Steel Propeller. Other subsidiaries added the following year included Stearman Aircraft Company of Wichita, the Sikorsky Manufacturing Corporation of College Point, Long Island, and John K. Northrop's first company, Avion Corporation, which was renamed Northrop Aircraft Corporation in the United family. (When United moved the Northrop operation, which was building the Alpha, from Burbank, California, to the Stearman facility in Wichita in 1931, "Jack" Northrop resigned and soon started another of his several Northrop companies.)

Finally, Boeing and Rentschler looked at their airline, renamed United Air Lines, and determined that it should fly all the way across the U.S. rather than terminate at Chicago. The problem was that National Air Transport possessed the mail contract for the Chicago-New York segment of the transcontinental route, and NAT had some financial heavyweights holding its stock.

Boeing and Rentschler bluffed a little by purchasing Stout Air Lines (Formerly Ford), which took United as far as Cleveland. Then, simultaneously, they began buying NAT common stock and wooing NAT's largest minority stockholder. It really wasn't much of a fight, because C.M. Keys, who was more than willing to tilt financial lances with United, could not muster the votes to block the merger after NAT's largest individual stockholder, Chicago banker Earle Reynolds, decided to cast his lot with United. NAT officially disappeared into United in March 1930, making United Air Lines the first of the great transcontinental operations. Meanwhile, in late November 1929, United had bought out Varney Speed Lines (originally, Varney Air Lines), which gave United a feeder into the Pacific Northeast connecting with the coast-to-coast route at Salt Lake City.

North American, Parent of Eastern and TWA

North American Aviation Company (which would evolve into North American Rockwell half a century later) began as a holding company organized by C.M. Keys of Curtiss-Wright and backed by GM, Hayden, Stone & Company, and Bankamerica-Blair. In the beginning, its sole purpose was the acquisition of promising aviation properties, among the first of which were the Fokker Aircraft Corporation (recently purchased by GM and renamed General Aircraft Corporation when it became a subsidiary of North American in May 1929); Pitcairn Aviation, Incorporated, and Transcontinental Air Transport (TAT).

Harold Pitcairn organized Pitcairn Aviation, Incorporated, in 1926, and received a mail contract in February 1927 for the New York-Atlanta route by way of Washington, D.C. and Richmond Virginia. North American acquired Pitcairn 10 July 1929 and changed its name to Eastern Air Transport in January 1930. EAT would be renamed Eastern Air Lines early in 1934.

Transcontinental Air Transport would become the second big coast-to-coast line across the middle of America when forcibly merged with Western Air Express and Pittsburgh Aviation Industries Corporation to form Transcontinental & Western Air, Incorporated (TWA; later Trans-World). But between 4 July 1929, when TAT began service without a mail contract, flying Ford Trimotors, and 25 October 1930, when it became part of TWA, TAT operated as an air-rail passenger service.

TAT reportedly lost almost $3 million during that time; at the beginning of the Great Depression, its 16¢ per mile fare was steep. It required 48 hours for the trip between New York and Los Angeles, while pure rail service took 78 hours. TAT did not fly at night, partly because the coast-to-coast middle route was not lighted at the time. Its passengers left New York at night in Pullman coaches of the Pennsylvania Railroad, thereby avoiding a flight over the Alleghenies. They switched to a TAT airliner the next morning at Columbus, Ohio, flew during the day to Waynoka, Oklahoma, then boarded a Santa Fe Pullman for the overnight run to Clovis, New Mexico. The last leg to Los Angeles was flown the following day.

Service was temporarily halted on 3 September 1929 after an eastbound Ford Trimotor crashed near Clovis, then resumed on 20 September with Curtiss Condors replacing the Fords. A year later, just a month before TAT's acquisition by North American, the first Fokker Trimotor entered service with TAT, suggesting that the deal with North American had been agreed upon by that time, and that General Motors' (which owned Fokker) influence was being felt.

AVCO, The Confused Parent of American Airlines

The Aviation Corporation—AVCO—was organized in March 1929, several months before the Lindbergh Boom began to show signs of expiring. AVCO was a holding company and, like United and North American, was formed to gather a system of satellite aviation properties. But AVCO, with $35 million in working capital, gained control—usually, with an exchange of stock—of an incredible collection of short feeder lines, aircraft engine companies, and other aviation businesses. It was said that no officer of the company could name more than half of AVCO's 80 subsidiaries, or knew their financial status.

AVCO was put together by Graham Grosvenor

of Fairchild Airplane Manufacturing Corporation, initially with the intention of financing Embry-Riddle Services, a successful Fairchild dealer that won the Chicago-Cincinnati air mail contract. But Grosvenor and his fellow incorporators appear to have become intoxicated with the possibilities presented to them when they found that the banks would underwrite so much stock for their venture. They should not have been surprised, because their little group contained some very impressive names: David Bruce, the son-in-law of Andrew Mellon; Robert Lehman of the investment banking firm bearing that name; Robert Dollar, the steamship magnate; Harry S. New, former Postmaster General who had left office a few days before AVCO organization (Presidential inaugurations were on March 4th until changed by Congress in 1932); Gen. Mason Patrick, former Air Service chief, and several lesser known multimillionaire types. So a proposal that began with a modest goal quickly snowballed as this group expanded its purpose and issued two million shares of stock underwritten at $17.50 per share.

For a while there appeared to be no cohesive plan, no discernible policy, to AVCO's seemingly reckless acquisitions, especially in the hodgepodge of air mail routes it collected. Among the companies AVCO controlled were four *other* holding companies, each with its *own* subsidiaries. Inevitably, some of the companies in the AVCO orbit directly competed with one another.

In an effort to get a handle on their holdings, AVCO directors formed a new super holding company on 25 January 1930, chartered as American Airways. AA issued 10,000 shares of preferred stock at $100 per share, which was the price for admission to all AVCO's subsidiary holding companies, proportionate to the amount of common stock assigned to each. Clearly, there is a way to do most anything in the world of finance. And although the placing of all AVCO's eggs in the AA basket did bring order to the bookkeeping and relief to horror-stricken auditors, AA's physical assets were as disjointed as ever.

Now, in fairness to AVCO, it should be mentioned that what seemed to be an indiscriminate grab of a number of air operations could have been planned from the start as the acquisition of necessary segments of a third great transcontinental airline—because that is exactly what happened to American Airways. It happened, however, at the direction of the new postmaster general in the Hoover Administration, Walter Folger Brown, and since Brown never made a penny out of it, it is unlikely that he acted as a conspirator with AA officials. He simply used AA's resources to benefit the public interest. At least, that is what he claimed when his day of reckoning came. If that also happened to enrich American Airways—well, so be it. Neither result can be denied.

Walter Folger Brown, Boss of the Airways

Walter F. Brown was a Toledo attorney and a power in the National Republican Party. He had faithfully toiled in the political vineyards for Republican candidates for 15 years, and had served as an assistant secretary of commerce under Hoover in the previous Coolidge Administration. In short, Brown was an able politician who knew his way around the nation's Capitol. He was also President Hoover's choice for the office of postmaster general.

Postmaster General Brown knew little about aviation matters when he assumed office, but he could hardly help learning because he was immediately caught up in the machinations of the newly formed aviation complexes and the direction of the air mail routes. To his credit, he assiduously studied the industry for many months before committing himself to a course of action. Then he fostered another amendment to the Air Mail Act, the provisions of which empowered him to put together the greatest domestic airline system in the world.

That amendment was known as the McNary-Watres Bill, sponsored by Senator Charles McNary of Oregon and Representative Laurence Watres of Pennsylvania (H.R. 11704), although it was said to have been written by Brown and his Second Assistant Postmaster General Warren Glover. It was approved by the Congress on 29 April 1930, and its two key provisions were that, first, air mail contractors were to be paid up to $1.25 per mile depending on the "space available" in their airplanes for the transportation of mail, and, second, that the postmaster general might "extend or consolidate" routes "when in his judgment the public interest will be promoted thereby."

The "space available" bit was Brown's way of encouraging the air mail contractors to place in service large, passenger-carrying airplanes, the theory being that the excess capacities would be utilized to garner the added profits that passengers would provide. Up to that time, most operators had no interest in carrying passengers, but flew the mail in open-cockpit biplanes. The "space available" basis for payment changed that. And the postmaster general's authority to "extend or consolidate" routes meant that he could extend the routes of selected operators without subjecting the "extensions" to the competitive bidding process.

Brown encountered a lot of opposition to his interpretation of the right to extend routes, and the comptroller general ruled that extensions must not exceed the length of the original route. Brown countered that by allowing certain lines to abandon unprofitable routes and add that mileage to the desired route.

The postmaster general sought the power to negotiate all air mail contracts without competitive bidding, but the Congress struck out that clause, although there remained in the new amendment a provision that required bidders to have flown routes of 250 miles or more. That effectively eliminated many of the small, poorly-financed bidders, which was what Brown wanted. He later explained that he was not against the "little guys" *per se,* but felt that he was forced to turn his back on them because they did not have the resources to buy multi-engine aircraft and operate, with maximum safety, long-haul passenger routes. In Brown's view, the transport of mail in single-engine biplanes was not advancing the industry and would always need subsidy. He was after well-managed and financially strong air carriers that would develop dependable passenger services. This would in turn allow gradual reductions in subsidies until the airlines were self-supporting.

Any authroity that Brown may not have specifically possessed under the law he assumed by way of his interpretation of the law. For example, in order to eliminate some "unqualified" bidders for air mail routes, he made a rule that bidders must have night flying experience, pointing to a clause in the original Air Mail Act which held that the postmaster general could impose such rules as were necessary to the safe operation of the routes. And while Brown dealt only with the financial fat cats in building the major transcontinental lines, he did not hesitate to crack his mail-pay whip about their ears if any failed to cooperate in the realization of that grand endeavor. That is how TWA was born.

Forced Merger of TAT and WAE Results in TWA

Within days after the McNary-Watres Bill became law, the Postmaster General invited the air operators to Washington for a series of meetings designed to plan two transcontinental routes. Brown was pleased with United's service between New York and San Francisco via Chicago; Northwest Airways would in time extend from Minneapolis through Billings, Montana, to Seattle, but in the meantime Brown wanted a middle route between New York and Los Angeles through Kansas City and Amarillo, as well as a southern route between Atlanta and Los Angeles by way of Dallas/Ft. Worth and El Paso.

American Airways was awarded the southern route after buying out C.E. Woolman's Delta Air Service, which was flying between Birmingham and Dallas, and Standard Airlines, a passenger service between Los Angeles and El Paso operated by Jack Frye, Paul Richter, and W.A. Hamilton with an eight-place Fokker Super Universal. Woolman would return in 1934 to start Delta Air Lines, while Frye would become president—and Richter vice president—of TWA by that time.

Frye had learned to fly in the Air Service during WWI, and operated a flying school in Los Angeles before starting Standard Airlines. Richter, taught to fly by Frye, had been a stunt pilot for the movies. They ended up working for TWA as a result of an involved deal fostered by the postmaster general.

Postmaster General Brown encouraged Pop

Hanshue of Western Air Express to buy Standard Airlines early in 1930, then immediately insisted that Hanshue sell Standard to American Airways so that AA would possess that vital segment of the southern coast-to-coast route. Hanshue gave $1 million in WAE stock for Standard, and received $1 million in AA stock for it. There would seem to be no point in the transaction since, presumably, AA simply could have purchased Standard without it passing through the hands of WAE. The difference was that WAE held $1 million in AA stock.

Meanwhile, Hanshue and his directors were forced to acknowledge that they had no choice but to merge with TAT (the well-financed though highly unprofitable air-train coast-to-coast service) and bid on the middle transcontinental route. Brown wanted it that way, but in any case others were ready to bid on the route, and once in operation it was certain to drain business from WAE since it would be a direct service, whereas WAE connected with United at Salt Lake City and Chicago.

The merger, however, was contingent upon a successful bid for the mail contract, and although the postmaster general was on their side, a couple of rival bidders had to be eliminated. Therefore, formation of the new company—Transcontinental and Western Air—was a paper company until receiving its contract from the Post Office Department in September 1930.

American Airways was in much the same position, and it was up to American to make the overt moves that solved the problems for both AA and TWA.

The main problem was Erle Halliburton, an Oklahoman who had made millions with his oil well-servicing company and then started Southwest Air Fast Express (SAFEway), a passenger airline operating Ford Trimotors in Missouri, Oklahoma, and Texas. SAFEway had a good record and was operating close to the break-even point, needing only a mail contract to become profitable. But SAFEway did not fit into Brown's scheme of things, and Halliburton had not been invited to the Washington meetings when the new coast-to-coast routes were planned.

Halliburton decided to fight. When the new routes were advertised, he offered to fly the mail coast-to-coast for about half the amount of the AA and TWA bids. To back up his bid, Halliburton ordered twelve new Ford Trimotors, the order to become effective when SAFEway was awarded the transcontinental mail contract.

The postmaster general rejected Halliburton's bid, saying that SAFEway did not have the required six months' scheduled night flying experience. But Halliburton replied that there was no such provision in the Air Mail Act or any of its subsequent amendments, and he was ready to take that issue as far as the Supreme Court if necessary.

Brown was unwilling to see his rule tested in court, or perhaps felt that too much time would be lost with the inevitable appeals. The way to silence Halliburton was to buy him out. Accordingly, American Airways purchased SAFEway for $1.4 million—about twice its book value.

The other stumbling block was United Avigation, which bid 30 percent below TWA for the middle route. UA was a paper company consisting of Pittsburgh Airways, which had been flying passengers over the Alleghenies since 1928; United States Airways, flying between Kansas City and Denver; and Ohio Transport, Youngstown-Dayton. These independents would merge to form United Avigation Company if awarded the mail contract.

United Avigation was disposed of when the postmaster general granted AA an extension to fly the Kansas City-Denver route. AA then sublet the route to United States Airways at maximum mail pay, whereupon United States Airways pulled out of United Avigation. That cleared the way for the award of the mail contract to TWA for the middle transcontinental route. Without United States Airways, United Avigation could not muster the backing to bid on the route.

American Airways got back the $1.4 million paid to Halliburton. That is exactly the amount TWA paid to AA for some ill-defined "property" and $1 million in WAE stock held by AA. Apparently, $1 million of the $1.4 million paid by TWA to AA was in the form of AA stock that WAE had received for

Standard Airlines. At first glance it would seem that AA got the best of the deal, but if one counts the profitable Kansas City-Denver contract that AA relinquished to buy off United States Airways, TWA got its money's worth.

In the end, TWA was forced to take in Pittsburgh Aviation Industries Corporation (PAIC). Little PAIC was a flying service organized in a fit of civic pride by a group of prominent Pittsburgh citizens (including three of the Mellons), and their claim to a "pioneering equity" across the State of Pennsylvania was worth a five percent interest in TWA. *That* transaction was never satisfactorily explained.

With the three great transcontinental lines established, plus Eastern Air Transport (a kissin' cousin to TWA because of its North American parentage) operating up and down the East Coast, United flying the West Coast, and Northwest expanding across the Dakotas toward Seattle, Postmaster General Brown spent the balance of his tenure in office piecing out the lesser routes with "responsible" bidders, often by way of route extensions to the major air carriers.

This latter course squeezed out most of the small independents because few could survive without a mail contract. And it was the elimination of one such independent that started the chain of events that broke up some of the aviation empires and almost destroyed the entire domestic airline system so efficiently put together by Brown and his chosen operators.

Ludington Lines, Handmaiden to Fate

Ludington Lines was an all-passenger airline that began service between Washington, D.C., and New York City in September 1930. The line was largely financed by the socially prominent Ludington brothers (Charles and Nicholas) of Philadelphia, and was ably managed by Gene Vidal (former Air Service pilot and recent general manager of TAT; later, director of the Bureau of Air Commerce, a director of Northeast Airlines, and father of Gore Vidal). Ludington Lines originated "Every Hour on the Hour" service over this short, high-density route, offering rates close to those of the railroads. Every economy was practiced; the equipment was ten-place Stinson Trimotors that, at $18,000 each, cost little more than one-third as much as the Ford Trimotor. Ludington captains flew alone, handled baggage, taxied on one engine, and cruised on automobile gasoline in flight.

Since Ludington's operating costs were about 35¢ per mile—an amount usually equalled or exceeded by passenger revenues—all Ludington needed for a reasonable profit was a minimal mail contract. Its bid of 25¢ per mile for flying the mail was ridiculous to the major carriers who were getting four to five times that much (with operating costs up to 80¢ per mile), but totally realistic to Ludington.

But Eastern Air Transport was awarded the mail contract for that route at 89¢ per mile. When a Ludington official bitterly complained to Washington reporter Fulton Lewis about Eastern's contract, Lewis sensed that he had a lead to the kind of sensational scoop that his boss, William Randolph Hearst, most appreciated.

Digging into Post Office Department records, and later interviewing a number of independent flight operators who felt that they had been treated unfairly by the postmaster general, Lewis excitedly documented his scoop. The only fault with his material was that it developed but *one* side of the issue—and that may have been the reason that Hearst himself sat on the story and refused to print it.

The Black Committee

The story would be front page news soon enough. Franklin Delano Roosevelt, campaigning with the promise of a "New Deal," won the presidential election in November 1932. After twelve years out of office, the Democrats were back in power, eager to correct the mistakes—and expose the misdeeds—of the Republicans during those dozen years.

Among the Congressional investigations instituted by the New Dealers during the summer of 1933 was a probe of the large banking firms. Next door, in Room 312 of the old Senate Office Building, a Special Committee headed by Alabama Senator Hugo L. Black, and including Senator Pat McCarran of Nevada, listened to testimony about the ocean mail subsidies.

The ocean mail hearings drug on into September—not that anything new was being revealed, but because all the news reporters were next door where J.P. Morgan and other financial barons were appearing. Senator Black had a pretty sensational show of his own waiting in the wings, and was unwilling to share the billing with other newsmakers.

The banking hearings ran out of gas early in October, so Black ran up the curtain on the "Air Mail Scandals," and Act One was a lulu.

Black ordered Interstate Commerce investigators to make surprise raids on the files of airline executives across the country. As expected this attracted the attention of the news media. Nothing much was revealed in those records, but the dramatic seizure was indeed newsworthy, especially when the attorney for Pop Hanshue (WAE/TWA) and Col. L.H. Brittin (Northwest Airways) provided some comic relief for the drama.

That attorney was William P. MacCracken, Jr. He had earlier contributed much to the drafting of the 1926 Air Commerce Act, had served in the Coolidge Administration as assistant secretary of commerce administering the first air regulations, was a member of the NACA, and had, it was said, aided Postmaster General Brown in writing the 1930 McNary-Watres amendment to the Air Mail Act. MacCracken appeared center stage in Black's production when he refused to produce certain papers belonging to Hanshue and Brittin, saying that he could not violate the confidential relationship between client and attorney without his client's permission.

Black responded that MacCracken was not serving as an attorney but as an airline lobbiest (which appears to be the case; MacCracken also represented Pan Am and Goodyear in Washington); therefore, when MacCracken held out, Black secured from the Senate a warrant for MacCracken's arrest, charging contempt of the Senate.

MacCracken then obtained his clients' permission to deliver the files in question, but Black insisted on MacCracken's arrest and appropriate punishment, whereupon MacCracken hid until the Senate recessed at noon on a Saturday. He then presented himself at the home of Senate Sergeant-at-Arms Chesley Jurney, saying that he was ready to be arrested. In the meantime, McCracken had obtained a writ of *habeas corpus* from the District of Columbia civil court. That meant that, as soon as the sergeant-at-arms arrested him, he would be released, and any subsequent legal action would have to be in civil court rather than in the heavily Democratic Senate. Since the files had been delivered, MacCracken was sure that the civil court would dismiss the case.

However, the suspicious Sergeant-at-Arms Jurney refused to arrest MacCracken, saying that the warrant was locked in a Senate safe for the weekend.

MacCracken insisted that his voluntary surrender *ipso facto* waived the need for a formal presentation of the warrant. He turned to the group of grinning newsmen and uttered the classic words "I have as much right to be arrested as anyone!" When Jurney remained adamant, MacCracken sat down on Jurney's doorstep, declared himself Jurney's prisoner, and refused to move.

Throughout the night and all day Sunday, MacCracken stuck close to Jurney, while Jurney sought to escape from his "prisoner," and an increasing herd of newsmen trailed along behind. Then, late Sunday evening, as Jurney strolled down a Washington street, MacCracken happily dogging his heels, Jurney suddenly leaped to the running board of a passing car (later identified as belonging to the district attorney) and sped away.

In court the following morning, Jurney testified that MacCracken had never been arrested. The judge agreed, and fined MacCracken $100 for obtaining a writ on speculation. Jurney then served his warrant, and MacCracken ended up in jail for ten days on the contempt-of-Senate charge.

The whole episode was ridiculous, of course,

but it helped raise an obscure Alabama senator to national prominence, and directed attention to his inquisition against the former postmaster general and the airline rascals who conspired with him to loot the U.S. Treasury.

The trouble with the Black Committee's investigation was that, proceeding in the beginning mostly on the evidence gathered by newsman Fulton Lewis, it assumed the guilt of its targets, and then set out to establish the degree of culpability. In other words, give 'em a fair trial and then hang 'em. By the time testimony from witnesses began to reveal both sides of the story, Black's extravaganza had become too much of a media event for him to back down or modify his righteous position. Black, a small man with a large ambition, had come to Washington with a mail order law degree, a soft-spoken manner, and a flair for old-fashioned, gallus-snappin' politics. Roosevelt would later appoint Black to the Supreme Court. But as a result of Black's intransigence, the "Air Mail Scandals" eventually proved more of an embarrassment for the New Dealers than for the Republicans.

Under oath, Brown cheerfully agreed that he had, in effect, been czar of the airways, pointing out that the power he possessed had been bestowed upon him by the Congress of the United States.

Brown explained why it had been necessary, as he saw it, to deal only with those who had the financial means and management ability to give the nation an air carrier network that would steadily grow and eventually become self-sufficient—and he had the figures to back his position.

When Brown left office in March 1933, there were 34 air mail routes operating 27,062 miles of airway. The first modern airliner—the Boeing 247—had appeared, its development fostered by Post Office Department policy. Passenger traffic was increasing at the rate of half a million boardings per year despite the Depression, and while it had cost the government $1.10 per mile for the air mail service in 1929, payment to the operators was down to an average of 54¢ per mile by 1933 as Brown had negotiated mail payments downward in return for ten-year Route Certificates to the contractors.

In retrospect, it is not difficult to see what the New Dealers should have done about the Air Mail Scandals at that time. They *would* do it by way of the Civil Aeronautics Act of 1938 (co-authored by Black Committee member Senator Pat McCarran), but meanwhile it was judged politically necessary to justify the work of the Black Committee.

Air Mail Contracts Cancelled

At four o'clock in the afternoon on Friday, 9 February 1934, President Roosevelt issued the following order:

> Whereas by an order of the Postmaster General of the United States, all domestic mail contracts for carrying the mails have been annulled; and
>
> Whereas the public interest requires that air mail continue to be afforded and the cancellation of said contracts has created an emergency in this respect;
>
> Now, therefore, I, Franklin D. Roosevelt, President of the United States under and by virtue of the authority in me vested, do hereby order and direct that the Postmaster General, Secretary of War, and Secretary of Commerce, together with other officers of their respective departments, cooperate to the end that necessary air mail service be afforded.
>
> It is further ordered and directed that the Secretary of War place at the disposal of the Postmaster General such airplanes, landing fields, pilots, and other employees and equipment of the Army of the United States needed or required for the transportation of mail during the present emergency over routes and schedules prescribed by the Postmaster General.
>
> Franklin D. Roosevelt

The airlines were given just ten days' notice. The Army Air Corps would take over the air mail routes on the 19th. During the night of the 18th, in a final gesture that spoke louder for Brown's policies than anything he may have said before the committee, TWA Vice President Jack Frye, with Eastern Air Transport's Vice President Eddie Rickenbacker riding as copilot, flew the last load of transcontinental air mail in a new Douglas DC-2 from Los Angeles to Newark in 13 hours 4 minutes. Part of the trip had been above a winter blizzard over the Midwest.

Their demonstration of the newest U.S. airliner, a machine that could cruise at 180 mph with 20 passengers for 50¢ per mile, also got the attention of news reporters, and would, without meaning to, soon emphasize what years of stagnant military planning had done to the nation's air defenses.

The Army Air Corps had no modern transport airplanes. It had no program for training pilots to fly in adverse weather. It would have to fly the mail in open-cockpit biplanes—observation craft and obsolete bombers—sometimes without so much as an ordinary magnetic compass. Few Army pilots had experience in cross-country night flying, and they were charged with the need to meet schedules during a period of the worst winter weather in years.

In view of the Army fliers' many handicaps, 40 percent of the 27,000-mile airway system was lopped off, leaving 11,000 miles of air mail routes. Nevertheless, it was an impossible assignment. No amount of courage and dedication to duty could make up for their lack of training and poor equipment. The result was predictable.

At the end of the first week, five Army pilots had died and six were critically injured in eight crashes. Schedules were further pared, but by the second week in March the mounting toll shocked the nation. Night flying was discontinued, and daylight mail flights were limited to what passed for reasonably good weather conditions. All flights were cancelled on 10 March for a week to give the Army time to regroup. Nevertheless, when the flights resumed, the crashes continued. By April a total of twelve pilots had died and the President had had enough. He announced that the air mail would be returned to the airlines as soon as possible. Not counting the tragic loss of life, the army's bill for flying the mail came to $3.76 million, or about $2.20 per mile. The average of 54¢ per mile paid to the airlines at the time of the cancellations, therefore, looked a bit less like the thievery Farley had charged.

Frontier Justice—the Black-McKellar Act

Actually, there is a good bit of evidence that Postmaster General Farley had little to do with the cancellation of the mail contracts, and had no sympathy for the Black Committee. He suffered much of the blame for it all like the good and faithful Democratic wheelhorse that he was. But when it came time to award new contracts he appeared to follow almost the same guidelines that Brown had used. True, as a face-saving measure for the Administration, he ruled that none of the airlines that had attended Brown's meetings in the spring of 1930 (called the "spoils conferences" by the Black Committee) would be allowed to re-bid for its former route. But then he made a rule that required only "responsible" bidders with multi-engine aircraft as bidders on the major routes, which effectively eliminated almost all of the independents, and saw nothing wrong with the awarding of contracts to the same big operators under different names. Thus, American Airways became American Airlines; Eastern Air Transport became Eastern Air Lines; TWA, a bit bolder, merely added "Inc." to its name, while United Air Lines remained unchanged because its original contract had been in the name of Boeing Air Transport.

Had the major airlines actually been guilty of collusion, as the new Administration had charged, they could have extracted some good contracts from Farley because the New Dealers were anxious to sweep the whole embarrassing mess under the rug. But the operators, then as now, were so distrustful of one another that they all bid their former routes at ridiculously low rates to make sure that they got them back, confident that a grateful government would negotiate fair adjustments later.

The New Dealers, of course, piously pointed to those low bids—as low as 19¢ per mile—as proof positive that the airlines had been grossly overpaid for flying the mail, as Senator Black had said all along.

Those adjustments may have indeed been made had it not been for Senator Black. He was determined to see to it that his position be fully vindicated and that the airlines be punished. He ignored the fact, made clear by testimony before his committee, that the airlines had no choice but to do as Brown directed if they wanted to stay in business. And he was surely frustrated by the fact that he could find no specific basis upon which to prosecute Brown under the law. (On 14 July 1941, Commissioner Richard H. Akers of the U.S. Court of Claims would rule that Brown had acted within the law in the awarding of mail contracts, but the public had long since forgotten the Air Mail Scandals by then.)

Senator Black, however, was to have his "victory." One way or another, it had to be established that the Roosevelt Administration had acted properly and responsibly in cancelling the air mail contracts. Therefore, Black teamed up with Tennessee Senator Kenneth McKellar to write a punitive law that was known as the Black-McKellar Bill (The Air Mail Act of 1934). Although more than 40 aviation bills dealing with the airlines were introduced in the Congress after the Army began flying the mail, the President put his weight behind the Black-McKellar Bill, which was all the endorsement any federal legislation needed at that time.

The mail was returned to the airlines on 8 May 1934, and the Black-McKellar Bill was enacted into law 12 June. Its provisions were harsh. It limited airline executive salaries to $17,500 per year, required the resignations of all airline officials who had attended Brown's "spoils conferences," ordered the separation of all airlines from manufacturing affiliates, and provided for re-bidding of all routes each year.

The new law confirmed the temporary contracts at the low rates the lines had bid the month before, an impossible situation for most, but the Interstate Commerce Commission, empowered to set the rates under the new law, later began to revise them upward after United and TWA threatened to cease operations. All of the airlines operated at little or often no profit during the following four years. Long-range planning was impossible as long as the routes could be threatened every year, while separation from manufacturing affiliates meant not only large auditing and legal fees, but reduced capitalization at a time when airline stocks were down to a fraction of their pre-cancellation values.

The airlines survived partly on faith during the middle 1930s, undoubtedly aware that nothing is very permanent in politics. The appearance of the Douglas DC series airliners, efficient enough that they promised profits from passenger fares alone, was a bright spot (by the end of 1941, 80 percent of all U.S. domestic airliners were Douglas DC-3s). An amendment to the Black-McKellar Bill extended the one-year contracts to three years in 1935, and Senators Pat McCarran and Clarence Lea each proposed constructive legislation designed to solve the industry's problems.

Eventually, the McCarran and Lea Bills were combined into a single measure that kept the best features of each. It took three years and the support of the President's son, Elliot, to obtain Roosevelt's backing, but the McCarran-Lea Bill became law on 22 August 1938 as the Civil Aeronautics Act. At last, sanity and stability had come to America's airways.

The Civil Aeronautics Act of 1938

The new law established the Civil Aeronautics Authority (CAA) and authorized the President to choose its administrator with the approval of the Senate. This five-member board could have no more than three members of a single political party, and none could be dismissed from the six-year term without cause. Since the Supreme Court had previously held that federal regulatory bodies were beyond the overt control of the President, the CAA was reasonably autonomous.

Concentrated in the hands of these five men was total power over civil aviation and the airlines. It formulated policy and controlled passenger rates. Its decrees were law with regard to civil aviation. It established an Air Carrier Economic Regulation Division to control the airlines, which immediately moved to protect the established air carriers by decreeing that only those lines to which it issued

Certificates of Convenience and Necessity would be allowed to offer scheduled air services—a practice that would remain in effect until the industry was deregulated 40 years later. No one could open new routes or extend or abandon existing ones without CAA approval, and it abolished the practice of competitive bidding for routes in favor of negotiated agreements (W.F. Brown was entitled to a chuckle of satisfaction when he heard of that).

In 1940, Roosevelt forced reorganization of the CAA and created the Civil Aeronautics Board (CAB) to regulate the airlines and investigate air accidents. The CAB was essentially independent (although President Truman would not hesitate to overrule it), while the CAA was tucked into the Department of Commerce. The President never explained why he demanded that amendment to the CAA Act, and, in practice, regulation of civil aviation appeared little affected by the change. The CAB would survive into the '80s, boss of the airlines under seven subsequent Presidents.

Independents and the New Deal

Some independent airlines squeezed through Postmaster General Farley's door when the mail contracts were re-bid following cancellation. One was Tom Braniff's Braniff Airways, which won United's Chicago-Dallas route. C.E. Woolman was back and his new Delta Air Lines was awarded a portion of American Air Lines' southern transcontinental route—the segment from Charleston to Dallas by way of Atlanta, Birmingham, and Shreveport; although that split up Brown's southern coast-to-coast route, American remained a transcontinental airline because it retained its New York-Los Angeles service by way of Nashville and Dallas.

A new company, Chicago and Southern Air Lines, received the Chicago-New Orleans route flying by way of St. Louis and Memphis. The route between Washington D.C. and Milwaukee through Norfolk and Chicago went to Pennsylvania-Central Airlines, which was formed by the merger of Pennsylvania Airlines and Central Airlines after they tired of draining one another's profits over most of the same territory.

Pop Hanshue's General Air Lines, which was the old Los Angeles-Salt Lake City route flown by Western Air Express and which was not included in the WAE-TAT merger, was successful in re-bidding that route as Western Air Express, and in 1941 was renamed Western Air Lines.

The old Varney Speed Lines reappeared as Varney Air Transport with the El Paso-Pueblo contract. This line would become Continental Air Lines in 1937 (and Varney next turned up operating an airline in Mexico).

Altogether, there were about 17 or 18 feeder lines—some new, some that had been struggling along a while without mail contracts, and some old hands with new identities. Most eventually disappeared into regional or trunk air carriers after WWII. Some expanded and changed their names; others merged into existing lines.

Pan Am, the Chosen Instrument

Concurrent with development of the U.S. domestic airline system was the extension of American wings to foreign shores, and one airline was destined—indeed, chosen by the U.S. Government—to fly the oceans: Pan American Airways.

Juan Terry Trippe built Pan Am with influence, money, ruthlessness, and ability. He was 28 years old when he started his airline in 1927, and he possessed all the necessary means. Born to wealth, educated in the best schools, and possessing as friends other young men of influential families, Trippe was assured of a hearing before those holding both financial and political power. He had muscle, and he used it.

Trippe had learned to fly in the Navy during WWI, and gained some experience between 1923 and 1925 operating a one-plane charter service under the name Long Island Airways. That part-time venture expired quietly and Trippe joined Colonial Airways as a minor executive. He left Colonial when it began scheduled service in 1926 on CAM-1, worked briefly for the banking firm of Lee Higginson & Company, and then tried to decide whether or not he should accept the security (and boredom) of a position in his family's business,

Trippe & Company, Brokers and Investment Bankers.

With aviation in his blood and avarice in his heart, the decision was not difficult. The time was right—the Lindbergh Boom was in its early days—and Trippe looked to the Caribbean and South America where the distances were great and the opportunity obvious.

It appeared from the beginning that Trippe was proceeding on something more than hope when he obtained the contract to fly the mail between Key West and Havana. That 90-mile stretch held little value in itself, but as one of two gateways to the Caribbean and Latin America, its potential was great.

Trippe's company—entirely on paper when he submitted his bid to Coolidge's Postmaster General New—was the Aviation Corporation of the Americas (not to be confused with AVCO); rival bidders were Florida Airways and Pan American, Inc.

Florida Airways, organized by John Harding and war aces Eddie Rickenbacker and Reed Chambers, wanted to extend its Atlanta-Jacksonville route, CAM-10, through Miami to Havana. Pan American, Inc. was headed by John K. Montgomery, a former Navy pilot, and G. Grant Mason. Their major backer was Lewis Pierson of the Irving Trust Company.

Pan American, Inc. seemed in a favored position to gain the mail contract for the Florida-Havana route, since Mason had obtained exclusive landing rights in Cuba from Dictator Gerardo Machado. However, it may have been that Trippe had already successfully argued his case for a monopoly on America's overseas airline routes because he went to Cuba and somehow convinced Machado that the agreement with Pan American, Inc. should be torn up and replaced with one granting Trippe's company exclusive scheduled landing rights in that country.

Thus armed, Trippe received the mail contract at the maximum rate allowed by law ($2 per mile for overseas routes). He then purchased—reportedly at bargain prices—both Pan American, Inc. and Florida Airways, renamed his company "Pan American Airways," and began scheduled operations 28 October 1927, when Capt. Hugh Wells took off from Key West with a handful of mail in the Fokker Trimotor NC-53. Passenger service to Havana began the following January.

From that 90-mile base route, Trippe reached out to the Dominican Republic and, in July 1928, gobbled up West Indian Aerial Express, which had been in operation just seven months, flying between Haiti and Puerto Rico. Pan Am got the mail contract and WIAE's pioneering equity was ignored, and therefore it seems safe to say that the Post Office Department and the Department of State had agreed with Trippe by that time that Pan Am should be America's "chosen instrument" to carry the U.S. flag on overseas airline routes.

In January 1929, Pan Am acquired control of *Compania Mexicana de Aviacion,* the Mexican airline that had been in operation since 1924, and that gave Trippe a route from Brownsville, at the tip of Texas, along the Gulf of Mexico through Guatemala to Nicaragua. Nine months later, service was extended through the Canal Zone to Barranquilla, Colombia, then continued eastward via Caracas to Port of Spain, Trinidad, where the system turned back northward to complete its circle of the Caribbean at San Juan, Puerto Rico. That locked up for Pan Am the two primary air routes to South America from the U.S., with one approach through Texas and Central America, the other by way of Florida and the Antilles.

Meanwhile, Miami replaced Key West as the Florida terminus, and Trippe began measuring the distances to Buenos Aires and Santiago, where some very formidable competition was firmly entrenched.

Anticipating continued expansion, Pan Am representatives fanned out across South America to obtain landing and operating rights while none other than Charles Lindbergh made the survey flights.

Trippe's toughest early antagonist was W.R. Grace & Company, a shipping, banking, trading monopoly that had for years been a mighty power on South America's west coast. Grace appears to have possessed as much favor in Washington as did Trippe, and in the end they formed a holding company in Peru, each taking 50 percent of the stock. The airline that resulted was called Pan American Grace Airways (PANAGRA), which flew the route from Buenos Aires up the South American west coast to Panama. Not until 1965 would PANAGRA's name drop from the list of international air carriers as Pan Am took over the routes and Pan Am subsidiaries were limited to domestic operations.

Trippe faced another determined opponent on South America's east coast, the New York, Rio & Buenos Aires Line. NYRBA (often referred to as "Near Beer" by its crews) was organized in 1929 by Ralph O'Neill, formerly associated with Bill Boeing, and was backed by James Rand of Remington-Rand (later Sperry-Rand, then the Rand Corp.), along with Lewis Pierson of the Irving Trust Company who, as a backer of Pan American, Inc., had jousted with Trippe before. J. K. Montgomery was also back, as an NYRBA officer.

NYRBA was well financed, and its fleet of 14 Consolidated Commodore flying boats was superior to Pan Am's assortment of Ford and Fokker Trimotors and Fairchild 71s. But NYRBA never had a chance; it lacked the U.S. Government backing that Pan Am enjoyed. The mail contract went to Pan Am, as usual, at the maximum rate, and on 15 September 1930 Trippe acquired NYRBA and added the entire east coast of South America to his routes. That closed the big circle, from Miami all the way around the South American Continent and back to Brownsville. In order to facilitate the takeover of NYRBA's Brazilian agreements, Trippe formed *Panair do Brasil* replacing *NYRBA do Brasil*. It, too, disappeared as an international carrier in the mid-'60s.

In 1932, foreseeing the need for future Pacific terminals, Trippe purchased Alaskan Airways from North American Aviation and the following year bought North American's 45 percent interest in Chiang Kai-shek's China National Aviation Corporation (CNAC). He also obtained a mail contract for a Boston-Halifax route in anticipation of the day Pan Am would fly the Atlantic.

As things turned out, Pan Am did not need the Alaskan bases to fly the Pacific because Glenn Martin, by then building airplanes in Baltimore, provided Pan Am with new, long-range flying boats that made possible a route to the Orient from California via Hawaii, Midway, Wake, Guam, and the Philippines.

That was a route that very much pleased the U.S. Navy, which lent Pan Am all possible assistance in establishing those bases, for the Navy had long been looking for an excuse to get such facilities on the stepping-stone islands of Midway, Wake, and Guam without upsetting the Department of State's delicate relations with Japan.

Pan Am began scheduled flights across the Pacific to Hong Kong on 22 November 1935. Capt. Ed Musick was in command of the Martin M-130 China Clipper on her initial flight.

Regularly scheduled Atlantic flights were delayed for several years while the British stalled negotiations on a reciprocal agreement until they possessed transport airplanes capable of flying the route. Finally, on 8 July 1939, Capt. Arthur LaPorte lifted his Boeing 314 flying boat Yankee Clipper from the channel at Port Washington, Long Island, and headed northeast on a Great Circle course for Southhampton, England.

By that time, Pan Am was confronted by a serious threat of competition at home as Hitler's aggressions in Europe at last prodded America into looking with concern at her deficient air defenses, and it became clear that the U.S. would be well-advised to build up its aerial lifeline to Britain in case of war and an attendant submarine menace in the Atlantic.

The last thing on Earth (or above it) that Juan Trippe wanted was to share one of his oceans with a competitor. But the exigencies of national security were about to change the rules.

A company formed as American Export Airlines, Inc. (AMEX) had been quietly preparing to challenge Pan Am's Atlantic monopoly for two years. It was headed by John E. Slater and James M. Eaton. Slater had earlier rescued the steamship company American Export Lines and put it on a paying basis through good management practices; Eaton was a former official of Ludington Lines and

Pan Am. AMEX was conceived as a subsidiary of the steamship company, but separated itself from its parent during the ensuing two-year battle with Trippe forces after filing for a route certificate in May 1939.

Trippe bitterly fought the AMEX application, with Nevada's Senator Pat McCarran serving as his point man in the Congress, and McCarran appeared to have blocked AMEX when he convinced friends in the House of Representatives that no money should be included in the Post Office Department's appropriation for an AMEX air mail contract. However, on the eve of America's entry into WWII, AMEX found so much traffic waiting that it didn't need the mail contract to survive. Then, when the military took over all U.S. airlines after the U.S. became a combatant, the Pan Am-AMEX battle was more or less suspended for the duration.

The precedent was established, however, and Juan Trippe knew that when peace returned he would be confronted with many AMEXes. The Department of State, too, was anticipating the problems of postwar global air travel.

Assuming Allied victory, it was clear that the U.S. alone would be in position to fly international civil air routes at war's end. The British had been forced to abandon transport airplane development in the late '30s in favor of desperately needed military aircraft. The U.S., its own terrible trial two years delayed, meanwhile produced such excellent transport craft as the Lockheed Constellation and Douglas DC-4.

The British had no intention of allowing U.S. airlines to monopolize postwar global air transportation, and they were not without effective bargaining power in the matter, because Britain controlled within her (then) far-flung empire many of the overseas terminals essential to such operations. Therefore, a conference was called in Chicago during November 1944 to divvy-up the postwar international air routes.

That conference wasn't as presumptuous as it may sound. No other country, except Australia and Canada, could hope to have overseas airlines soon after the war. But agreement between the U.S. and Britain did not come easily, because President Roosevelt had decided to discontinue the chosen instrument policy and encourage the operation of as many U.S. overseas airlines as the traffic would support.

The British looked with horror on that wasteful Yankee concept because their flag carrier, British Overseas Airways Corporation (BOAC), which had temporarily disappeared into war service, was a state-owned enterprise that Parliament regarded as a prestigious symbol of Britain's global influence. Really, old boy, one could hardly expect such an instrument of the Crown to grub about in a competitive market with a gaggle of freebooters, now could one?

However, after much haggling, British-American differences were settled when the airlines themselves organized the International Air Transportation Association (IATA), offered membership to all nations expecting to possess an overseas airline, and proclaimed that IATA would establish international air fares.

That allayed British fear of rate-slashing by the Americans and promoted an optimistic setting for a new conference at Bermuda early in 1946, from which delegates from most of the Free World emerged in happy agreement.

The single dissenter was Pan American. From the time of the Chicago conference, Trippe, aided by his most visible field general, Senator McCarran, along with a recent recruit, Senator Owen Brewster, had been fighting in Congress to re-establish the chosen instrument principle. They repeatedly introduced bills (S 1790, S 326, and S 1814) that would, in effect, return America to its airline policies of the early '30s. It was an interesting effort (McCarran never explained why the overseas air routes were of such importance to the desert state of Nevada), but all it engendered in the end was some resentment for Pan Am, especially in the White House. And since Harry S. Truman had occupied the White House since April 1945, resentment felt there was seldom borne in silence.

There is evidence that President Truman was ready to substantially chastize Pan Am. He reversed decisions of his own CAB and arbitrarily awarded to Braniff a route from Houston to Argentina and Brazil via the South American west coast—right through the heart of Pan Am's hitherto-private preserve. The President also announced that Eastern Air Lines would be granted a route from New Orleans to Mexico City, although Eastern had not asked for it. But then CAB Chairman L. Welch Pogue managed to calm the President, and the CAB was allowed to proceed with its task of parceling out America's portion of the world air routes to the 18 U.S. airlines that wanted to share Mr. Trippe's air.

On 1 June 1945 the CAB ruled that three U.S. air carriers would fly the Atlantic to European terminals: TWA, Pan Am, and American Airlines. Five months later, AA bought a 60 percent interest in AMEX and renamed that company American Overseas Airlines. United, alone among major U.S. domestic carriers, sought no international routes at the time. United's president, William Patterson (the California banker who had originally backed Pacific Air Transport), believed that not more than two dozen airplanes would be needed to serve the postwar Atlantic traffic.

The CAB's decisions on Latin America were handed down in May 1946. In addition to the routes given to Braniff and Eastern by the President, National Air Lines (begun in 1934 flying between Tampa and Daytona Beach), American, and Chicago & Southern (Chicago-New Orleans since 1934) were awarded certificates to operate south of the Border.

Then, 15 months later, the CAB decided that Pan Am's competitors in the Pacific would be United and Northwest Airlines (later, Northwest Orient), the latter flying from Seattle to Hawaii, soon extended to the Orient where it connected with TWA to effect round-the-world service. Pan Am, retaining its prewar route, would be the other round-the-world U.S. airline. United was authorized to fly between San Francisco and Hawaii.

These decisions, all carefully monitored by the President, resulted in the basic structure from which present-day U.S. overseas air routes evolved. There have been many additions, including foreign flag carriers (all government-owned or heavily subsidized), which have the right to serve the United States in return for U.S. airline service to and across their countries, and there have been route extensions and duplications as traffic steadily increased. Introduction of the jets brought a marked increase, and Pan Am was first with jet service when its new Boeing 707 began flying from New York to Paris in October 1958.

Juan T. Trippe stepped down as Pan Am's board chairman and chief executive officer early in 1968, and Pan Am clearly lost much of its elegant arrogance. Under Trippe, Pan Am was the haughty *grande dame* of the international airways. It set the standards.

Scheduled Airlines 1926-1931

	1926	1927	1928	1929	1930	1931
Airliners	NA	NA	325	525	600	590
Airway mileage .. (domestic)	8,252	8,865	15,590	24,874	29,887	30,451
Airway mileage .. (foreign)	152	257	1,077	11,456	19,662	19,949
Number of accidents	NA	25	86	137	91	126
Miles per fatal accident (thousands)	NA	NA	889.4	1,047.5	4,105	3,384
Fatal accidents	NA	4	12	24	9	14
Passenger fatalities	NA	1	15	18	24	26
Miles flown per passenger fatality (thousands)					4,322	4,614
Express/freight Domestic (thousands lbs)	3.5	45.8	210	249.6	359.5	788
Express/freight foreign (thousands lbs)	0	0	6.2	7.8	109	363
Mail carried (thousands lbs) Domestic	269	1,065	3,545	7,099	7,985	9,097
Mail carried (thousands lbs) Foreign	107	204.8	517.6	672.4	528.6	545.8
Mail payment to airlines Domestic (thousands)	$710	$2,561	$7,205	$13,873	$14,702	$19,900
Mail payment to airlines foreign (thousands)	$55	$82	$277	$3,168	$5,313	$6,983
Operators, domestic	11	16	31	34	38	35
Operators, foreign routes	2	3	5	6	7	7
Passengers carried, domestic (thousands)	5.7	8.6	47.8	159.7	374.9	469.9
Passengers carried, foreign routes (thousands)	0	(18 only)	1.8	13.6	42.5	52.3
Pilots, domestic & foreign	NA	107	308	562	675	690

Source: U.S. Department of Commerce

Scheduled Airlines 1932-1937

	1932	1933	1934	1935	1936	1937
Airliners	1,300	1,300	1,400	1,500	1,600	1,700
Airway mileage (domestic)	28,550	27,812	28,084	28,267	28,874	31,084
Airway mileage (foreign)	19,980	19,875	22,717	32,184	32,658	32,572
Number of accidents	115	101	73	62	70	50
Miles flown per fatal accident (thousands)	2,996	6,071	4,878	7,942	7,330	12,832
Fatal accidents	17	9	10	8	10	6
Passenger fatalities	25	8	21	15	46	51
Miles flown per passenger fatality (thousands)	5,862	24,850	10,727	24,037	10,690	10,777
Express/freight Domestic (thousands lbs)	1,033	1,510	2,133	3,822	6,958	7,127
Express/freight Foreign (thousands lbs)	566	942	1,316	1,689	1,391	1,786
Mail carried (thousands lbs) Domestic	7,393	7,362	7,411	13,268	17,706	NA
Mail carried (thousands lbs) Foreign	515	454	460	503	617	714
Mail payment to airlines Domestic (thousands)	$19,249	$16,467	$8,804	$10,662	$12,433	$13,100
Mail payment to airlines Foreign (thousands)	$6,939	$6,946	$6,917	$6,603	$7,290	$8,194
Operators, domestic	29	24	22	23	21	17
Operators, foreign routes	6	7	4	7	7	7
Passengers carried, domestic (thousands)	474	493	461	746	1,020	1,102
Passengers carried, foreign routes (thousands)	66	75	99	113	127	164
Pilots, domestic & foreign	566	543	503	652	690	749
Copilots, domestic & foreign	143	206	248	335	543	596

Source: U.S. Department of Commerce

This German airliner of 1930 was built by Focke-Wulf Flugzeugbau A.G., formed in Bremen in 1923. Its full cantilever wing and general lines, as well as the FW A-17 that preceded it, were strikingly similar to the Fokker Super Universals built in the U.S. at the time. Fokker's Atlantic Aircraft Corporation in New Jersey was formed at the same time Focke-Wulf appeared. Fokker's main plant was in Amsterdam, Holland. (Bill Selikoff)

A National Air Transport Curtiss Carrier Pigeon arrives in Kansas City with the mail as barnstormer Ben Gregory and airport kids watch from Gregory's OX-5 Standard. The New Swallow at left has a Curtiss-Reed twisted-metal prop. Gregory continued to barnstorm, with Ford TriMotors, until the eve of WWII. (Earl Reed)

Pop Hanshue's Western Air Express operated Liberty-powered Douglas M-2s over Contract Air Mail Route 4, Los Angeles to Salt Lake City. Profitable WAE was later forced to merge with unprofitable TAT to form TWA. (McDonnell-Douglas)

This Whirlwind-powered WACO had the same airframe as the OX-5 and Hisso-powered WACO 10s. Maximum speed was 125 mph with 220 hp. This airplane, flown by Charlie Meyers, won the 6300-mile 1928 National Air Tour, an annual reliability and efficiency event flown across the Central, Southern, and Western U.S. (Charles W. Meyers)

The Boeing 40A, powered with the 420-hp P&W Wasp, appeared in 1927 to fly Boeing Air Transport's Contract Air Mail Route 18, San Francisco-Chicago. BAT began operations with 25 of these machines, later fitted with 525-hp Hornets to make the 40B model. BAT became United Air Lines after Bill Boeing, Fred Rentschler of P&W, and Chance Vought got together to form the United Aircraft empire in 1928. (Boeing)

Clement Melville Keys put together the Curtiss-Wright family of companies, as well as North American Aviation. Keys was unsurpassed as a corporate wheeler-dealer during a period of aviation industry mergers when the foundations were established for important present-day companies. (Francis Dean)

Varney Speed Lanes began operation in 1931 with six Lockheed Orions flying between Glendale and Oakland, California, after Varney sold his original route in the Pacific Northwest to United. A total of 36 Orions was built 1931-1934, most fitted with 450-hp Wasps. Each seated six plus pilot and had a top speed of 225 mph. Walter Varney was another of the freewheeling second generation air pioneers, who made and spent money with equal gusto. He started several airlines, and was a founder of the present-day Lockheed Corporation.

The 1928 Lockheed Air Express, aimed at the emerging airline market, was a parasol-wing Vega with aft-positioned pilot's cockpit. Wasp-powered, an Air Express flown by Frank Hawks established a nonstop transcontinental record, but it was not ordered by the airlines; eight were built. (Texas Company)

The Fokker F-14A, a seven-to-nine passenger craft with a P&W Hornet engine of 575 hp had a maximum speed of 145 mph. The Air Corps purchased a few as the C-14, but it was not successful as an airliner. (Bill Selikoff)

The Cierva Autogiro, developed by Juan de la Cierva in Spain beginning in 1923, is pictured at the 1929 National Air Races in Cleveland. Pitcairn produced improved models in America during the '30s, as did Kellett. Rotor blades turned freely, and received no power from the engine except initial starting boost. The autogiros did not find a place in commercial aviation, but they did point the way for the helicopter. (Charles W. Meyers)

Walter Folger Brown gave the United States the best domestic airline system in the world employing his unique power as postmaster general. His methods were legal if high-handed, and the small independents were largely eliminated during Brown's reign as boss of the airways. (U.S. Postal Service)

John K. "Jack" Northrop's Alpha seated six passengers. Fitted with the 420-hp Wasp, the Alpha cruised at 145 mph and six were put in service by TWA in 1931. Northrop's Alpha, Beta, Gamma series featured all-metal stressed-skin construction, a concept developed in the U.S. by Northrop. (Don Dwiggins)

One of Erle Halliburton's Ford TriMotors of Southwest Air Fast Express. Halliburton was forced out of the airline business by W.F. Brown, but Halliburton did not go down without a fight and profited when his line became part of American Airways. (Burrell Tibbs)

This Stinson Model U airliner was in service with American Airways in 1932. These ten-passenger machines were powered with 240-hp Lycomings. Sixteen were built.

135

Interior of the Stinson Model U airliner.

Ludington Lines operated Stinson 6000s, which could be purchased for about ⅓ the price of Ford TriMotors. Ludington was the original "no frills" air carrier. Ludington captains sold tickets and handled baggage, taxied on one engine, cruised on automobile gasoline, and used the station facilities of the Pennsylvania Railroad. The Stinson 6000s engines were 215-hp Lycomings.

An Eastern Air Transport Stinson 6000B. Normal cruise was 115 mph with ten passengers; engines were 215-hp Lycomings. At least 40 were built, serving on nine airlines. Originally priced in 1931 at $25,900, the last ones sold for $19,500.

The Boeing 247, first of the "modern" airliners, appeared in 1933 with the first ones going to United; its cruising speed of 160 mph cut seven hours from the transcontinental schedule. Its only weakness was its size. Carrying but ten passengers, it was soon eclipsed by the larger Douglas DC-2. (Boeing)

The Fokker F-32 had four engines, two tractors and two pushers, Hornets of 575 hp each. It was a 30-passenger airliner that entered service with Western Air Express in April 1930. It cruised at 120 mph. Only seven were built.

The Stinson Model A was an attempt by that company to continue its line of minimum-cost airliners. The Model A appeared in 1934 and was operated by American and Central (the first of at least three Central Airlines)

The Douglas DC-1 made its maiden flight 1 July 1933. Only one was built, but it established the basic airframe for some 169 DC-2s and 10,000 DC-3s to follow. (McDonnell-Douglas)

The pilot of this Eastern Air Lines DC-2 apparently found the runway too short at Teterboro. A total of 131 DC-2s were delivered to the airlines in 1934; an additional 38 of the DST "sleeper" version were produced in 1936. The Navy took delivery of five DC-2s as R2Ds, and the Air Corps purchased 56 as the C-33, 34, 38, and C-39. (Charles W. Meyers)

Juan T. Trippe built Pan American Airways with ability, money, and the blessing of the U.S. Government. (Pan Am)

Pan Am's fleet of 14 Consolidated Commodores was acquired when Pan Am absorbed the New York, Rio, and Buenos Aires Line (NYRBA, commonly called "Near Beer"). The Commodores had a span of 100 feet; two P&W Hornets provided a cruising speed of 100 mph with a typical passenger load of 22. (Pan Am)

Pan Am owned 73 Douglas DC-3s, taking delivery of the first one in September 1937. The DC-3, with a pair of Wright Cyclones of 1000 hp each, carried 21 passengers at a normal cruising speed of 180 mph. (Pan Am)

The Sikorsky S-42 entered service with Pan Am in South America 16 August 1934. The S-42's P&W Hornets were rated at 700 hp each, giving this 32-passenger flying boat a cruising speed of 140 mph. Pan Am owned ten S-42s. (Pan Am)

Pan Am operated 28 Lockheed 049 Model Constellations, taking delivery of the first one 5 January 1946, and placing them in trans-Atlantic service 15 days later. These pressurized aircraft had a normal seating capacity of 45, were powered with four Wright R-3350s of 2200 hp each, and cruised at 250 mph. (Pan Am)

Chapter 6

General Aviation During the '30s

General aviation advanced sluggishly during the 1930s. Airplanes owned by individuals were mostly the open-cockpit biplane type, although WACO offered a series of cabin biplanes, while Stinson, Curtis-Robertson, Fairchild, and Lockheed produced cabin monoplanes. All except the Lockheeds were covered with cotton cloth that was filled, tautened, and finished with celulose nitrate "dope." Fuselages were framed with welded chrome molybdenum steel tubing; wings were built of spruce. The Lockheeds of the early '30s had plywood-covered wings (similar to those of the Fokker Trimotors), and fuselages built up of spruce strips bonded in concrete molds.

The most notable advances in aviation technology made during the '30s did not filter down to general aviation (all civil flying except airlines) before WWII. These were: all-metal stressed-skin airframes (introduced by John K. "Jack" Northrop with his single-engine Alpha transport of 1930), a successful controllable-pitch propeller (designed by Frank W. Caldwell at Hamilton-Standard Propeller Company in 1932), the first constant-speed propeller (in 1934), and the development of alloys for aircraft engine bearings (at P&W) that could withstand much greater loads than previous bearing materials.

The development that most influenced private flying probably was the appearance of the Continental four-cylinder engines of 50 to 65 hp. The Piper Cub airframe was literally waiting for such an engine, having been around as the J-2 Taylor Cub for several years. The low-horsepower Continentals—soon joined by Lycomings and Franklins in the same power range—encouraged production of the Taylorcraft immediately, the Stinson 105 and Luscombe a little later. The Cub and T-Craft could be operated for less than $3 per hour, and ensured profits to fixed base flight operators (FBOs) while making it possible for many thousands to learn to fly who otherwise could not afford it.

Private flying was (still is, and growing more so) a stratified activity, socially segregated in accordance with one's economic status. Young people were flying the Cubs and T-Crafts and older OX-5 biplanes. The local small businessman was more likely to own a late model cabin WACO, while the well-to-do flew Beechcraft Staggerwings and Stinson Reliants in the late '30s.

Private Airplanes of the '30s

A number of civil aircraft manufacturers, most of them small, managed to enter and/or remain in business as the Great Depression slowly eased during the middle and late '30s. Labor costs were low, even for skilled workers, and so many manufacturing plants had failed that finding a low-rent facility in which to build airplanes was no problem.

Edward E. Porterfield was a 35-year-old Kansas City Ford automobile dealer in 1925 when he traded a used Model T to barnstormer Blaine Tuxhorn for six hours of dual flight instruction. That led to establishment of the Porterfield Flying School, which in turn led to the formation of the American Eagle Aircraft Corporation in September 1925.

Lloyd Stearman's brother, Waverly, was hired to design the American Eagle biplane, and if that machine turned out to look remarkably like the Laird Swallow, it is understandable; after Matty Laird left Wichita, Lloyd Stearman became chief engineer at Swallow, and Waverly followed Lloyd in that position when Lloyd and Walter Beech quit Swallow to form Travel Air.

The first American Eagle proved to possess a pronounced proclivity for flat spins, however, so Porterfield hired Guiseppe Bellanca to redesign the machine. Approximately 400 of this model, the American Eagle A-129, were built between December 1928 and January 1931. Most were powered by the 90-hp OX-5, and were sold for $2495 at the factory.

When the American Eagle biplane expired with the depression, Porterfield brought in Noel Hockaday to design a $1000 low-hp airplane that, hopefully, would be more in tune with the tough times. Hockaday's creation, called the Eaglet, was a 470-pound parasol, open-cockpit monoplane powered with the 40-hp air-cooled Szekely engine.

About 70 Eaglets were sold, priced at $995, during 1933 and 1934.

Meanwhile, Hockaday redesigned the Eaglet airframe into a tandem-seat two-placer with enclosed cabin, planning it for the 70-hp Velie or LeBlond radial engine. This machine was certificated in 1935 with the LeBlond, and in 1937 with the 90-hp Warner radial. About 200 were sold.

In 1938, when the 65-hp Continental, Lycoming, and Franklin opposed-type engines became available, The Porterfield Collegiate, powered with a choice of the 65-hp engines, successfully competed with the Cubs and T-Crafts—especially after famed racing pilot Roscoe Turner became an officer of the company in 1939. More than 800 of the 65-hp Porterfields were sold 1938-1942. The deluxe model was priced at $2895.

During WWII Porterfield built WACO GC-4A troop gliders, but did not return to lightplane manufacturing after the war.

Porterfield was one of the more successful small firms building private airplanes during the '30s. Dozens of others appeared, produced a prototype little different from many already in the market, and quietly folded. Some prevailed for months, or even a year or two, turning out a handful of airplanes before they, too, disappeared. Almost every design could be classified as one of six types: two- or three-place open-cockpit biplane, open cockpit parasol monoplane, cabin biplane, four- to six-place cabin monoplane, two- to three-place cabin monoplane, or minimum-horsepower fun mono-plane. WACO, Stinson, and Fairchild (plus Piper late in the decade) could easily have satisfied the limited market for all types, but it was a time for dreamers, and they left us the Moonbeam, Kari Keen, Chic, and Red Arrow (all of which were certificated), along with many others possessing less intriguing names.

Spartan was another example of a marginally successful civil aircraft manufacturer. The Spartan Aircraft Company of Tulsa, Oklahoma, was organized in January 1928 by oilman William G. Skelly. He purchased the assets of the Mid-Continent Company, those assets mostly consisting of the prototype of a three-place open-cockpit biplane, and the services of its designer/builder, one Willis C. Brown.

Skelly built a new factory near Tulsa's municipal airport, opened a flying school (which would become one of the largest and best in the country), and put Brown's airplane, designated the Spartan C3, into production. Early models were powered with the imported (from Czechoslovakia) 120-hp Walter radial, but Skelly soon switched to the Wright Whirlwind J6-5 of 165 hp, which occasioned a redesign of the C3 airframe by George Hammond and Brown's early exit from Spartan.

Approximately 120 Spartan C3s were built 1928-1930 inclusive, most of them Whirlwind-powered. They were priced at $5975.

In 1930, Spartan introduced a pair of low-wing, open-cockpit airplanes—the C2-60, a side-by-side two-placer with a 55-hp three-cylinder Jacobs, and the C2-165 with tandem cockpits and the 165-hp Whirlwind. Neither was sold in significant quantity.

The Spartan C4 followed in 1931; except for the shape of its rudder, it could have passed for a Travel Air 6000. It was offered with a choice of engines ranging from 225 to 300 hp. The C4-300, with a Wright J6-7, was priced at $12,350, and eleven were sold.

The Spartan 7W Executive was the company's best achievement. It was a low-wing, four- or five-place all-metal cabin airplane powered by the 450-hp P&W Wasp Junior (R-985). The Executive was far enough ahead of its time that it would appear quite at home on any airport today. It first flew in 1935, and cruised at 190 mph. It was tested with a tricycle landing gear and, as the 8W Zeus, became a tandem two-placer with greenhouse canopy, several of which were sold to Mexico.

WACO built a bewildering array of biplanes during the '30s, both open-cockpit and cabin types. By 1930, WACO's home was the former factory of the Pioneer Pole & Shaft Company (buggy builders) in Troy, Ohio, and the WACO system of three-letter identifiers for each model began at that time. There were so many, however, that they will be meaningless to all but the most avid WACO fan. For example, the first letter of the model designation indicated the engine installation; the second letter, the wing type, and third letter, the type of airplane (open cockpit, cabin, etc.). Thus the WACO RNF Model was powered by the 125-hp Warner radial engine (R); "N" was type of wing, and "F" indicated open-cockpit sport/trainer. Since the WACOs used 16 different engines, just learning which letter represented which engine was too much for most people. In case that is exactly what you want to know, here they are:

Q . . . . . Continental 165-hp radial
R . . . . . Warner 110/125-hp radial
U . . . . . Continental 210-hp radial
I . . . . . Kinner 125-hp radial
K . . . . . Kinner 100-hp radial
M . . . . Menasco 125-hp in-line
P . . . . . Jacobs 170-hp radial
Y . . . . . Jacobs 225-hp radial
Z . . . . . Jacobs 285-hp radial
A . . . . . Jacobs 330-hp radial
C . . . . . Wright 250-hp radial
E . . . . . Wright 320-hp radial
W . . . . Wright 420-hp radial
H . . . . . Lycoming 300-hp radial
S . . . . . P&W 450-hp radial
V . . . . . Continental 240-hp radial

The Civilian Pilot Training Program

The proliferation of the two-place, low-horsepower trainers was due, in part at least, to the Civilian Pilot Training Program (CPTP), later called the War Training Service. The CPTP was the idea of CAA Member Robert H. Hinckley, a former FBO in Ogden, Utah. It was sponsored by the CAA and was initially conceived as a stimulus for general aviation by providing 35 to 50 hours of flight training to college and university students. The plan was to turn out 20,000 private pilots per year, and the initial funding would come from the National Youth Administration, one of the New Deal's economic recovery programs. (Although the country began a slow recovery from the Great Depression in 1934, jobs were less than plentiful and pay was low throughout the remainder of the '30s. At P&W, for example, the average wage was 81¢ per hour in 1939. Two years earlier, Charles Taylor, who had helped the Wright brothers build the world's first successful airplane, was discovered running a lathe for Dutch Kindelberger at North American Aviation for 37.5¢ per hour.)

So the CPTP was strictly a civilian project, and no military training was to be given, although the military implications could not be ignored. A vastly increased pool of young private pilots was an obvious asset to the nation's defense, and was almost certainly the reason President Roosevelt supported it. The clear need to build up the Army Air Corps was evident to the President by the late summer of 1938; while Britain's Prime Minister Chamberlain was in Munich agreeing to Hitler's demand for the Sudetenland, Roosevelt dramatically called for an aircraft production rate in the U.S. of 10,000 annually (the total for 1938 was 1800 military, 1823 civil).

In the President's message to Congress on 12 January 1939, Roosevelt asked for $10 million in extra NYA monies to fund the CPTP. Congressional doves argued against the project, but the CPTP Act became law on 27 June 1939 and the first 330 students begain training in 13 colleges. When the program ended in the summer of 1944, 1132 colleges and universities had been involved and 435,165 student pilots, including several hundred women, received Private Pilot Certificates after training with 1460 FBOs.

The CPTP was open to graduate and undergraduate students between the ages of 18 and 25 who were U.S. citizens and could pass the student pilot's physical examination. One in ten could be female, although that ratio was never remotely approached.

The CAA administered the project with amazing efficiency, and tried—successfully in most cases—to hold payments to the contracting FBOs to $6 per hour. That alone encouraged the contractors to obtain the economical Cubs and T-Crafts, which met the CAA's requirement that the trainers possess at least 50 hp.

Gradually, the CPTP felt the pressure of the developing world crisis and its secondary purpose became its primary one. In the summer of 1941 girls were no longer accepted, and a military pledge was required. Then, less than a week after the U.S. was plunged into WWII with the surprise attack on Pearl Harbor, the President, by executive order, declared the CPTP to be exclusively devoted to the training of men for ultimate service as military pilots. Shortly thereafter its name was changed to the CAA War Training Service.

More than 10,000 airplanes were employed as trainers in the CPTP/WTS, and approximately half of those were an assortment of civilian machines purchased from private individuals by CAA field men and leased to the contract flight schools. By early 1943 the operators could not replace their training planes and spare parts were almost nonexistant for essential maintenance and overhaul. Therefore, their training fleets had to be somehow augmented. In December 1944, after all contracts between the CAA and the flight operators had terminated, these airplanes were offered for sale in the surplus market.

While training more than 435,000 student pilots, the CAA's program resulted in fewer than 100 fatalities—a record so unbelievable at the time that the CAA was accused of withholding accident information.

As so many developments and innovations in aviation, the contribution of the CPTP have been widely misinterpreted. Much has been made of the fact that many American aces of WWII (Richard Bong, Robert Johnson, Joe Foss, et al), were products of the CPTP, but the CPTP trainees were simply civilian private pilots, and they had to pass the military's tougher physical exams and then begin their Air Force or Navy flight training at the same level as those who had not participated in the CAA program. (Many, as War Service Pilots, were civilians who wore military khakis, without military insignias or rank, and served as flight instructors and transport pilots.)

The greatest impact on the CPTP was less direct. It provided a significant infusion of cash into a faltering industry. It brought profits to many hundreds of small FBOs who logically should have gone out of business long before, but who had stuck it out because they had already invested too much of their lives in aviation. It precipitated the sale of thousands of lightplanes (and engines), which in turn increased employment and helped to expand manufacturing facilities that would soon be critically important to the defense of our freedoms. And then when meaningful military flight training programs were begun in 1940, the thousands of CPTP-trained people were identifiable and largely available—mostly as volunteers—as prime recruits. That was a circumstance that saved a great deal of time, and time was the most precious commodity of all once the nation's awful peril was belatedly faced by our leaders.

Never before or since has American aviation's controlling authority initiated such a bold and useful project; seldom has it spent the taxpayers' money so wisely. The total cost worked out to $400 per student.

General Aviation 1932-1937*

	1932	1933	1934	1935	1936	1937
Airplanes (licensed & unlicensed)...	9,760	8,780	7,752	8,613	8,849	10,446
Total accidents	1,951	1,603	1,504	1,517	1,698	1,917
Miles flown per accident	40,071	44,431	50,267	55,871	54,959	53,728
Number of fatal accidents	208	182	186	164	159	185
Miles per fatal accident (thousands)	375.8	391.3	406.4	516.8	586.9	556.7
Airports	2,117	2,188	2,297	2,368	2,342	2,299
Repair stations	127	139	148	174	181	193
Flight schools	22	19	20	24	27	29
Airplanes certified	7,330	6,896	6,339	7,371	7,424	9,152
Mechanics certificated	8,373	8,226	8,156	8,432	8,738	9,314
Pilots certificated	18,594	13,960	13,949	14,805	15,952	17,681
Student pilots certificated	11,325	12,752	11,994	14,572	17,675	21,779
Aircraft produced (civil; includes airliners)	803	858	1,178	1,251	1,869	2,824
Miles flown (thousands)	78,178	71,222	75,602	84,755	93,320	102,996

*Source: U.S. Department of Commerce

Charles Lindbergh purchased a Lockheed Sirius in 1930 and used it to survey future world airline routes planned by Pan Am, taking wife Anne along as a passenger and radio operator. Originally Wasp-powered, Lindbergh installed a 575-hp Cyclone, along with a sliding canopy. The standard version with a 420-hp Wasp cruised at 150 mph and sold for $18,985. (Lockheed-California)

The Lockheed Altair of 1931 was the sport version of the Orion, with canopied, tandem seating, and the first retractable gear offered on a civilian production airplane. (Lockheed-California)

The gull-wing Stinson Reliants, SR-7 through SR-10 models, were built during the late '30s with Lycoming engines of 225 hp through 300 hp. More than 1000 were produced, the last ones going to the USAAF as C-81s.

George Pollard's Thomas-Morse Scout advertises an air movie parked in the street before a theater in downtown Lawton, Oklahoma, in the early '30s.

The WACO S3HD of 1934 was an attempt at a low-cost fighter-bomber for the Latin American market. (Warren Shipp)

The WACO Model E Aristocrat had a 300-hp Lycoming; was four-place with a 135 mph cruising speed.

The Spartan 7W Executive was the most advanced private airplane to appear prior to WWII. Powered with the P&W R-985 Wasp Junior of 450 hp, it cruised in the 150 mph range. This one was photographed at New Garden, Pennsylvania, in 1975. (Francis Dean)

Beechcraft C17B with a 285-hp Jacobs engine as it appeared in February 1936. Maximum speed was 177 mph. (Beech)

The hybrid Lockheed Orion 9E Explorer flown by Wiley Post, with famed humorist Will Rogers as a passenger, crashed near Point Barrow, Alaska, taking the lives of both, 15 August 1935. (DuPont Eleutherian Library)

The Lockheed Model 12 followed the Model 10 Electra in 1936. A scaled-down Electra, seating six passengers and a crew of two, the Model 12 cruised at 206 mph. (Lockheed-California)

Chapter 7

Air Racing

There was something very special about American air racing between 1929 and 1939. Each machine was Depression-born, conceived by men of modest means and built with limited private funds. Common sense and hope largely substituted for slide rules and engineering degrees. These wondrous airplanes were normally operated mostly on guts and on credit, and everyone understood that. These were the vehicles of the air frontiersmen—and therein lay their enormous appeal. The world's landplane seed record could be claimed by those who were, essentially, amateur aircraft builders.

It began in 1929 when Walter Beech's Travel Air racer, the Model R, or "Mystery Ship," as news reporters called the red-and-black low-winger, outsped the Army's fastest pursuit plane, a Curtiss P-3A Hawk, in Event 26 at the Cleveland National Air Races.

The National Air Races, begun as an annual event in 1924 at Dayton, Ohio, and in ensuing years held at New York, Philadelphia, Spokane, and Los Angeles, had been dominated by the military until Beech's Mystery Ship appeared. The civilians had to be content with cross-country races ("derbies") to the NAR's host city, in which speed was less important than good dead reckoning navigation.

Beech's flair for showmanship drew attention to his racer. Designed by Travel Air engineers Herb Rawdon and Walt Burnham, it was built in secret, and Walter allowed the press but a brief look at it when it arrived in Cleveland prior to the races, then moved it into a hangar and covered it with a canvas. He did nothing to discourage the impression that his "little" racer was sort of a sheep among wolves, a decided underdog pitted against the powerful military fighters. As a former barnstormer, Walter appreciated the value of a good hype.

Actually, the P-3A Hawk was fitted with a P&W R-1340 rated at 450 hp at that time, and had a gross weight of about 2600 pounds. The Travel Air was powered with a Wright R-975, normally rated at 300 hp, but souped up with an increased impeller ratio and higher compression to 420 hp. Taken together with the Travel Air's 1940-pound gross weight and much cleaner aerodynamics, that should have made the P-3A the underdog; Walter could not have been surprised when his racer, flown by airline pilot Doug Davis, easily won the 50-mile main event at an average speed of 194.9 mph, despite a turn in mid-course to circle a missed pylon. The P-3A Hawk, flown by Army Lt. R. G. Breene, was second at 186.8 mph. Trailing among the five other entries was Roscoe Turner in a Lockheed Vega.

The fact remained, however, that for the first time a civilian airplane had beaten the best the military had to offer, and that encouraged a lot of others to try. Of course, the deciding factor was meaningful prize money, and the sources of that money should be recognized for their role in fostering the golden age of air racing.

Event 26 of the 1929 NARs was sponsored by the Thompson Products Company, which produced, among other things, the sodium-filled exhaust valve for aircraft engines invented by Sam Heron and Ed Jones. Thompson was clearly impressed by the event and established the Thompson Trophy that would go to the winner of an unlimited class air race each year, and which carried with it a nice cash prize. For the 1930 Thompson, the cash prizes would be $5000 for first place, $3000 for second place, and $2000 for third. Those rewards would grow until by 1938 the Thompson paid $22,000, $9000, $4500, $2500, and $1800 for the first five places—and those were Depression-era dollars, equal to five times or more those amounts today.

The transcontinental was sponsored by Bendix Aviation Corporation beginning in 1931, and its purses ranged from $7500, $4500, and $3000 for the first three places that year to $13,000, $5000, $5000, $2000, and $1000 for the first five places in 1938.

The major lower-horsepower contest was the Louis W. Greve Trophy Race, sponsored by the Cleveland Pneumatic Tool Company, for airplanes with engines of 549 cubic inch displacement or less. It was held as part of the NAR 1934-1939 inclusive. In 1934-35 the Greve Trophy went to the pilot who collected the most points in three events; the purses were small. After that, the Greve became a single race that paid $4900, $2125, and $1190 in 1936; this was upped to $12,000, $5000, and $2000 in 1938.

A host of other closed-course races provided many thousands more in prize money, largely to the contestants in the limited horsepower categories. Some of those pilots took home more money during the early '30s than did the winners of the unlimited races, because the smaller one's engine, the more races one was eligible to enter.

The Thompson Trophy Race was for 100 miles 1930-1934 inclusive—either 20 five-mile laps, or 10 ten-mile laps. In 1935-36 it was lengthened to 150 miles with ten-mile laps, and in 1937 the Thompson consisted of 20 ten-mile laps. Finally, in 1938-39 it was stretched to 30 ten-mile laps.

The only significant rule in flying the transcontinental Bendix Trophy Race was that each contestant must arrive at the finish line in Cleveland between noon and 6:00 P.M. on race day—which meant before-dawn takeoffs from Los Angeles.

The Greve Trophy Race was a 200-mile 20-lap event.

The National Air Races were managed, quite ably, by brothers Clifford W. and Phil T. Henderson 1929 through 1939 inclusive, by which time the seven- to nine-day programs had shrunk to three days and half as many events, with a corresponding drop in the number of entries. Although spectators continued to number between 70,000 and 100,000, and prize money was bigger than ever, it was clear that the 1939 races marked the end of an era. The 1930 Thompson was won with a P&W Wasp producing 450 hp; Roscoe Turner's winning aircraft in the 1939 Thompson was fitted wth an R-1830 Twin Wasp rated at 1200 hp. While the horsepower almost tripled during those ten years (or perhaps quadrupled; Turner claimed that he was getting 2000 hp from his engine in the 1939 race), the winning speeds in the Thompson increased by 81 mph. The privately built racers had clearly reached their limit. With very few exceptions, the Thompson Trophy racers were, essentially, homebuilts, and to significantly improve upon the performances of existing racers would require more power—and a *lot* more money.

In any case, America was to have few more holiday weekends in which to enjoy such a spectacle. While the 1939 air races were being flown, Hitler's panzers were rolling across Poland to signal the beginning of WWII.

Air Racing After WWII

The National Air Races were resumed at Cleveland over the Labor Day weekend in 1946, and nothing was the same as before. There was a Thompson and Bendix race flown by war surplus fighters, and one for military jets. The transcontinental race for propeller-driven airplanes was notable for the speeds attained—Paul Mantz averaged 435.5 mph between Van Nuys, California, and Cleveland to win in a P-51C-10—but one could get that figure from the evening newspaper; cross-country speed derbies were never spectator events.

The Thompson race for propeller airplanes wasn't much better. True, the P-51 Mustangs, P-39 Airacobras, and P-38 Lightnings circled the race course at speeds 100 mph faster than their predecessors, but it was 30 miles around each lap and the spectators didn't see much of the racers.

The 1947 NARs saw the 300-mile Thompson race flown in 15-mile laps, which gave the crowd much better viewing. A pair of Goodyear-built F2G-1 Corsairs fitted with 4000-hp P&W R-4360 engines edged out a P-39Q powered with a 2000-hp Allison V-1710-135 in that still-popular contest. Cook Cleland flew the first-place Corsair, averaging 396.1 mph.

The Bendix was an all-Mustang affair, and Mantz bettered his mark of the previous year to win with an average speed of 460.42 mph.

Additional events introduced that year included the Kendall Oil Trophy which was limited to P-51s, the Tinnerman Trophy for Bell P-63 Kingcobras, the Halle Trophy for women flying AT-6s (redesignated T-6 by the Air Force in 1948), and the Sohio Trophy for Lockheed P-38s.

The most notable of the new events was the Goodyear Trophy Race, limited to racers fitted with the 190-cubic inch continental engine. It was a serious attempt to promote design innovation and encourage individual builders. The speeds would (presumably) be well below 200 mph, but that would allow two-mile laps for good spectator viewing. (The 190-cubic inch continental was the C85, normally rated at 85 hp, although undoubtedly souped up to as much as 125 hp for racing.)

William Brennand flew a Steve Wittman racer to victory in the first Goodyear race at a speed of 165.85 mph. By 1949 the winning speed was up to 177.3 mph, and two dozen of the little racers had appeared by that time to share in the $25,000 Goodyear prize money.

Meanwhile, Continental Motors Corporation decided to sponsor a race in 1948 for the Goodyear-class racers. The Goodyear event was planned for only three years, enough to get the movement off to a good start; therefore, Continental stepped in to sponsor the little racers through 1951, offering purses at the mid-winter Miami Air Maneuvers for the first three years, and at Detroit in 1950 and 1951. By that time the winning speed was up to 197.2 mph for a 15-lap 37½-mile race, achieved by John Paul Jones in *Shoestring*.

The National Air Races were suspended between 1949 and 1964 (Detroit called the 1951 190-cubic inch races "Nationals" but that was stretching things a bit). The Cleveland NARs were discontinued after Bill Odom's modified P-51 apparently entered a high-speed stall in a tight turn on the second lap of the 1949 Thompson and crashed into a house. Odom was killed, along with a mother and her young son.

The pilots of the little Goodyear-class racers flew the Continental contest two more years, but after 1951 were effectively grounded (except for an occasional exhibition race at a local airshow) until "big time" racing was revived at Reno in 1964.

Modern Air Racing

The National Air Races were resuscitated by Bill Stead, a Nevada rancher, pilot, and champion powerboat racer. Stead's choice of a site was Sky Ranch, a mesquite-covered area in Spanish Springs Valley near Reno. He proposed four classes of racers: Unlimited, 190-cubic inch (soon to be called

Formula I), stock production airplanes for women pilots, and the Sport Biplane Class for homebuilt two-wingers.

The 1964 event established that both pilots and spectators in significant numbers would support air racing if the "big iron" competed. (Later, races in other cities without the Unlimiteds failed to attract meaningful crowds.)

The Unlimited machines were mostly unmodified WWII-vintage fighters, but not the early-model Mustangs and heavily-muscled Corsairs of the 1946-1949 Thompsons, most of which had not been preserved because such machines represented relatively small investments during the late '40s and there was almost no market for them during the '50s. Meanwhile, such airplanes had become very scarce. America's WWII combat airplanes were available as surplus for a very short time and apparently only on the West Coast. Not many passed into civilian hands before the sale was halted, and the overwhelming majority of them were sold only to be melted down for the metal they contained. Had it not been for the fighters diverted from surplus to the Canadian and Latin American Air Forces in the late '40s, and later available to U.S. civilian buyers, it it unlikely that any would still be racing into the '80s.

Along with the Mustangs came several Grumman F8F Bearcats to the 1964 races, and the first unlimited race at Reno was won by Mira Slovak in a Smirnoff-sponsored F8F-2 fitted with a P&W R-2800-34W engine normally rated at 2300 hp. Just two seconds behind Slovak was Clay Lacy in a P-51D, although Darryl Greenamyer would have beaten them both in a modified F8F-2 had he not been disqualified for refusing to fly from the dirt airstrip, flying instead from Reno Municipal Airport. Greenamyer would be back, however, and would take six championships in seven years at Reno. In 1969, he would set a new world's speed record of 480 mph for propeller-driven airplanes.

In the '70s, the Mustangs began to dominate the Unlimited Class, and a highly-modified P-51 fitted with a Rolls-Royce Griffon engine turning a pair of contrarotating propellers established a new world's speed record for prop-driven airplanes of 499 mph. (The Griffon, a development of the famed WWII Merlin, has a normal rating of 2375 hp at 2750 rpm.) The record-setting Mustang, owned by Ed Browning of Idaho Falls, Idaho (sponsor of the *Red Baron* racing team, which also included a T-6 Class racer), flown by Steve Hinton, was completely destroyed at the conclusion of the 1979 Championship Unlimited Race at Reno when its engine seized, out of oil. Hinton survived.

Formula I Racers

The little Formula I racers have found a home at Reno, although their pilots bicker a lot among themselves. Sadly, Bill Stead was one of a half-dozen Formula I pilots who were killed during the '60s and '70s — about the same number who died in crashes of the Unlimiteds.

These machines, formerly the 190-cubic inch racers, became Formula I airplanes on 1 January 1968, and at that time the rules were changed to allow an increase in engine size to 201 cubic inches. This, in effect, meant the Continental 0-200 engine, normally rated at 100 hp (the standard Cessna 150 engine for many years), and that powerplant has offered many possibilities to gifted enginemen for uprating its advertised output.

True, the rules say no souping up, but the nature of Mankind in general, and of those who race in particular, makes such a rule a challenge rather than a curb. One may be assured that the Formula I machines all have more than 100 horses beneath their tight cowlings. That has resulted in a lot of finger-pointing and some temporarily derated engines, one court injunction, and the suspension of Formula I champion Ray Cote. Ray's problem was that he won the championship every year 1968 through 1975, and that was just too much. His official transgression was an illegal camshaft. Cote was back in 1979 to establish a record qualifying speed of 246 mph in *Shoestring*, but placed second behind John Parker in Parker's new *American Special* in the championship race.

In 1980, the contending factions split into two groups, Formula I and IMF, and although 25 Formula I-type racers appeared for the 1983 Reno races, the dissention continued. This racing class will probably survive.

The only female Formula I racer was Judy (Mrs. Les) Wagner, a serious competitor who usually placed among the top four finishers. Unfortunately, Judy and her husband were killed in the crash of a private airplane in 1982.

Sport and Racing Biplanes

The Sport Biplane Class of racers was in some turmoil from 1968, when the first of four specially designed pure racers turned up at Reno. It was Don Beck's all-metal *Sorceress*, and she cast her spell over everything in her racing class. She was 50 mph faster than the other little bipes, and *Sorceress* ruled until 1973 when Sidney White showed up with another super bipe, *Sundancer*. The sport biplanes raced for third place. But after 1982 and the appearance of a *third* pure racer in the homebuilt biplane class, the Racing Biplane Class was created to separate the pure racers from the gentler fun machines.

A Rutan biplane was prepared for the Racing Biplane Class in 1983, but crashed at the beginning of that event. Another pilot lost interest and offered his airplane for sale. Since only three such racers were built during the 15 years following the first appearance of *Sorceress*, it seems that the Racing Biplane Class has a limited future.

The Reno races were moved to the deactivated Stead AFB (named for Bill Stead's brother) after their dusty beginning at Sky Ranch, and after two decades are well-organized with many added attractions, including a program (differing each day) featuring the best aerobatic pilots, the Air Force or Navy air demonstration teams, the Army's Golden Knights, and numerous ground activities designed to entertain the children and wives of air racing buffs.

The most remarkable thing about the Reno National Air Races is that this air extravaganza continues to grow into the '80s. Many industry observers predicted years earlier that the principal attraction, the Unlimited Class racers, would expire by the end of the '70s as parts for the big engines became unavailable. Production of the high-horsepower reciprocating powerplants had ceased in the early '50s. But in 1983 more unlimiteds (32) were registered at Reno than ever before.

James Haizlip with the Shell-owned Travel Air Mystery that he flew to second place in the 1930 Thompson Trophy race, and to first place in the 1000 cu/in contest.

Speed pilot Frank Hawks with the Laird Solution. This aircraft was owned by Goodrich Rubber Company, and flown to victory in the 1930 Thompson by Charles "Speed" Holman. (E.M. Laird)

The highly modified Curtiss Hawk, the XFC6-6 Navy racer in which Marine Capt. Arthur Page crashed to his death during the 1930 Thompson race, apparently overcome by carbon monoxide gas from his engine exhausts. (NASM)

Jimmy Wedell in his Wedell-Williams Number 44, winner of many air races during the early '30s. Wedell established a landplane speed record of 305.33 mph in this machine in 1933. He was killed 24 June 1934 while instructing a student pilot.

Roscoe Turner and his Wedell-Williams racer as it appeared for the 1934 races. Turner won the Thompson at 248.13 mph.

Lowell Bayles and the Gee Bee Model Z, the first of what has since become known as the "Killer Bees." Bayles won the 1931 Thompson in this airplane, then died in it on December 5th that year during an attempt to establish a new world's speed record. The best official speed of the Model Z was 286 mph. Its engine was a 525-hp P&W Wasp.

The Gee Bee Model Y designed as a two-place sport plane, it and a sister aircraft were used for little except racing. The Model Y's best closed course speed was 187.5 mph. Florence Klingensmith died in one when it shed wing fabric during a race.

Ben O. Howard's *Ike,* a sister craft to Howard's *Mike* ("they look alike," the whimsical Howard explained as his reason for so naming them). Ike and Mike were perennial winners at the National Air Races. Howard was aided in the design and construction of his racers by Gordon Israel, who later contributed to the design of the Learjet, among other things. Ike and Mike had dash speeds in the 240-mph range with engines of 225 hp.

John Livingston, the inspiration for Bach's *Johnathon Livingston Seagull,* and his Monocoupe racer. Air racing was highly profitable for Livingston because he took advantage of the "cubic inch rule" and entered many races against more powerful machines. He did not expect to win, but often placed in the money while winning more than his share in his racing class. His Monocoupe had a normal top speed of 135 mph; he increased that by 40 mph with detail improvements. (Dave Jameson)

A pair of deHavilland Comets were prepared for the great 1934 London-to-Melbourne air race, which was won by G-ACSS. American entries included Jackie Cochran flying a Gee Bee Q.E.D., who dropped out in Bucharest, and Clyde Pangborn and Roscoe Turner in a Boeing 247 airliner, who placed third.

Flown by Joe Jacobson in the 1936 Greve Trophy race, Howard's *Mike* is damaged in a landing accident.

The Miles and Atwood *Miss Tulsa* was built in 1933 and had a dash speed of 234 mph with a 200-hp Menasco Pirate engine. Lee Miles died in this airplane when a wing flying-wire fitting failed in flight and the racer shed its wings at low altitude.

Howard Hughes' H-1 racer, in which he established a world landplane speed record of 352.4 mph 13 September 1935. On 19 January 1937 he averaged 327.1 mph between Burbank and Newark for a transcontinental record. (Hudeck Collection)

The 85-hp midget racers built for the postwar Goodyear-sponsored events were the parent aircraft of today's Formula I machines. Some of these original airframes, modified, are flying in the Formula I class in the '80s. (Peter M. Bowers)

165

This clipped-wing Corsair F2G-1, with a 4000-hp P&W Wasp Major engine, was flown to victory by Navy pilot Cook Cleland in the 1947 Thompson race at a speed of 396.1 mph. Ten F2G-1 and -2s were built near war's end by Goodyear as Kamikaze killers. None saw combat. (Peter M. Bowers)

Today's National Air Races have been held at Reno, Nevada, since the mid '60s. The WWII-era fighters continue to dominate the unlimited events. (Jim Larsen)

"Unlimited" has a broad interpretation at the National Air Races; Clay Lacy once showed up with a DC-7 airliner. This A-26 was flown by both Wally McDonald and Dwight Reimer, at Mojave and Reno, but was not competitive. (Jim Larsen)

Howie Keefe placed third in the Unlimited Class in 1976; is pictured at the Mojave races in 1974. (Dustin Carter)

The T-6 events are popular at the National Air Races, but have suffered more than their share of fatal accidents. Pictured is Bob Metcalf at Reno. (Dustin Carter)

The late Bob Downey's Formula I racer *Ole Tiger* at Reno. Downey usually finished in the money during a 20-year racing career. He was a paint manufacturer in real life. (Jim Larsen)

A stock P-38 at the Reno races in 1971. The Lightning did not prove competitive. (Roger Besecker)

One of the winningest Formula I pilots was the late Bill Falk (R), shown here with wife, crew, and his racer, *Rivets*. Today's Formula I machines are capable of dash speeds in excess of 240 mph. Engines are 0-200 Continentals, normally rated at 100 hp. (Jim Larsen)

The Mace R-2 *Snark,* built and flown by Harvey Mace; also flown by Ralph Twomby, at Reno during the early '70s. (Dustin Carter)

A new approach to the Formula I Class racer is Jim Miller's *Texas Gem,* pictured at Reno in 1974. The *Gem* has not sparkled with speed. (Jim Larsen)

Don Fairbanks' Knight Twister has shown closed course racing speeds in the 170-mph range in the sport Biplane Class at Reno. The Knight Twister was originally a 1930s design. (Dustin Carter)

Frank Sanders' Hawker Sea Fury, raced in Unlimited Class events by Sanders, Bob Metcalf, and Lyle Shelton. (Dustin Carter)

Pitts Specials are always entered in the Sport Biplane Class contests. This one was flown by Don Beck during the '70s. Beck's radically-configured, all-metal *Sorceress* was fastest. (Dustin Carter)

The Salina-Hoffman homebuilt biplane racer as seen at the 1974 Reno and Mojave races had a canopy that enclosed the entire center-section struts. (Dustin Carter)

Mike Smith's Bell P-63 *Kingcobra* at Reno. (Jim Larsen)

Pat Palmer's T-6 racer, first place in its class '74, '75, and '76. Since 1967, the T-6s have also raced at Mojave. (Jim Larsen)

Chapter 8

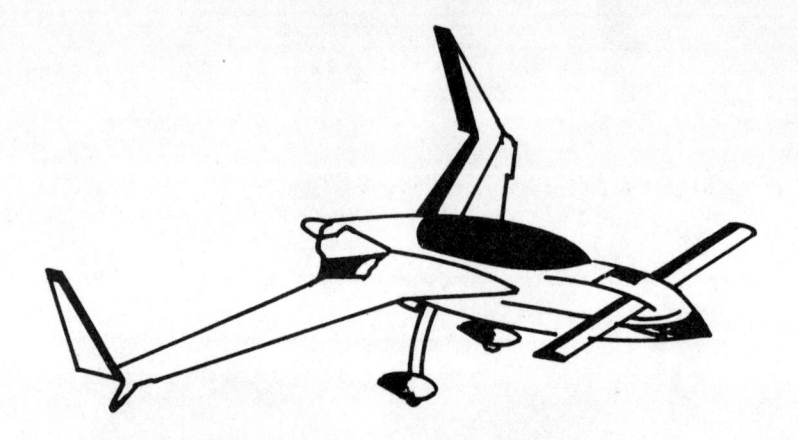

The Homebuilts

Amateur aircraft builders have been constructing personal airplanes at home since the beginning of manned flight, and their biggest problem—aside from the disapproval of alarmed spouses—has usually been the acquisition of a proper engine. During the 1920s, homebuilt airplanes were often powered with motorcycle engines. Automobile engines were tried, especially the Ford Model A, which was acceptable in the Pietenpol Air Camper. There were some WWI surplus Lawrance A-3 engines available, 28-hp air-cooled opposed type (which had powered the Breese Penquin Trainer; it did not fly, but taxied rapidly about the airfield in flight attitude). Small, French-built Anzanis were also to be found secondhand.

The Heath Parasol, designed by Ed Heath, was a single-place monoplane planned for the Henderson motorcycle engine. It was built in some numbers because it was offered as a kit, less engine, for $200. Bernard Pietenpol's wood-framed Air Camper was popular at the end of the '20s after its plans were published in *Popular Mechanics* magazine. A Model A Ford engine could be purchased new for about $100, and it would produce 35 hp at 1600 rpm. It was a heavy but simple engine, and reasonably reliable.

The early homebuilt airplane movement had its wings clipped by the Air Commerce Act of 1926, which required the certification and licensing of commercially-produced airplanes but contained no provision for the licensing of amateur-built aircraft. That meant that the homebuilts could not be legally flown across state lines. There was spotty enforcement of the new law for a year or so because the Department of Commerce inspectors were new to their jobs and there weren't many of them. Also, many homebuilt pilots clearly took the position that, as far as the new law was concerned, catchin' came before hangin', and few were hung because few were caught.

By 1934, however, all the states except Oregon had adopted laws that were rubber stamps of the Air Commerce Act. That effectively destroyed the homebuilt movement.

A small group of Oregonians continued to build airplanes in their barns and backyards right up to WWII, when all private flying was suspended in the U.S.

After the war, that same group of Oregon plane builders returned to their workshops and were organized into the American Airmen's Association with the hope of attracting enough like-minded people in other parts of the country to make their voices heard in Washington. Then, early in 1946, the leader of this determined little band, George Bogardus, of Troutdale, Oregon, went to the nation's capitol to plead the cause of the amateur plane builder directly to the CAA. Remarkably, the CAA officials not only listened to Bogardus, but cautiously agreed with him.

Another 18 months passed before CAA Safety Regulation Number 236 proclaimed that amateur-built airplanes could be certified. Each project would be judged by a local CAA safety agent, although no guidelines for the builders or safety agents were issued at that time. It was a start, and by the early '50s the WWII training plane bargains were gone (they were expensive to fly and maintain in any case), and interest in amateur-built personal airplanes was growing once again.

The Builders and Their Organization

The potential amateur plane builder is where you find him—or her; several women have built their own airplanes. These people are scattered throughout the country, and they seem to have no common denominators except for a strong desire to fly and the pleasure they take from creating things with their own hands. Among them are represented all professions, trades, and age groups, although most appear to be men past 30, with a family, and with average or above-average incomes. They do not build in order to save money; they could buy good used commercially-produced two-placers for the amount they will invest in engines and materials for their dream airplanes. The labor invested—up to 2000 hours—is part of the pleasure.

However, the homebuilt airplane scene is not static; new materials and new building techniques are resulting in better airplanes constructed with less labor, while the number of amateur builders has grown to the point that some very gifted aircraft engineers are designing airplanes for this still-growing segment of civil aviation. The advent of molded airframe components, made from epoxy-impregnated fiberglass and foam, fostered the appearance of airplane kits, which are legal as long as the builder accomplishes at least 51 percent of the construction.

Due to its superior strength-to-weight ratio, minimum maintenance, accelerated building time, and smoother finish, treated fiberglass is becoming ever more popular with amateur plane builders, particularly after the homebuilt airplane designs and construction methods of Burt Rutan became so fabulously successful in the early '80s.

The homebuilt airplane constructors' official organization is the non-profit Experimental Aircraft Association; "Experimental" because homebuilts have, since August 1947, been licensed in the Experimental Category by the FAA. In 1984, the EAA had approximately 80,000 members, up from less than 70,000 in 1980.

Today the EAA is headquartered on 500 acres of land directly adjacent to Wittman Field in Oshkosh, Wisconsin. Everything was new in 1984, and a mite elegant: headquarters building, shops, museum, and all of it a tribute to the dedication of one man who, with the support of the membership, built this organization over a period of 30 years.

The EAA was formed in January 1953 by half a dozen homebuilt airplane enthusiasts in the Milwaukee area led by Paul Poberezny. Writing their charter, Paul and his friends did not limit membership to amateur plane builders, but welcomed everyone interested in sport aviation. That has resulted in several separate divisions within the EAA, including the Warbirds of America, for those members who refurbish and fly ex-military airplanes; an Antique/Classic Division, for those who restore and fly the airplanes of yesteryear; the International Aerobatic Club; and, since 1980, the Ultralight Division. Each has its own elected officers, its own publication, and its own staff at headquarters. However, the amateur plane builders are the hardcore of the EAA.

The EAA Museum Foundation is a separate entity, and is self-supporting from donations and paid admissions.

The EAA also has a research facility and has led the way in testing—and gaining FAA approval for—the use of unleaded automobile gasoline in a number of lightplane engines.

Poberezny has had an enormous amount of volunteer help (and cash donations) from EAA members in making this organization a power in civil aviation. That, of course, reflects his leadership ability and his own dedication to the task of preserving the rights of aviation's "little guys."

Paul Poberezny is a former Air Force pilot. He has flown everything from jet fighters to military transports, and was on full-time duty with the Wisconsin Air Guard when the EAA was organized. He has designed and built several personal airplanes. Poberezny has been confronted by but one serious challenge to his leadership. It came in the early '60s when other founding members became fearful of Paul's "big ideas." But he had enough support among the membership—about 20,000 then—to remain in control. He began drawing a modest salary in the middle '60s and left the Air Guard in 1969 to devote full time to the EAA. By 1975, membership passed the 44,000 mark and the EAA was grossing over $1 million per year, not counting profits from its annual fly-in/convention. In 1984, dues from 80,000 members put $2 million into the EAA treasury, while the fly-in at Oshkosh added almost that much more.

The Oshkosh Fly-In

The EAA Fly-In at Oshkosh is a week-long event that begins on the first Saturday in August each year. It is *big*. It is easily the biggest airshow anywhere in the world. A half-million people attended in 1983. They came in automobiles, campers (*acres* of campers), and airplanes—more than 15,000 airplanes—and the miracle of it all is how smoothly Poberezny's horde of volunteers stage-manage this multifaceted happening. In addition to the flying, seemingly endless fly-bys of unusual airplanes, aerobatic demonstrations, and warbird formations, several hundred commercial exhibitors man booths in the permanent buildings while rows of colorful tents contain workshops where one may learn aircraft construction techniques, design theory, and related subjects. No liquor is allowed, and *everyone* is responsible for keeping the airport and surrounding grounds completely free of litter.

International Aerobatic Club

Modern aerobatics is not "stunt flying." It is *precision* flying, a demonstration of the pilot's ability (or lack of it) to maintain total control of his/her aircraft in a series of intricate maneuvers that require a high level of skill and extreme mental concentration.

It is fair to say that Wilbur Wright started it all with his "graceful figure-eights" at LeMans, France, in 1908. The Frenchman Adolphe Pegoud performed the first inside loop in 1913; Germany's Max Immelmann added the Immelmann Turn two years later as a combat maneuver, and Eddie Stinson is generally credited with being the first to deliberately spin an airplane after discovering spin recovery technique in 1916. Jimmy Doolittle performed the first outside loop in 1927, and Leonard Povey added the Cuban Eight in 1936.

During the '20s, freestyle aerobatic routines became common at air meets, and in 1929 "Fearless" Freddie Lund, who had been a pilot and wingwalker with the Gates Flying Circus, became the first to equip his airplane with a smoke generator for airshow aerobatics. Lund was also the first civilian to perform the outside loop—1928, in a WACO Taperwing. Later, Tex Rankin, a Northwest Great Lakes dealer, performed 131 outside loops in 131 minutes.

Airshows proliferated in the '50s and '60s, and some really great aerobatic pilots appeared: Merle Torrance, Roy Timm, Frank Price, Lindsey Parsons, Rodney Jocelyn, Charlie Hillard, Bevo Howard, Harold Krier, and Duane Cole, to name a few.

Mary Gaffaney and Margaret Ritchie were the outstanding women aerobatic pilots from the mid '60s into the early '70s, although Margaret's career was cut short before she reached her peak when she was killed practicing in a new aerobatic airplane. In 1972, Gaffaney capped her career by winning the

World's Championship for women. Preceding Gaffaney and Ritchie was Joyce Case of Wichita, a three-time national champion; earlier, during the '50s, it was Caro Bayley and Betty Skelton. All flew Pitts Specials except Ritchie, who had a clipped-wing T-Craft similar to the one Duane Cole has flown so well for so many years.

With spotty sponsorship, the U.S. National Aerobatic Championships were flown at Reno as part of the National Air Racing programs during the late '60s, the Antique Airplane Association having dropped the event following the 1960 contest in favor of an aerobatic "championship" for amateurs only.

Sponsors are difficult to come by because championship aerobatics (as opposed to airshow aerobatics) do not attract a lot of paying spectators. They are not flown with smoke, only one contestant at a time is in the air, and since the object is precision rather than scare-the-pants-off-the-crowd maneuvers, these contests are fully appreciated only by other aerobatic pilots. With a meager gate, little publicity, and no prize money, the attraction for the participants is, besides prestige and a desire to excel, a chance at a world title. The World Aerobatic Championships are held every two years, usually in Europe, and the U.S. team is picked from the high scorers in the U.S. national contests.

It has been charged that some of America's best akro pilots do not go to the Nationals because they cannot afford it. That may be a cop-out. Desire, dedication, and sacrifice are essential ingredients to every championship performance, whatever the activity. If you want to be the best, you must pay your dues.

It is true that other world championship teams (except the British) are supported by their respective governments, but the fact that the U.S. team has no government support is in the American tradition—it is the same with our Olympic athletes—and it is, in fact, a source of pride. A great deal of satisfaction may be taken from the fact that the U.S. team competes in individually-owned homebuilt airplanes against government-subsidized pilots flying the best specially-designed akro airplanes their governments can provide for them.

The Americans participated as a team for the first time in the 1966 World Aerobatic Championships in Moscow and did poorly. Upon his return to Wichita, team member, the late Harold Krier, had this to say: "Our perception of this event is wrong. American akro pilots tend to firewall everything, bend the wings, and go like hell. You don't get any points for that. Precision is the key, along with a good selection of maneuvers that blend well together and have high point values. Vertical maneuvers accumulate the most points, and the Czech Zlinn and Soviet Yak-18 have the edge there. Our little biplanes have got to have more power or we'll have to go to aerodynamically-cleaner monoplanes to match the East Europeans in vertical maneuvers."

In 1968, in East Germany, the U.S. team placed five members in the top ten spots, with Bob Herendeen claiming third place in the men's division, and Mary Gaffaney placing third among the women. The Americans were learning. Their Pitts Specials had new symmetrical airfoils and 180 hp. Their Pitts would get even larger engines later, and some would switch to monoplanes.

In 1970, in England, the U.S. male team took the world team title with the points amassed by Herendeen, Charlie Hillard, and Gene Soucy. Then, at the next world meet in France in 1972, the Americans won it all. Hillard took possession of the individual men's title, Gaffaney won the world's championship for women, while Hillard, Soucy, and Tom Poberenzy (Paul's son) collected the world's team championship for men.

The event was not held in 1974. In 1976 the site was Russia again, where the Eastern bloc judges effectively eliminated the Americans. The favoritism was so obvious that at least one Soviet could not stomach it; he apologized to a supporting crew member of the U.S. team.

That judging bias was clear to others as well. Therefore, the rules committee of the *Federation Aeronautique Internationale,* the French-based international sactioning organization for all official world and national aviation records since 1905 (the

U.S. representative of the FAI is the National Aeronautic Association), adopted a computer-assisted system of analyzing judges' performance at such events.

Encouraged by this attempt to ensure fair play, an American men's team returned to the world championship contests in 1978, which were held in Czechoslovakia. U.S. team member Kermit Weeks won the silver, a Russian took the gold, and the Czech male team collected team honors in contests that were honestly judged.

The United States hosted the 1980 World Aerobatic Championships and ten nations were represented. The communist bloc nations boycotted the contests, apparently in retaliation for the U.S. boycott of the 1980 Moscow Olympics following the Russian invasion of Afghanistan. The Americans took 28 of the 30 medals awarded including the men's individual world championship, Leo Loudenslager; women's individual championship, Betty Stewart; and men's team championship, Loudenslager, Kermit Weeks, and Henry Haigh. The Swiss won the silver team award, and the Australians the bronze.

This event, held in Oshkosh, may or may not have lent substance to the one remaining charge of "unfair advantage" leveled against the conduct of the World Aerobatic Championships—that of the "home field" edge. More often than not, team honors have gone to the host nation. Questionable scoring aside, the complaint was that the home team had ample practice time over the contest site, while visitors—particularly the Americans with very limited funds—had no chance to become familiar with the contest area and firmly fix ground reference points in their minds. Both Krier and Loudenslager had, at different times, blown their chances by becoming disoriented.

If there is substance to this complaint, the obvious remedy is simply more financial backing for the U.S. team. This would allow it to arrive in the host country a couple of weeks in advance of the contests for on-site practice.

However, the true competitive "edge," or lack of it, may be more subtle than that—and it may be shared, more or less, by all contestants. It may be an indefinable thing within, made of will, determination, confidence, and total control—not only of the airplane, but of oneself. Sometimes it all comes together; sometimes a flicker of doubt creeps in unbidden. As Harold Krier once remarked, there are probably 100 akro pilots in the world who, all outside forces being equal, are capable of winning a world's championship on any given day. Not all of them are Americans.

The 1982 world championships were held in Austria, and in the women's division America's Betty Stewart won the World Championship for the second time, a feat never before accomplished by any female pilot (Betty then announced her retirement from competition aerobatics—too expensive, she said). Henry Haigh took the silver, and a Soviet pilot won the gold.

As this is written (summer '84), the 1984 world event is yet to be flown. The site is Hungary, and the U.S. team is counting Haigh and Kermit Weeks among its five-member men's team, with three rookies. The women are to be Brigitte de Saint Phalle and Linda Meyers, who participated in 1982, along with two rookies.

Where does America find its akro pilots? The International Aerobatic Club sanctions 25 to 30 contests around the country every year. There are four levels of competition: Sportsman, Intermediate, Advanced, and Unlimited. The category that draws the greatest number of entries is the Sportsman. The talented ones ("talented" meaning all those traits previously mentioned) may work up from Sportsman through the other categories in a couple of years. Assuming good instruction, dedication, and practice, practice, practice, they may then be counted among those "100" that Krier mentioned.

Aresti Aerocryptographics

The Aresti Aerocryptographic System, or the "Aresti Key," as it is sometimes called, is the written, universal language of the serious aerobatic pilot. It is standard throughout the world, and all official contests employ it.

The Aresti Key makes it possible to diagram any aerobatic sequence, and allows a positive frame of reference within which any precision performance may be accurately judged. It was devised by Count José L. Aresti, a colonel in the Spanish Air Force and a famous akro pilot himself. It was adopted by the *Federation Aeronautique Internationale* (FAI) in 1958.

It has been called a form of shorthand, because it employs simple symbols to represent all basic maneuvers. Combinations of these are put together to diagram complex maneuvers.

Aresti divided all maneuvers into nine "families:"

- ☐ Lines and lines plus angles.
- ☐ Horizontal turns.
- ☐ Vertical turns.
- ☐ Spins.
- ☐ Wing slides.
- ☐ Tail slides.
- ☐ Loops.
- ☐ Rolls.
- ☐ Half-loops and half-rolls; and half-rolls plus half-loops.

Each Aresti figure is numbered, and each is assigned a coefficient of difficulty (K). For example, the inside loop, in the seventh family, is figure 7.1.1, and has a K-value of 12.

In scoring a performance, judges award a grade of one to ten for each maneuver, then this grade is multiplied by the maneuver's coefficient of difficulty. With few exceptions, however, a maneuver repeated suffers a decreased K-value.

Also, the Aresti System assigns K-values to the harmony, rhythm, and diversity of an aerobatic program, its framing within an ideal zone, and provides penalties for violation of time limits.

A complex maneuver will have a K-value equal to the sum of the K-values possessed by the basic figures it contains.

This is why, in championship aerobatics, pilots select compound maneuvers that string together well and with no penalizing repeats. Allowed but ten minutes in a freestyle competition, a contestant must select a routine composed of maneuvers that will total as near the 700-point maximum as possible. This will result in 25 to 30 compound maneuvers as a rule.

In competition, one must furnish to the judges, before takeoff, an Aresti diagram of the sequence to be flown. Most pilots attach copy to their instrument panels.

National and international contests normally call for each pilot to fly a "Known Obligatory" akro sequence, a set of maneuvers designed by the host country and made known to each contestant weeks, or even months, before the event. After the known obligatories are flown at the competition, each pilot is required to fly an "Unknown Obligatory" sequence, a routine put together the night before by team captains and judges. Finally, each contestant flies his/her "Freestyle" program, one the pilot designs (and, presumably, one he/she has practiced for months). In major contests, there may be a fly-off for individual honors between the top scorers.

The International Aerobatic Club is a division of the EAA, with headquarters on Wittman Field, Oshkosh, Wisconsin.

The Pietenpol Air Camper of 1930 was wood-framed and fabric-covered; fitted with the Model A Ford automobile engine of 40 hp, hundreds were constructed and flown from the nation's cow pastures during the Great Depression by amateur builders who spent as little as $200 for materials and engine. Performance was roughly equal to that of the WWI Jenny. (Bernard Pietenpol)

Pietenpol construction was simple but structurally sound although its designer was not an aeronautical engineer, but an automobile mechanic. Ordinary hand tools and a modicum of patience were all that was required for the successful completion of such a project. (Howard Levy)

The Bowers Fly Baby is a modern embodiment of Pietenpol concepts—all-wood, easy-to-build, and cheap. Designed and built by aviation writer Pete Bowers, who describes it as a "scaled-up model airplane," Fly Baby is a minimum-cost personal airplane that may be built with a minimum of skill with simple tools. Hundreds have been built during the past 20 years. That is Pete flying this one. (Peter M. Bowers)

The Corben Baby Ace is another ageless homebuilt airplane design, dating back to the '30s and popular into the '60s. While the Pietenpol did well to cruise at 55 mph, the Baby Ace could do 75 mph with 65 hp.

The Spezio Tuholer (two cockpits, two holes) is an amateur design by Tony Spezio that appeared in the early '60s and which was built by a number of others. It has a 125-hp Lycoming 0-290 and a cruising speed of 120 mph. (Dorothy Spezio)

Some wood and metal homebuilts are still being constructed, but the future appears to belong to the composites—fiberglass and foam and carbon-fiber materials bonded with epoxies. The Rutan airplanes are so constructed, with procedures that markedly reduce building time. (Rutan Aircraft Factory)

The warbirds—restored classic military aircraft—are flown to the EAA Fly-In at Oshkosh, Wisconsin, each year. Martin Caidin's Junkers Ju 52 is framed here by a soft-nose B-25 Mitchell and a Corsair, all WWII airplanes. (Don Downie)

Warbird owners usually try to restore their craft to original configuration and finish. This Ryan PT-22 WWII primary trainer looks like new. (Bob Dean)

An aerial view of just some of the 15,000 airplanes that showed up for the 1980 Oshkosh Fly-In; easily the biggest airshow on Earth, the event was still attracting more aircraft than ever in 1984. (Done Downie)

Sailplane flying is part of sport flying although the sailplane adherents are generally a clannish lot, of the club-soda persuasion. Their quietly-superior air is, however, more easily borne than the assertive arrogance of many warbird owners. People will be people. (Don Downie)

A Schweizer two-place sport and training glider on tow over the Plains. Sailplaning is exhilarating but surprisingly expensive because of the ground support required. Serious sailplane pilots seek areas of lifting air to remain aloft, usually thermals and orographic waves. The cross-country distance record is near 700 miles.

Modern aerobatics is a form of sport flying that polishes pilot skills, and is the basis for commercial airshows. It all began in 1913 when Adolph Pegoud performed the first "loop-the-loop" in France. Here, akro artist and Hollywood film pilot Art Scholl cuts a ribbon with his rudder, inverted at 20 feet. There is no margin for error in this kind of airshow work. (Don Downie)

Scholl's aerobatic airplane began life as a two-place trainer for the RCAF—a deHavilland Chipmunk with fixed landing gear, Ranger engine, and fixed-pitch prop. Scholl designed a retractable gear, installed a modern engine of much greater power, an Aeromatic propeller, added rudder area, and fabricated a new canopy to get his agile Super Chipmunk. (Don Downie)

The late Harold Krier, a three-time national champion, with Betty Skelton. Betty performed aerobatics in her Pitts Special *Li'l Stinker* before there was an officially sanctioned contest for female akro pilots. (Krier Collection)

Former airline stewardess Joyce Case of Wichita was a national akro champion and airshow performer in the '60s flying a Pitts, *Joy's Toy*, she later worked for Beech and Cessna. (Cessna)

A master aerobatic pilot for almost four decades, the late Beverly "Bevo" Howard flew a Bucker Jungmeister, a Luftwaffe trainer of the late '30s, originally brought to the U.S. by Italian Count Alex Papana. Bevo is pictured at a Detroit airshow in 1950. (Don Wigton)

Charles Hillard, Jr., one of the most skillful akro pilots of all time, began his career as an airshow pilot flying a clipped-wing Cub; he flew a Pitts during the '70s, and a Christen Eagle in the early '80s. A Ft. Worth Ford dealer, Hillard no longer competes in sanctioned contests.

Another old master on the aerobatic scene is Duane Cole, pictured here in his clipped-wing Taylorcraft. The Cole family operated Cole Brothers Airshow for many years. Duane continues to instruct aerobatic students these days and is a successful author.

F. Don Pittman, flying a Pitts, was the National Aerobatic Champion when that event was sponsored by the Antique Airplane Association at Ottumwa, Iowa, in the '60s. Pittman also appeared in airshows, then disappeared into the cockpit of a TWA airliner where, presumably, his most spectacular maneuver is a 30-degree banked turn.

The Soviet YAK 18P, powered with the Ivtchenko 260-hp radial, was the Russian aerobatic team's akro mount during the '60s and, with more power, during the '70s. (Julio Toledo)

Chapter 9

Modern Civil Aviation

At the end of World War II, Douglas and Lockheed were well-positioned to supply the pent-up demand for large civil airliners created by the exigencies of war. Both the Douglas DC-4 and the Lockheed Constellation had been proven as the C-54 and C-69 on long-haul military operations. Both had originally been designed as airliners just before America was thrust into the war. Then, during the war, Douglas had built 14 C-74s for the Army Air Forces, the prototypes of the DC-7, while Boeing built the pressurized Stratocruiser near the end of the war as a military air transport by giving the B-29 a different fuselage. Nearly 900 were delivered as the KC-97 (tanker) and C-97. In civilian livery it became the 100-passenger Model 377 Stratocruiser, flying for Pan Am, American Overseas, United, BOAC, and the Scandanavian flag carrier, SAS.

The plethora of C-47s released as surplus went on short-haul and low-density airline routes in America and around the world. (In 1947 a brand new C-47 could be purchased for as little as $2500. When the Korean War started, the USAF was trying to buy them back again for $50,000.) In the meantime it soon became clear that a 40-passenger airliner was needed—something between the 28-passenger DC-3 (*nee* C-47) and the 60-passenger DC-6 and the first postwar Constellation, the L-649 model—and that prompted the appearance of the 40-passenger Martin 202 and Convair 240, both of which were twin-engine transports. These airliners dominated America's domestic and overseas airways—in fact, the airways over much of the rest of the world—until Pan Am introduced the first jet service with the Boeing 707 in October 1958.

Actually, the British, in a bold attempt to grab a commanding lead in the world's jet airliner market, began jet passenger service in May 1952 from London to Johannesburg with their new 40-passenger deHavilland Comet I. Service to Colombo, Singapore, and Tokyo followed that same year. But this effort was premature. Five Comets crashed between October 1952 and April 1954 due to metal fatigue problems. The Comets were grounded after the fifth crash and did not return to airline service until September 1957, following extensive re-work.

Postwar Lightplanes

Meanwhile, in the U.S., the major lightplane makers reverted to their established design philosophies. The first Beechcraft Bonanza made its maiden flight in December 1945, and was clearly viewed as a high-performance personal airplane in the Staggerwing tradition. Cessna's first postwar models were the two-place 120 and 140 (production of which reached 30 units per day in the summer of 1946), and the five-place 190/195, which appeared in 1947. The 190/195 (same airframe, different engines), essentially an all-metal version of the prewar fabric-covered Airmaster, appears to have been a bit too much airplane for this market. It did not sell well, and the Cessna 170, resembling nothing more than an oversized 140, was introduced in 1948.

At war's end Piper was ready with a three-place Cub called the Super Cruiser (100-hp O-235 Lycoming engine, the same basic engine used today in the Piper Tomahawk and Beechcraft Skipper), while Aeronca, Luscombe, Ercoupe, and Taylorcraft stuck with two-place designs that were, except for a little more power, practically the same as their prewar models. Stinson scaled up a prewar design.

Two manufacturers, North American and Republic, which had grown in size to rival Douglas and Lockheed during the war, took note of the roughly 250,000 pilots trained by the military during the previous six years and, anticipating a boom in private flying, entered the lightplane field with two all-new designs, the North American Navion and the Republic Seabee.

Republic's Seabee was a small amphibian, seating four, designed as the result of replies to 10,000 questionnaires mailed to licensed civilian pilots in 1945. This direct market survey established that the average private pilot's dream plane was an all-metal four-place amphibian with 6-7 gph fuel consumption and a price tag below $10,000. The Seabee fulfilled all these requirements—including price, provided that enough were sold.

Enough were not. So much for market surveys.

The Navion, with lines faintly resembling those of the famed North American Mustang, was also offered below $10,000, as was the first Bonanza. But the expected boom did not develop, and it soon became clear that these all-new machines were markedly underpriced in relation to the numbers that could be sold.

Both North American Aviation and Republic soon pulled out of the lightplane market. Ryan acquired manufacturing rights to the Navion and produced 1240 of them 1949-1951 inclusive. North American built 1109. During the '60s and early '70s, the Navion was twice revived by a couple of Texas companies and sold in limited numbers as the Navion Rangemaster. Originally fitted with the E-185 Continental engine, subsequent models were powered with the 260-hp GO-435 Continental, the O-470 and IO-470 of 240-260 hp and, finally, the IO-520 of 285 hp.

The Seabee, which was powered with the 6A-215-G8F Franklin engine of 215 hp, did not find a foster home after Republic abandoned it in 1948.

The Last Stinson

The Stinson Detroiter had evolved into the 225-hp Reliant by 1933 which, in turn, developed into the gull-winged SR-7 through SR-10 series in the late '30s.

In 1939 the company, controlled by AVCO, entered the low-horsepower lightplane field with the 75-hp Model 105, followed by the 90-hp Model 90 in 1940. The postwar 108 series were the Voyagers and Station Wagons, fitted with 150-hp and 165-hp Franklin engines respectively. At least 5000 were built before Piper purchased the design in 1948, but only a few Piper-Stinsons were sold before production stopped in mid-1949. The last 108 sold for $6484 at the factory. Eddie Stinson did not live to witness his company's demise. He was killed in a plane crash in 1932.

The Ragwings of the '40s and '50s

There are those who claim that the best time ever for pleasure flying was the period between 1945 and 1960. That was the era of the fabric-covered Pipers, T-Crafts, and Aeroncas, and those who have never flown a Piper Vagabond from a grass airport on a dewey-fresh spring morning are in no position to argue about it. Of course, the air was different then, and so were we.

One of the nicest things about the Pipers of those years (and the Aeroncas and T-Crafts) was that you didn't have to plot with your tax accountant to plan the acquisition of one. The first Vagabond appeared in 1948. It was the PA-15 model, a side-by-side two-placer fitted with the 0-145 Lycoming of 65 hp. It was priced at $1995 and was equipped with nothing that did not directly contribute to sustained, manned flight; it was a mite spartan. The best Vagabond, which cost a couple of hundred dollars more, was the PA-17; it followed a month or so later with shock absorbers (the rubber cord type), floor mats, and other luxuries including the Continental A-65 engine. The Continental was preferred because it took its 65 hp from 171 cubic inches, while the Lyc, working harder, sought comparable power from only 144.5 cubic inches. Both engines would operate on four gallons of gasoline per hour, giving the compact (compared to the J-3) Vagabond a normal cruise of 90 mph.

At the beginning of the '80s there was still a lot of uncontrolled airspace out there, mostly west of the Mississippi, and still some Vagabonds around that could be purchased for no more than three or four times their original price. That probably is a bargain if you can find one. The Vagabond is a real airplane, with a gross weight of 1100 pounds, 480 pounds of which is useful load. If you insist, you can install a secondhand radio in it and fly into controlled airports.

As mentioned earlier, the first post-War Two Piper was the J-5C Super Cruiser of 100 hp. It was followed almost immediately by the PA-12, the Super Cruiser with 108 hp. This same airframe had been offered in 1941 with a 75-hp Continental engine, but had barely entered production before the war intervened. The PA-12 had a cruising speed of 100 mph and a cross-country range of about 400 miles with its 25-gallon fuel capacity.

The PA-11 Cub Special appeared in 1947, employing the J-3 airframe with a fully cowled engine and an 18-gallon wing tank. It was intended as a replacement for the J-3, and was available with either the A-65 or C-90 Continental engine. PA-11 production halted in 1949 when the Super Cub was introduced.

The PA-18 Super Cub was first offered with the 90-hp Continental or Lycoming 0-235 of 108 hp, but it soon became available with the 125-hp 0-290D, the 135-hp 0-290D-2 Lycoming, and, finally, the 150-hp Lycoming 0-320 engine. The Super Cub remained in production until 1982, when Piper sold the manufacturing rights to a West Texas firm.

The Piper Family Cruiser PA-14 was four-place with flaps, metal propeller, and the 0-235C-1 engine of 115 hp. It immediately preceded the Vagabonds, and sold poorly.

Department of Commerce figures reveal what happened to the civil aircraft market in the postwar years: In 1946, 35,000 civil aircraft were delivered; the following year, 15,617; in 1948, 7302 units were sold, and in 1949 total production was down to 3545. About the same number was sold in 1950, and the low point was reached in 1951, when only 2447 civilian airplanes were sold.

In an effort to spark sales and consume a large inventory of materials that had gone into the Cub Special and Family Cruiser, Piper returned to the basics and brought out the no-frills Vagabonds in 1948 as sales slipped. Then, in 1949, Piper edged back into the four-place market with the PA-16 Clipper, essentially an enlarged Vagabond with the 115-hp 0-235C-1 Lycoming engine. The first Super Cub was announced at the same time.

The Clipper had a short production life, but it spawned the Piper PA-20 Pacer in 1950. The Pacer was powered with the 125-hp 0-290D Lycoming engine as well as the 0-290D-2 of 135 hp, the latter coming equipped with an Aeromatic or Sensenich controllable-pitch propeller.

The last fabric-covered Piper (besides the Super Cub) was the popular PA-22 Tri-Pacer, which entered production in 1952 and was built throughout the remainder of that decade and into 1961, when the all-metal Cherokee was introduced. The Tri-Pacer was originally offered with the same choice of engines that powered the Pacer. Indeed, it was simply the Pacer with its tailwheel under the nose.

Veteran civilian pilots did not embrace the tricycle landing gear with enthusiasm. The nosewheel was derisively referred to as a "training wheel" or "idiot wheel." The fact that the concept dated back to some of the earliest flying machines and had again been proven during WWII on both fighter and transport airplanes did not mitigate the fear that operation of such an airplane suggested that one lacked competence in the art of safely landing a "properly" configured aircraft. The tri-gear airplanes did prove easier to land and to operate on the ground, and when Cessna announced the tri-gear 172 in 1956, the trend was well-established. The prewar Ercoupe had been the first modern lightplane with a tricycle landing gear, followed by the postwar Navion and Bonanza. Stinson had tried one in 1930 but it didn't sell.

The last Tri-Pacers had 150 hp (Caribbean model) and 165 hp, powered by the Lycoming 0-320 and 0-320B.

The Piper Colt, PA-22-108, is a two-place Tri-Pacer powered with the Lycoming 0-235C-1B engine of 108 hp. The Colt was built from 1960 into early 1963. More than 7600 Colts and Tri-Pacers were produced.

By that time, the Piper Commanche was well-established as a high-performance single-engine private airplane and an alternative to the V-tail Bonanza. Subsequent Pipers were all-metal.

The first Taylorcraft was the 40-hp B-12 model of 1937, produced by E. Gilbert Taylor in Alliance, Ohio, following Taylor's split with William Piper. The T-Craft had 65 hp by 1941, and when production was resumed after the war with the BC-12D Model, nearly 3000 of those were built before the company went out of business in 1947. During the war, 1949 T-Craft L-2s (at first called 0-57) were produced for the USAAF.

In 1974, a new company in Alliance announced production of the Taylorcraft F-19 Sportsman, a re-engineered version of the BC-12D fitted with the 100-hp Continental 0-200.

The 1945 plywood and fabric-covered Culver V was a two-place, low-wing cabin plane with retractable landing gear and was developed from the prewar Cadets. During the war, 1259 were built as the PQ-8 through PQ-14 target aircraft, but the Culvers were not commercially successful, despite their good performance.

The Culvers were designed by Al Mooney, who began his career at age 19 with Eaglerock in Denver. It is interesting to note that Mooney was able to maintain his design numbers in an unbroken string from the 1925 Eaglerock to the 1954 Mooney M20, while most of those designs were for someone else. His M4 was the Alexander Eaglerock Bullet, a low-wing cabin airplane of advanced lines that first flew in 1928. The M7 through M9 were Bellancas. The M10 was the 1940 Dart designed while Mooney was at Monocoupe, then the Culvers filled in intervening model designations until, at last, the M18 Mooney Mite appeared in 1947 as the first offering of Mooney Aircraft, Incorporated, of Wichita, Kansas.

The Mite was a single-place low-wing retractable, plywood covered, and originally offered with a 26-hp Crosley automobile engine. Later Mites were powered with the 65-hp Lycoming and Continental. Believe it or not, the M19 was a military version of the Mite, fitted with .30-caliber machine guns. Not surprisingly, it was known as the "Cub Killer." It did not go into production.

The M20, the basis for all Mooneys built since, was Al Mooney's last design for the company that bears his name. He left the company in 1955 to work for Lockheed.

The Engineering Research Corporation of Riverdale, Maryland, produced the original Ercoupe in 1940. Designed by Fred Weick (who had accomplished important work at the NACA, and would later design the Piper Cherokee, among other things), the Ercoupe was certified non-spinnable. With limited control travel and no rudder pedals it was easy to fly, although difficult to land in much of a crosswind. The first Ercoupe was the 415-C with 65 hp. It reappeared after the war as the 415-D with the Continental C75-12, and in 1948 the E Model was fitted with the 85-hp Continental. It was priced at $3995.

No Ercoupes were built during the first half of

the '50s; then the Forney Aircraft Company of Fort Collins, Colorado, acquired manufacturing rights and began production of the Fornaire F-1 and F-2 "Aircoupe." Fornaire built Aircoupes, powered with the 90-hp Continental, through 1960. Next, in 1963, Alon, Incorporated, of McPherson, Kansas, revived the design once again, offering the 90-hp Alon Aircoupe at $7825.

Finally, Alon sold out to Mooney in 1967, and Mooney redesigned it and called it the Mooney Cadet. That didn't matter; it was no longer recognizable as an Ercoupe/Aircoupe. The Cadet did not sell. However, ERCO, Forney, and Alon sold more than 6000 Ercoupes and Aircoupes. The Ercoupes had fabric-covered wings.

The Aeronca and its offspring, the Citabria, stuck with tube-and-fabric construction (as did the Cub and Super Cub) into the '80s. (So did the Great Lakes and Pitts, but the Lakes was out of production from 1933 to 1973 when Doug Champlin resurrected it, and both, especially the Pitts, appeal to a very small and unique segment of the civil aircraft market. What about the Maule? Yes, the four-place STOL Maule has a steel tube fuselage frame covered with fiberglass cloth to go with its all-metal wings, but the Maules are modern airplanes that first appeared in 1962, as are the Bellanca Vikings, which we'll recognize momentarily).

The Aeronca Champion series was introduced in 1945, directly descended from the WWII Aeronca L-3 liaison aircraft. The first of the series was the Model 7AC, built in large numbers with the 65-hp Continental, Lycoming, or Franklin engines. Affectionately referred to as "Airknockers," the Champion series added the 7BCM Model in 1947 with the 85-hp Continental, and a year later the 7CCM Champ appeared with an enlarged fin (vertical stabilizer) and some minor structural changes, followed by the 7DC, produced in 1949 and '50 fitted with the 85-hp Continental which provided for installation of a metal propeller.

The Aeronca Champions have wood spars with aluminum wing ribs; the fuselage is welded steel tubing, and the entire structure is fabric covered. Seating is tandem. None of this series was originally equipped with starter or generator/alternator, and these accessories may not be installed without switching to a dash-12 engine of 85 or 90 hp.

More than 7000 Aeronca Champions were built 1945-1951 inclusive. In 1942, 1474 Aeronca L-3s were delivered to the Army Air Forces, the first 405 of which were originally designated 0-58s. Almost all of the L-3s were powered with the A-65 Continental, which the Army called the 0-170.

Aeronca production stopped in 1951 with the industry-wide drop in civil aircraft sales, but the Champion Aircraft Company of Osceola, Wisconsin, a new firm, took over the design in 1955. They offered essentially the same airframe, with 90 hp, as the Champion Traveler Model 7EC. Also offered were several tri-gear versions—the Tri-Traveler, Tri-Con, Sky-Trac, DXer, and the Challenger 7GCB, powered with 0-290D and 0-320 Lycomings ranging from 125 to 150 hp. None was notably successful.

During the early '60s, Champion introduced a twin-engine two-place craft called the Lancer powered with two Continental 0-200 engines of 100 hp each. It was meant as a multi-engine trainer, but this high-wing, fabric-covered craft found few buyers.

The Citabrias are modern airplanes in that production began in 1964, although their Aeronca heritage is obvious to anyone who has ever seen an Aeronca Champion. The first Citabria (so named because of limited aerobatic certification; spell it backwards), the 7ELA, had 100 hp, but few were built before the 7ECA appeared with the 115-hp Lycoming 0-235.

In 1966, Champion added the 7GCAA fitted with the 150-hp Lycoming 0-320-A2B engine. Later, the 7KCAB was added with fuel injection. The 7GCBC, with the carbureted dash-A2B Lyc, has flaps and an extra foot of wingspan. Late models of the Citabria all have spring-steel landing gear legs.

Champion Aircraft Company became the Champion Division of Bellanca in 1970, and the company attempted to revive the old Aeronca 7AC with an inadequate engine—the 60-hp two-cylinder Franklin 2A 120-B—and priced this combination at $4995. It did not sell.

The Champion Scout 8GCBC, with the Lycoming 0-360 of 180 hp, 2½ ft of additional wingspan, and oversize wheels, was intended for back-country operation. It entered the market in 1974 and enjoyed very limited success. Another specialized Champion was the Decathlon 8GCBC, which also first appeared in 1974. It is a beefed-up 150-hp Citabria with constant-speed propeller and a special wing designed for advanced aerobatics.

By 1979, Champion Aircraft and its parent company, Bellanca, had fallen upon evil times, and the Champion inventory was purchased by B&B Aviation of Tomball, Texas, in 1982. At this writing, B&B is offering complete aircraft as well as parts for the 12,000 Citabrias in service.

The first Cessnas produced after WWII were the two-place 120 and 140 Models, which had all-metal fuselages and cloth-covered wings. Actually, these represented two versions of the same airplane, the 120 being the economy model without flaps or electrical system. Both were originally powered with the C85 Continental. A total of 2164 Cessna 120s were built 1946-1949 inclusive, while a total of 5560 140s were produced 1946-1951 inclusive (strangely, two were built in 1953, according to Cessna records). The last 521 were 140As, which had metal wings and the Continental C90 engine. The 140A appeared in mid-1949 and was priced at $3695. The fabric-wing 140 sold for $3495.

The first four-place Cessna 170 was almost identical in appearance to the 140 except for its size. It was introduced in 1948 priced at $7245, powered with the 145-hp Continental. Like the 120s and early 140s, it had an all-metal fuselage and fabric-covered wings.

The 170s were in production from 1948 through 1957, with a total of 5136 sold. The 170A, which appeared in 1949, had a new all-metal wing, a new dorsal fin, and a single lift strut on each side. The 170B came along in 1952; it was similar to the 170A but adopted the flaps used on the Army's Cessna L-19 Bird Dog (later 0-1), and its price was up to $8295.

Although production officially ended in 1955, another 72 170Bs were sold in 1956, and 36 in 1957, perhaps from inventory. In December of 1955, 173 of the new 1956 Model Cessna 172s were delivered to begin the apparently never-ending production of this airplane.

There were other postwar lightplanes with fabric covering. These included the Funk (which resembled a fat Piper Pacer, offered in 1946,) and the four-place, low-wing Johnson Rocket (which was also called the Sky Gem by a successor company). Another was the Hockaday Comet, also a Piper lookalike, and the Fairchild 24, a high-wing cabin monoplane that had first appeared in 1932 as the Model 24C-8 side-by-side two-seater powered with a 95-hp air-cooled, in-line Cirrus engine.

As a two-place machine the Fairchild 24 was also fitted with a 125-hp Warner (24C-8A), and as a three-place airplane, the 145-hp inverted Ranger (24C-8D). By 1937, it was the 24G with the 145-hp Warner radial and a four-placer, and the 24H with the 200-hp Ranger. The 1937 G Model was priced at $5390.

During WWII, 995 Fairchild 24s were delivered to the USAAF as UC-61s, and the Navy purchased a few prewar models and designated them JK-1 and J2K-1s.

Civil production was resumed for a short time after the war in a short-lived Fairchild "Personal Airplane Division" at Dallas, Texas. The parent plant was in Hagerstown, Maryland, where the last A-10 Thunderbolt II tank buster was coming off the line at the time of this writing.

The CallAirs were built in Afton, Wyoming, by the brothers Ruell and Spencer Call, along with cousin Ivan. Their first machine was the 1940 three-place low-wing cabin design of 90 hp planned for operation from unimproved mountain airstrips. The CallAirs A-2 through A-4 appeared after WWII and had 135 and 150-hp Lycomings (0-290 and 0-320). Only a few hundred were built, but these airplanes were regionally popular.

In 1953 the CallAir A-5 sprayer/duster was

offered, and evolved through the A-9 Model with 235 hp. The design was purchased by Aero Commander in 1968 and briefly marketed as the Ag Commander.

The Temco Swift and Luscombe Silvaire were significant postwar lightplanes and prewar designs.

The Swift GC-1 was first produced by the Globe Aircraft Corporation of Fort Worth, Texas, in 1940. That first Swift had plywood-covered wings and fuselage with fabric-covered control surfaces, and was powered with the C85 Continental engine. After the war, Globe redesigned the Swift as an all-metal airplane, and when it proved markedly underpowered with 85 hp, went to the 125-hp Continental engine. That proved to be a very good combination.

Texas Engineering and Manufacturing Company (TEMCO) built the Globe Swifts under contract until the Globe company failed late in 1946. Then TEMCO bought the design and continued production into 1950. The Swift is two-place, side-by-side, and has a retractable landing gear. The all-metal version is the GC-1B and it sold new for $4995.

The first Luscombe Silvaires entered production in 1937, powered with engines ranging from 50 to 75 hp. Designed by Don Luscombe (designer at Monocoupe), the two-place Silvaire originally had fabric-covered wings and all-metal fuselage.

The first postwar Silvaire was the 65-hp 8A Model. The 85-hp 8E followed, and then the 8F with the Continental C90. Production stopped late in 1949, but during the early '50s TEMCO took over the design and built a few Silvaires until, in 1955, Silvaire Aircraft Corporation was formed in Fort Collins, Colorado, to produce the 8F Model. That company suspended production in 1960. The last ones sold for $4950.

G.I. Bill

A significant percentage of the civilian students during the '50s consisted of WWII veterans whose training was paid for by the government under provisions of the law popularly known as the "G.I. Bill." Vets could receive up to 36 months' training, along with $110 per month sustenance aid, and many thousands elected to seek a Commercial Pilot's Certificate. Eligibility requirements for WWII veterans began to expire in 1962. Although Korean War vets had more or less similar benefits, the Veterans Administration generally made it more difficult for them to get flight training under the G.I. Bill.

Apparently, no useful figures are available from the VA listing the actual number of vets who have completed flight training under this program. Therefore, we can only generalize by stating that it did launch a lot of young men and women into aviation careers, while providing a hefty financial boost to a large number of FBOs, continuing to a degree throughout the '70s and into the '80s.

Federal Aviation Agency/Administration

In August 1957, the Airways Modernization Act established the Airways Modernization Board following a two-year study of future airline needs. This was the prelude to the Federal Aviation Act of 1958, which went into effect on 23 August of that year.

The Federal Aviation Act created the Federal Aviation Agency. It took over the functions and personnel of the preceding Civil Aeronautics Authority (CAA, which had controlled American aviation since 1938) on 1 January 1959. The Civil Aeronautics Board (CAB), however, continued to act as the independent governing body responsible for the economic regulation of the airlines, as well as for determining the probable cause of aircraft accidents. The CAB, in other words, retained its absolute power over the airlines, although overseas routes were subject to presidential approval.

The FAA maintained its agency status for less than ten years. Such autonomy was too much for Lyndon Johnson and his lieutenants of the "Great Society." Therefore, on 1 April 1967 new legislation reduced the Federal Aviation Agency to an "administration" and stuffed it into Johnson's new Department of Transportation. Thus the FAA fat cats became pussycats, and much more responsive to political pressures from Capitol Hill.

Whether or not it was so intended, a number of President Johnson's top aides later found high-paying positions with airlines that benefited from White House decisions made while they were there. Referring to the lucrative trans-Pacific routes awarded by Johnson (contrary to CAB recommendations), syndicated columnists Rowland Evans and Robert Novak wrote, on 20 January 1969, that "The list of rainmakers . . . benefiting from CAB decisions . . . reads like a Who's Who of the Great Society." In Washington, a "rainmaker" is a person of political influence in the pay, or expecting to be, of a vested interest. After President Nixon took office, he rescinded the route awards, and followed the CAB's recommendations.

The effect on general aviation of Johnson's subordination of the FAA was not immediately apparent, but in time it has become clear that it may be best described as "taxation without representation."

Deregulation

The CAB chairman under President Carter was one Alfred Kahn, who has since been described as "an instant expert on airline operations, once someone explained to him which was the front, and which was the back, of an airplane." According to some industry observers, Mr. Kahn appears to possess a similar background in economics. However, in the interest of objective reporting, it should be noted that others—including some in the airline industry—regard Kahn as an economic genius.

In either event, Kahn decided that the airlines should be totally unregulated (along with the bus and trucking industries), and that the CAB should be abolished. That proposition was endorsed by President Carter, and legislation to deregulate the airlines was pushed through the Congress in 1978 by Senator Ted Kennedy of Massachusetts, with strong support from Nevada's democratic Senator Howard Cannon.

The new law required that the CAB gradually give up its regulatory role and go out of existence altogether no later than 1 January 1985. After that, the Federal Trade Commission could arbitrate any charges of unfair trade practices among the airlines. The Federal Aviation Administration would continue with its traditional duties of writing and enforcing FARs, certifying civil aircraft and airmen, etc., and would continue under the direction of the Secretary of Transportation, a political appointee with plenty of patronage jobs to hand out to the party faithful. From their ranks will come the current and future rainmakers.

The chaos throughout the airline industry that immediately followed deregulation defied useful analysis of that factor alone, because other enervating influences intruded to confuse the issue—not the least of which was poor management of some of the largest air carriers. Other unsettling factors included high interest rates for new equipment, a recession, and escalating fuel costs. The controllers' strike, and their subsequent firing by President Reagan, probably helped more than it hurt the major air carriers, because it provided a legitimate opportunity for them to cut back on wasteful overscheduling, allowed a good excuse to eliminate unprofitable routes, and to substantially pare swollen work forces.

We will return to the airline situation of the '80s and how it evolved.

Modern General Aviation

Increasingly, during the late '70s, and especially during the recession of the early '80s, it became clear that American aviation's "little guys" were steadily being priced out of the sky. Throughout the previous quarter-century, people with median incomes could afford flying lessons and could, depending on the extent of their family obligations, fly for pleasure in new or late-model two or four-place private airplanes, either rented or individually owned. But the cost of lightplane rental or ownership soared beyond the means of those with median incomes in the early '80s. A measure of the increased costs may be seen by comparing the list prices of new two-place Cessna trainers of 1971 and 1983: $10,775 vs. $36,190. Those prices are for the *lowest-cost* production aircraft available. Meanwhile, fuel went from 40¢ per gallon to an average of $1.80 per gallon, and all fixed costs of airplane ownership—insurance, hangar or tiedown, maintenance, taxes, etc.—increased proportionately.

Since few people will easily give up flying once

they have been seduced by its unique satisfactions, private pilots sought affordable ways to remain airborne. That sent some to the used plane market looking for older lightplanes in the $8000 to $12,000 bracket. Others considered the construction of a homebuilt machine, while a relative few tried the ultralights—at least temporarily.

Ultralights and ARVs

Actually, the ultralight aircraft that evolved during the early '80s was no substitute for a *real* airplane—certainly not to experienced pilots. True, a flight in an ultralight is an unusual experience, but to one who knows airplanes the ultra-minimum flying machine poses some unsettling questions, beginning with its obviously flimsy construction that necessarily lacks any structural redundancy. Inspecting such a machine, the knowledgeable airman will spot a number of small bolts and fittings, the failure of any one of which would spell disaster in flight. Rigging is another question mark; wings and tail must be attached at exactly the correct angles, a task that, by law, may be performed only by licensed aircraft mechanics on real airplanes. The two-stroke, two-cycle engines used in ultralights because of weight considerations are inherently too unreliable for aircraft use. Perhaps the main drawback to ultralight flying, however, is its relatively high cost. These lightly constructed machines, with their sensitive, single-ignition engines, require too many hours of maintenance for each hour of very limited flying. Priced between $4000 and $6000, and with a cost per hour of operation estimated no lower than $20 by even their strongest supporters, it was inevitable that the ultralight would evolve into a more practical machine.

Although most ultralighters were not licensed pilots, they, too, recognized that they were expending too much effort and money for too little of a much diluted form of flying.

The immediate (and interim) solution was small, lightly constructed machines that resembled real airplanes, especially famous old biplanes. A few remained in the ultralight class—254 pounds empty weight or less; max speed of no more than 55 kts—while others, exceeding those limits, were technically airplanes and therefore required certification in the Experimental Category and, of course, 51 percent owner construction. These aircraft, weighing up to 300 to 400 pounds, were called Aerial Recreational Vehicles (ARVs), and occasioned a move to petition the FAA to establish an ARV Category providing for minimum materials and construction standards for such machines, as well as a special ARV Pilot's License. At this writing, the FAA has reached no decision on the proposal.

The ARV was certain to evolve into a reasonably substantial flying machine, particularly if it could be marketed as a kit with major components prefabricated for home assembly requiring a minimum of unskilled labor—and priced below $15,000. Production of the lowest-priced two- and four-place Cessnas was suspended for months at a time during the early '80s, and Cessna was the largest manufacturer of general aviation airplanes in the world.

Modern Cessna

Since its founding in 1927, Cessna had delivered more than 175,000 airplanes by 1983, including 24,000 twins, 2000 military jets (the "Tweety Bird" and its A-37 variant), and more than 1100 Citation bizjets. More than half of the aircraft flying in the Free World today are Cessnas, sold and serviced by some 600 Cessna dealers worldwide, plus 1100 flight training operations.

Cessna, as the other gen-av manufacturers, concentrated on development of corporate aircraft throughout the '70s, and by 1982 offered three Citation bizjets as well as the Conquest II project. Both the Citation and Conquest were "me too" airplanes, designed to share markets pioneered by the Learjet and Beechcraft King Air. Cessna has never been a leader in any area except sales—and, surely, there's a lesson in that somewhere.

Cessna did not have a two-place trainer between February 1951, when the last 140 left the factory, and November 1958, when the 150 was introduced. Cessna waited for the market—and then took the lion's share of it.

Modern Beech

Beech, meanwhile, had no training plane for entry-level pilots until the Model 23 Musketeer appeared in 1962. Beech didn't really want a low-cost, low-horsepower machine, but accepted the fact that "product loyalty" is an important factor in the marketplace. People who learned to fly in Cessnas or Pipers tend to *buy* Cessnas or Pipers later. And after buying, say, a Piper, tended to buy a Piper *next* time. Since most pilots did not—and should not—go directly from a training airplane to a Bonanza, Beech recognized the need for interim Beechcrafts. So the Musketeer soon gave its airframe to the Sundowner and Sierra. Eventually, Beech felt the need to build a for-real two-place trainer, the Skipper.

Beech has never been too successful in the low-horsepower lightplane market. For more than two decades that has been Cessna's turf, with Piper in second place. Actually, Piper had the "pioneering equity," but blew it in the '60s with poor public relations and inadequate marketing practices (the only real advantage the Cessna 150 had over the Cherokee trainer was that spins could be taught in the 150). The entire Beech organization has been, since Staggerwing days, oriented to the affluent customer with (in accordance with its self-image) quality products.

Mooney, long troubled by poor management, entered the uncertain '80s with a turbocharged version of its thoroughly proven high-performance single-engine retractable, and as a subsidiary of Republic Steel. With but a single product of a type that competes in a diminishing market, it would seem that it is time for some practical innovation at Mooney.

Unfortunately, there are formidable deterrents to innovation in civil aircraft manufacturing. The FAA's very expensive certification program for new designs is one of them. The desirability of regular dividends to shareholders is another. There are other factors, and together they resulted in 40 years of design stagnation for gen-av airplanes.

However, as fuel and other expenses associated with flying soared, and then a recession multiplied the impact of those ruinous forces, gen-av plane makers saw their sales and profits drain away. Backed against the wall during the early '80s, they were forced to accept the obvious need for more efficient airplanes, and the obvious place to start was with new designs aimed at the most viable segment of the market: corporate aircraft.

Rutan

One other fact was equally obvious: the one airplane designer in America who had demonstrated his ability to produce significantly more efficient airplanes was Burt Rutan, a 40-year-old individualist who had developed a series of canard ("tail first") designs since 1974 for the homebuilt airplane constructor. Built of fiberglass and foam, without a need for molds, the Rutan machines combined ease of construction with unusual strength and remarkable performance. The Rutan VariEze ("very easy") has a cruising speed of 196 mph with 100 hp. It is a two-place and, as the other Rutan canards, its wing cannot be stalled. The prototype of a light twin with centerline thrust, the Rutan Defiant, took to the air in 1979; Burt (Elbert L.) apparently intended it only as a proof-of-concept vehicle, but public demand—or maybe the times—finally persuaded him to release plans in 1984. The first plans-built prototype Defiant was constructed by Fred Keller of Anchorage, Alaska, in time for Oshkosh '83.

By 1983, a subsidiary secret development company in operation next door to Burt's homebuilt plans business on the sand-blown airport at Mojave, California, and one of its first known projects was an 85-percent scale flying prototype of the new Beechcraft Starship I. Its maiden flight was on 29 August 1983.

Burt's more-or-less secret facility also completed a military trainer prototype for Fairchild in 1983. Contracting such projects to Rutan is said to be the most economical way for a commercial aircraft manufacturer to develop new aircraft. It also allows the manufacturers to claim credit for the designs, even if each looks like a modified VariEze.

Burt went from California Polytechnic University, with a B.S. Degree in Aeronautical Engineering, to Edwards Air Force Base as a Flight Test Project Engineer, where he remained until 1972. That was a dangerous but rewarding experience. Then, planning his first canard, the VariViggen, and lacking access to a wind tunnel, Burt mounted a model on a pylon atop his 1966 Dodge Dart station wagon and drove it at night at 80 mph on a sparsely traveled highway to collect the data he needed. (He was never ticketed for speeding.) He did his own test flying, and found it necessary to make a number of modifications to his first airplanes, but the several aerodynamic principles he brought together in a single concept, and then fine-tuned, clearly pointed the way to safer and more fuel-efficient airplanes. Rutan's work could not be ignored, especially with Rutan-designed homebuilts proliferating throughout the Free World, their "21st century" configurations attracting admiring crowds at every stop.

Nowadays it is said that Burt communes only with God and girlfriend Pat Storch, but that is unfair. The guru of the canard regards his work as unfinished, and clearly would get nothing done if he made himself available to the steady and increasing stream of visitors to the Rutan Aircraft Factory. Actually, Burt is a pleasant man, and while not particularly loquacious, does have a sense of humor. (Asked how many people lived in the nearby community of Mojave, he replied, "About half of them.") And he does have some good and able people around him, including elder brother Dick, an Air Force pilot who has contributed much of his free time to Burt's projects, and the Melvills, Mike and Sally, who handle the expanding office force and public relations duties. All are pilots, including Pat. Burt's father, George, is also a pilot; sister Nellie is a stewardess for American Airlines.

It is too early at this writing to fully assess Burt Rutan's true impact on future aircraft designs, but it seemed obvious by the mid-'80s that it will be, as a Damon Runyon character might put it, somewhat more than somewhat.

The Airlines: Hand-Flying into Cloudy Skies

A central fact to be considered by observers of the U.S. scheduled airline industry is that for more than 50 years it was strictly controlled by one federal authority or another. Except for the four-year period following the Black Committee hearings in the middle '30s, Federal regulation was benevolent and supportive. It was the airlines' security blanket. It protected them from one another, because routes and rate structures required CAB approval. Such a system worked well. The air carriers could plan ahead with confidence, could go in debt for the best new equipment, and could compete on the basis of service and convenience to the public.

However, President Carter's CAB Chairman, Alfred Kahn, a Cornell University professor in real life, theorized that U.S. transportation systems should not be treated as public utilities, but should be deregulated and, in the case of the scheduled airlines, should be free to fly wherever they pleased, at rates dictated by the forces in a competitive market. Thereupon Senators Teddy Kennedy and Howard Cannon shepherded the airline deregulation bill through the Congress in 1978, and despite the disaster that followed, the Reagan Administration, committed to "getting the government off the backs of the people," supported airline deregulation. (The trucking industry was similarly deregulated in 1980 with the same result. By the end of 1982, 72 established trucking firms, including some of the largest, had gone broke, while thousands of new, small, non-union, cut-rate truckers fought for survival.)

Now, it's easy for those affected to become emotional over airline deregulation, but its direct and identifiable effects must be separated from other enervating forces that, unfortunately, happened to coincide with deregulation. The concurrent rising fuel and labor costs, along with a recession that markedly reduced airline boardings, were enough in themselves to cause financial problems for the air carriers. Deregulation intensified those factors while introducing others that placed the established airlines at a competitive disadvantage.

Between 1975 and 1983, the cost of jet fuel increased 500 percent while labor costs doubled (average employee compensation was $20,000 in 1975, $40,000 in 1983). Fuel amounts to one-third of total airline costs, and labor another third.

Another factor was declining revenues. Industry-wide, net earnings peaked in 1977 at $1.3 billion. A year later they were down to $800 million, and the slide continued unabated for the next five years, with industry losses totaling more than $800 million in 1982.

A portion of the falling revenues, which contributed to these losses, was chargeable to the dilution of the market by the appearance of dozens of new commuter airlines and the air fare wars that followed. Recurringly, for short periods on high-density routes, air fares went as low as 2½¢ per seat-mile, while costs were as much as 10¢ per seat-mile.

CAB Chairman Kahn, and his successor, Marvin Cohen, agreed that deregulation would result in the demise of some established airlines, but new ones would come along to replace the failures, they said. In a free market, it was survival of the fittest.

Under deregulation, the "fittest" were penalized during the critical adjustment period, and the newcomers aided. The CAB did not give up its authority at once. Under the new law, it stopped setting routes in 1981, and its right to approve domestic fares ended in 1982. After that, the CAB retained the authority to regulate overseas fares and routes, continue subsidies to "essential" local service airlines serving 88 small communities, and had the power to enforce regulations concerning smoking on airliners, baggage handling, the bumping of passengers, and airline advertising until 31 December 1984. At that time the CAB would be officially dissolved, and its residual duties given to a new office, presumably within the Department of Transportation. In mid-1984, leaders in both the industry and the Congress were quoted as saying that there was no chance that a law reinstating airline regulation would be enacted in the foreseeable future.

The Controllers Strike

During the second week in August 1981, the members of the Professional Air Traffic Controllers Organization (PATCO) voted to strike in violation of Federal law. Roughly 12,000 left their posts demanding a $10,000 pay increase, shorter work week, earlier retirement, and gentler treatment by superiors. Average annual pay at the time was $33,000 for an average of 25 hours per week; any could retire after five years of service with 40 percent of their pay if the job proved too stressful, and they were undoubtedly correct in charging that their supervisors were a demanding lot.

Secretary of Transportation Drew Lewis offered the controllers a 34 percent pay increase spread over a period of three years (about $11,220). The controllers turned it down, then ignored President Reagan's ultimatum to return to work within 48 hours or be fired.

The President did indeed dismiss them all, and FAA Administrator J. Lynn Helms kept the system alive with the 2000 or so non-strikers and supervisory personnel, aided by 500 military controllers, furloughed airline pilots, and recently-retired controllers. That allowed the system to function at about 65 percent of its normal capacity. The airliners were hurt the least, while private flying suffered the most. Airline overscheduling stopped, and some low-density routes were abandoned or cut back as the airlines were allocated a given number of "slots" per day at hub airports. Private pilots wishing to fly cross-country on an instrument flight plan had to file twelve hours in advance, and perhaps as many as half of those requests were denied during the fall of 1981—particularly, east of the Mississippi; the crunch was less severe over the western half of the nation.

Meanwhile, most pilots—and the general public—applauded the President's action, and the pared-down system, despite delays, was more pleasant to fly. Instead of the predicted confusion and short tempers, both pilots and controllers took the position that "We're in this together," and the radio contacts were friendly and revealed a desire to cooperate. Controller "chipping" (sarcastic replies) disappeared, while pilots began adding "Thank you" and "Have a good day" to their transmissions.

The predicted rash of accidents did not occur in

the wake of the controllers' dismissals. No air accident was charged to controller error from mid-August 1981 to the time of this writing in mid-1984.

The controllers who defied the law and were fired, however deservedly, paid heavily for their intransigence, because theirs is a professional skill that is not transferrable to other jobs. Most are working elsewhere today for a good deal less money.

By mid-1984, the system was fully manned once again, it having been discovered that approximately 8000 controllers were actually all that were needed. The newcomers had been trained at the FAA's Aeronautical Center in Oklahoma City during the preceding three years. The veterans remained at the high-density hub airports and in the 23 Air Route Traffic Control Centers, and "flow control" continued to an extent while the new controllers gained experience.

Slots and Flow Control

With the number of flights per terminal cut back due to the controller shortage of the early '80s, the CAB allocated a given number of slots to each airline at each terminal. That seemed reasonable at first, but the slot program was badly administered. Long-established airlines, already suffering from declining revenues, had slots taken from them by CAB edict and awarded to newly formed lines. Slots were sold—United bought twelve from Pioneer at Denver—and when Braniff ceased operation in May 1982, its slots at major air terminals were parcelled out by lottery, with the result that some good ol' boys with a secondhand airliner bought on credit suddenly found themselves in possession of a valuable asset.

Eventually, the slot system would fade away, and when a reconstituted (if emaciated) Braniff, backed by the Hyatt Corp., returned to the skies in March 1984, it could not be denied access to the terminals it chose to serve. After all, that is what deregulation is all about.

Meanwhile, flow control continued to an extent. It was a commonsense procedure first used by the FAA back in July 1968, when controllers in the New York area chose to slow the system rather than risk a strike for their demands. They increased the spacing between aircraft arrivals and also between aircraft departures at John F. Kennedy International Airport, and that very quickly stacked up dozens of airliners in holding patterns in the area.

The FAA stepped in and held on the ground, at their points of origin, all airliners bound for JFK, issuing clearances only as the slowdown could accommodate them at their destination. That was flow control, and it proved useful while the air controller force was being rebuilt.

Labor, the Airlines, and the Supreme Court

In February 1984, the U.S. Supreme Court held that bankruptcy judges have the authority to allow companies to abrogate union contracts when those companies seek to reorganize under provisions of Chapter 11 of the federal bankruptcy law, and that such relief is permissible even before the actual bankruptcy petition is approved.

That was exactly the position taken by Continental Airlines' boss, Frank Lorenzo, when he announced on 24 September 1983 that Continental's domestic system had filed for permission to reorganize under Chapter 11. Continental laid off two-thirds of its 12,000 employees, sharply cut the pay of the remainder, and reduced the number of U.S. cities it served from 78 to 25—this, after Continental had suffered repeated operating losses following deregulation, and had sought, unsuccessfully, to renegotiate its labor agreements.

In the meantime, Eastern's Frank Borman warned his 37,500 employees that Eastern would probably be forced into Chapter 11 proceedings if they did not accept 20 percent pay cuts. Eastern settled for a lot less for the time being, but has remained financially troubled into 1984.

Other major lines that faced uncertain futures as the midpoint of the decade approached included Pan Am, Republic, TWA, and Western. Retrenchment and reorganization under Chapter 11 might save some, at least temporarily, but a labor-oriented Congress could well pass new laws to nullify the 1984 Supreme Court ruling that allowed cancellation of burdensome union contracts.

Where would it all lead by, say, 1990? Some

industry observers believed that no more than three major airlines would completely dominate the U.S. domestic airways by the end of the '80s. If so, two of them would be United and American, both of which had large cash reserves in mid-decade, turned a profit in 1983, and were aggressively moving in on all the hub terminals while shoring up feeders that served them at those hubs. As if to underscore the above prediction, American announced in March 1984 the purchase of 67 new Douglas DC-9-180 airliners, the largest civil airplane purchase (in dollars) in history.

A day after American announced its historic purchase, Braniff returned to the air with less than half the number of (refurbished) airplanes (37) it possessed at the time of its bankruptcy, and one-third the number of employees—who seemed happy to be working for 40 percent less pay.

It was a time of great change in the U.S. airline industry.

The first Convair 240 was delivered to American Airlines in February 1948, and the 240 remained in production ten years, with 571 built including T-29s and C-131As for the Air Force. The series—340/440/540/600/640—followed, the 540 being the first turboprop. Substitution of the gas turbine engine for the Convairs' original P&W R-2800s has kept this ageless machine in airline operation into the '80s. Pictured is a Central Airlines 540.

The Beechcraft Model 35 Bonanza made its debut in 1947 fitted with a 165-hp (185-hp takeoff) engine, the E-185 Continental, and was priced at $8945. (Beech)

The Cessna 190/195 Models were introduced in 1947 with engines of 240 to 300 hp. A total of 204 190s was built, and 890 195s were produced, plus 83 195s for the Army as LC-126s. Prices ranged from $12,750 for the first 190s to $24,700 for the last 300-hp 195s. (Cessna)

The Piper PA-12 Super Cruiser with the Lycoming O-235 engine of 108 hp appeared early in 1946 and was the postwar version of the 100-hp 1941 J-5. Three-place, the PA-12 was followed by the PA-14 Family Cruiser.

The North American Navion, like the first Bonanza and Republic Seabee, was an attempt to market a modern family airplane priced under $10,000 to meet an anticipated demand for 50,000 such aircraft immediately after WWII. There proved to be no such market. The design was acquired by Ryan and, ultimately, by Texas-based Rangemaster, and the Navion, like the Bonanza, went from 185 hp to 285 hp in subsequent models. (Don Downie)

Olive Ann Mellor joined the Travel Air Company as a secretary in 1926, became Walter Beech's wife in 1930, and assumed the duties of chief executive officer in Beech Aircraft Corporation when Walter died in 1950. O.A. Beech served in that position until Beech's merger with Raytheon 30 years later. (Beech)

Arizona Senator (R) Barry Goldwater, Sr., one of America's best-known private pilots, was a fighter pilot in WWII. When this photo was made, aviation writer Don Downie—a WWII Hump pilot—was riding as copilot as Senator Goldwater made a nighttime instrument approach into Ontario, California. (Don Downie)

Three-niner-delta, a Piper Caribbean (150-hp) version of the popular PA-22 TriPacer series built 1952-1960 inclusive. Engines were Lycoming 0-290s and 0-320s, 125-165 hp. The 1959 Caribbean above sold new for $8890.

The Cessna 152, essentially a 150 with an engine change to accommodate LL-100 av- gas. The first 150 was the 1959 model, which had the 100-hp 0-200 Continental engine. As late as 1971 it was priced at $8895 for the Standard model, and the 150 accounted for an estimated 15 percent of all civil flying. In 1976 the price was up to $12,650, and in 1984 the Standard model was $29,700 at the factory. (Cessna)

The Cessna 185 Skywagon appeared in 1961 fitted with the Continental IO-470 of 260 hp—a little more airplane than the 180 with 230 hp and which had been around since 1953, originally with 145 hp. The 1984 185 Skywagon has the Continental IO-520 of 285 hp, and cruises at 154 kts with a 1600-pound payload. Basic price: $86,850. (Cessna)

The 1984 Cessna Stationairs, six and eight models, trace their ancestry back to the Cessna 205, 600 of which were produced 1962-1964 inclusive. The 205 could be described as a 210 with fixed gear. The 206 followed in 1965, at first called the Super Skywagon, but when the 207 appeared in 1970, rather than call it, logically, the Super Duper Skywagon, the Super Skywagon became the 206 Skywagon—which was replaced in 1971 by the Stationair; same airplane, new name. Meanwhile, the Stationair Eight has replaced the 207 Skywagon. In 1984, there were 15 Skywagon/Stationair versions, counting the turbo models. (Cessna)

The Cessna 172 was introduced in 1956 to replace the four-seat 170 taildragger. Both had the 145-hp Continental engine and essentially the same performance. The 172 and Skyhawk, the latter being the deluxe version, switched to the 150-hp Lycoming in 1968, and the present models are rated at 160 hp with the Lycoming 0-320-D2J. Standard model 1984 Skyhawk: $47,700.

The Cessna Turbo Stationair Six on amphibious floats is primarily a working airplane with a 970-pound payload; it requires a water run of 2400 feet for takeoff, and has a maximum speed of 150 kts, 24 kts slower than its landplane counterpart. (Cessna)

The Cessna model 210 Centurion has been in production since 1959 (although it was not originally called the Centurion). It has always been competitive with the Bonanza in performance, and the 1984 Centurion has a max cruise of 175 kts at 6500 feet. The turbo version does 204 kts at 17,000 feet. Prices with radios: $144,250, and $156,400, respectively. (Cessna)

The Cessna Crusader seats six and cruises at 196 kts at 20,000 feet. Engines are turbo Continental TSIO-520s of 250 shaft hp. Basic list is $278,450. (Cessna)

The Cessna 340 is a pressurized six-place businessliner with a maximum speed of 244 kts at 20,000 feet. Engines are turbocharged fuel-injected TSIO-520s of 310 hp, and price with radios is $473,000. (Cessna)

The pressurized Cessna Golden Eagle has eight seats, a maximum speed of 258 kts at 20,000 feet, and is powered with turbocharged fuel-injected GTSIO-520 Continentals of 375 hp. With basic radios, its price in 1984 dollars was $731,500. (Cessna)

The two-place Temco Swift was built 1946-1950 by Texas Engineering Manufacturing Company of Dallas, which took over the design from Globe Aircraft of Ft. Worth. This airplane, with 125 hp, has a cruising speed of 140 mph and sold new for $4995. A good one in the used market today will bring more than twice that. The tip plates and double-bent pitot tube on this one are not standard. (Francis Dean)

The first Luscombe Silvaires were built in 1937, fitted with 50-hp engines. After WWII, the 8A model appeared with 65 hp, and the 8E with 85 hp. Production ended in 1949 with the 90-hp 8F. Temco acquired the design in the early '50s and built a few 8Fs, and then the Silvaire Corporation of Ft. Collins, Colorado, purchased the rights and manufactured 8Fs until 1960. Maximum speed is 128 mph with 90 hp.

The non-military helicopters are almost exclusively utility machines; they must earn their keep. Relatively expensive to buy and maintain, the helicopter's place in the affairs of mankind is secure because it can do things that are not possible with any other vehicle we possess. The Bell Jet Ranger is popular with TV News departments and as an air taxi. (Bell)

The Mitchell P-38 meets ultralight criteria but is intended by its manufacturer to be licensed in the experimental category as a homebuilt. It cruises at 55 mph and lands at 30 mph. Construction time averages 150 hours. (Larry Collier)

Foldable wings and a trailer allow home storage of ultralights and homebuilt aircraft. One person may prepare the Mitchell P-38 for flight, or return it to its trailer for transport, in ten minutes. (Larry Collier)

If an ultralight is to remain unlicensed, and its pilot is unlicensed, it may weigh no more than 254 pounds, possess a speed of no more than 55 kts, must have a landing speed of no more than 24 kts, and a maximum fuel capacity of five gallons. (Don Downie)

The Beechcraft Skipper two-place trainer is powered by the time-tested 0-235 Lycoming, which has been in production since it was first fitted to the 1946 Piper PA-12. Beech operates a chain of 33 pilot-training facilities around the nation. (Beech)

The "entry level" Beechcrafts for new pilots/owners are the fixed-gear Sundowner (foreground) and the retractable-gear Sierra. Possessing essentially the same airframe, these craft are descended from the Model 23 Musketeer which first appeared in 1962 and was priced at $13,300. The '84 Sundowner was $53,550; Sierra $73,000. (Beech)

The Beechcraft Lightning originally scheduled for production in 1985, is a pressurized propjet that promises a max cruise of 281 kts (323 mph) at 25,000 feet with a 1000-mile range. Fully equipped the price is well over half a million dollars, but production is now uncertain. (Beech)

The V-tail Bonanza, offered with turbosupercharging as the V35B, had a 1983 basic price of $161,500. After 36 years of continuous production, the Mercedes of single-engine civil aircraft clearly faced an uncertain future. It had gained 115 hp, 1000 pounds in gross weight, and had fostered a loyal Bonanza cult. It also symbolized the end of an era. (Beech)

The Beechcraft Barons are the true "twin Bonanzas," and in the mid '80s remained in production, though struggling, while many other light twins were out of production or "temporarily" suspended. The Cessna Skymaster (the most sensible light twin) was gone, along with the 310, the Piper Seminole, Aztec, and Aerostar, the Rockwell Commander Shrike, and apparently the Beechcraft Duchess. The key to light twin (under 6000 pounds) survival is probably the minimum pressure altitude such a machine can maintain on one engine. (Beech)

The Beechcraft 1900 Executive is a corporate airliner version of the 1900 commuter airliner. Powered with P&W PT6A-65B turboprops, it has a max cruise of 263 kts (303 mph) with 15 passengers. Interior appointments include two beverage bars and a lavatory. Price: $2,832,750. Deliveries began in 1984. (Beech)

The Beechcraft Super King Air dominated its market from the time the first one appeared in 1964, that market being the corporate propjet. Almost 3800 had been produced by mid-1984. Seating is configured for eight to 15 passengers, and maximum cruise is 294 kts (338 mph) at FL220 (22,000 feet); maximum range is 2200 miles. The Super King Air is priced in the two million dollar range. (Beech)

Burt Rutan at the EAA's Oshkosh Fly-In/Convention in his twin-engine Defiant. Plans for this four-place homebuilt were being prepared for sale in mid-1984 according to a source close to Burt. (Don Downie)

The EAA's annual Fly-In/Convention attracts dozens of Rutan designs flown in by proud builders of these advanced technology homebuilts. (Don Downie)

The Beechcraft Starship I, an 85 percent scale flying prototype that created a sensation when unveiled at the National Business Aircraft Association convention in Dallas 4 October 1983. Twin propjets ("jetfans" is suppossed to be the new term) will give the Starship I a max cruise in excess of 400 mph. Projected price is $2.75 million. (Beech)

Beechcraft Starship instrument panel. (Beech)

The Lockheed L-1011 TriStar first flew 16 November 1970. Several versions were produced, but it entered stormy economic skies. Along with the McDonnell-Douglas DC-10 and the Boeing 747, the L-1011 must be judged an engineering triumph and an economic failure. (Lockheed-California)

Proliferation of the commuters following airline deregulation weakened some of the large air carriers but did not, during the first six years of that experiment, provide better service. In 1984 it appeared that two large airlines would emerge as super carriers while most of the rest would fight for survival.

The Ted Smith-designed Aero Commanders can trace their lineage back to the Douglas A-20. The prototype Commander flew 27 April 1948, and the model 520 left the Betheny, Oklahoma, plant in August 1951 with two 260-hp Lycomings and six seats. The series progressed through the 560- and the 240-mph Grand Commander in 1965 to the Rockwell Shrike Commander in 1974. Sales lagged after that, and the 1121 Jet Commander was taken over by Israel Aircraft Industries and offered as the Westwind. (Francis Dean)

Learjet has built more business jets than anyone else—almost 1400 by mid-1984, roughly 25 percent of all bizjets in the world. The first model 23 Learjets sold for approximately $500,000 in 1964. The 1984 model 25 was $2 million plus; the model 55, pictured, sells for over $5 million fully equipped. The first of 80 USAF C-21A Learjets was delivered in March 1984. (Gates Learjet)

Section II
U.S. Military Aviation

Chapter 10

The First U.S. Military Airplanes

The U.S. Army took delivery of its first airplanes at the end of July 1909. The Army paid the Wright brothers $30,000 for their Military Flyer following successful demonstration flights by Orville Wright at Ft. Myer, Virginia. The Army's advertised specifications—actually based on performance promised by the Wrights—called for a machine with a speed of at least 40 mph and capable of remaining aloft for a minimum of one hour with a passenger aboard.

With Lt. Benjamin Foulois as a passenger, Orville flew to Alexandria and back, a ten-mile round trip, at an average speed of 42.58 mph, and accompanied by Lt. Frank Lahm remained airborne for an hour and twelve minutes. Takeoffs were aided by the release of a 1600-pound weight from the top of a 50-foot tower; this catapulted the airplane down a 100-foot monorail on a wheeled dolly.

The Army would have had an airplane a year earlier, when Orville first took a Wright machine to Ft. Myer, but for an unfortunate accident (caused by a propeller failure) that took the life of passenger Lt. Thomas E. Selfridge and seriously injured Orville.

Between 1909 and America's entry into World War I on 6 April 1917, the Army ordered a total of 532 airplanes from 14 different manufacturers, 227 of which were actually delivered prior to the U.S. declaration of war. Contracts were canceled for 155 prewar orders, and 150 prewar orders were delivered later. The Army's airplanes were assigned to the Signal Corps, and when America went to war in April 1917, the Army Signal Corps had 55 flyable airplanes, the rest having been destroyed or junked.

The U.S. Navy purchased its first airplane 1 July 1911, when Glenn Curtiss demonstrated the Curtiss Triad in a couple of short flights, taking off and landing on Lake Keuka near Hammondsport, New York. Lt. T.G. Ellyson accompanied Curtiss as a passenger on one flight. Then Ellyson, who had learned to fly at the Curtiss school on North Island, San Diego, flew the Triad alone on two flights.

The Navy had ordered the Triad, serial A-1, on the 8th of May, 1911, and Curtiss had tested it as both a landplane and seaplane before the official tests were held two months later, because the Navy had contracted for an amphibian (which is why this machine was called the Triad—for land, sea, air).

The U.S. Navy took delivery of 129 airplanes prior to 6 April 1917, and 54 remained in service on that date, all of them seaplanes or flying boats, plus two captive ballons and one dirigible.

Three U.S. Marines, led by Lt. Alfred A. Cunningham, formed the nucleus of the Marine Corps' Aviation Branch in 1912 after Cunningham learned to fly at the Burgess school. The Marines, however, possessed no airplanes they could call their own until after America went to war in 1917.

When WWI began for America, the nation had seven Marine, 38 Navy, and 35 Army pilots assigned to aviation duty (the Army had trained a total of 66 pilots). A handful of other officers in each service were attached to aviation units in administrative posts. A total of 76 enlisted Marines were assigned to Marine aviation; 230 Navy noncoms supported Naval aviation, while the Aviation Section of the Army's Signal Corps could count almost 1000 enlisted men.

The first U.S. Coast Guard pilots were 3rd Lt. Elmer F. Stone and 2nd Lt. Charles E. Sugden, who were sent to the Navy's air training facility at Pensacola, Florida, in March 1916. These two would become Naval Aviators 38 and 43 respectively. Four additional Coast Guard pilots were trained by the Navy later that year.

Naval aviation's first base was near the Naval Academy on Greenbury Point at Annapolis. It was known as the Aviation Camp, and was moved to Pensacola on 20 January 1914. Nine officers, 23 men, and seven aircraft under Lt. J.H. Towers established a flying school there.

The Army's first flight training facility was at College Park, Maryland; it began operation in June 1911. It moved south to Augusta, Georgia during the cold months. After a second unpleasant winter in Georgia, the Army pilots received training at Curtiss' school on North Island, at San Diego, California, as did some Naval aviators. A few officers from both services attended the Wright school at Dayton as well as the Burgess school at Marblehead, Massachusetts. (Burgess, well known as a yacht designer, became involved with airplanes when he designed a pair of floats for a Wright machine in mid-1911, at approximately the same time the Navy Triad was in test by Curtiss).

Both Army and Navy pilots were subject to hostile action of a sort prior to WWI. Army fliers were sent to the Mexican Border in 1913, and then three years later were with Pershing as he chased Villa into Mexico. Navy pilots saw action over Vera Cruz in 1914.

With Pershing in Mexico

Army aviation was first to be summoned for possible action. Late in February 1913, while the school was at Augusta, a tense situation on the Mexican Border due to the political instability in that country resulted in a call for air assistance from the Second Division, stationed at Texas City, Texas. The school commander organized the First Provisional Aero Squadron with five pilots, 21 enlisted men, and seven Wright pushers, and this group reached Texas City a week later by train. By that time the crisis had passed, and the First Provisional Aero Squadron was absorbed by the new facility at North Island.

A court-martial of several Army pilots at North Island, fostered by a spit-and-polish commander, prompted establishment of a new training school away from the easy pleasures of Southern California. San Antonio was the new site selected, and the First Aero Squadron (no longer "provisional") had been there for four months when ordered to join the Pershing expedition into Mexico.

The First Aero Squadron, consisting of 15 Officers and 85 enlisted men, was commanded by Capt. Benjamin Foulois, and equipped with eight Curtiss JN-2 Jennies. Its first duty station was Ft. Sill, Oklahoma, where it arrived at the end of July 1915 to work with the Field Artillery School. The squadron was transferred to San Antonio in November 1915.

Early on the morning of 9 March 1916 a band of Mexican bandits (or rebels, depending upon where one's sympathies rested during the turmoil in Mexico at the time) numbering about 400 men, and under the command of Pancho Villa, raided the U.S. border town of Columbus, New Mexico, killing nine civilians and seven U.S. troopers. The surprise attack would have been more devastating had not Col. H. J. Slocum's 13th Cavalry charged into town.

Slocum's cavalry chased the bandits about 14 miles into Mexico but Villa escaped, whereupon President Wilson ordered ". . . an adequate force . . . be sent in pursuit of Villa . . ." And that resulted in the hasty organization of a 5000-man expeditionary force at Ft. Sam Houston under the command of Brig. Gen. John J. "Blackjack" Pershing. It was to be accompanied by the First Aero Squadron.

Pershing entered Mexico on 15 March, and his air contingent was expected to catch up with him five days later at the Mexican Pueblo of Casa Grande. Six of the Jennies made it; two went down enroute.

The airplanes were all underpowered, being fitted with the early Curtiss OX engine. Two "more powerful" engines were sent to the squadron, along with two Curtiss R-2 airplanes, as the Jennies were lost one by one until only two remained. Mechanical failures of the simplest kind resulted in forced landings. Lt. C.G. Chapman lost his plane at San Rosalia when Mexican Federal troops put him in jail; Lt. Ira Rader wrecked his machine near Ojito, and sabotage inflicted on Capt. Foulois' aircraft at Chihuahua City resulted in a crash-landing. Capt. Willis and Lt. Dargue crash-landed in the mountains following engine failure and walked 60 miles back to their base.

The airmen were expected to reconnoiter and, hopefully, locate Villa's forces, but during the month they served the Pershing expedition the inadequacy of their equipment largely limited them to the role of couriers and the transport of mail to forward elements of the expedition. By 22 April Pershing's cavalry was 300 miles into Mexico, and the First Aero Squadron's two remaining airplanes were ordered back to Columbus.

Villa was never captured. He remained under arms until, in 1920, the Mexican Government bribed him into retirement with money and the gift of a large estate. On 20 July 1923 his automobile was ambushed by "unidentified" assassins and he and his three companions were killed.

Military Air—1914

When World War I began in Europe in August 1914, Germany had 218 airplanes in military service in 49 units; France could count 142 in 25 *escadrilles* (squadrons). Britain's Royal Flying Corps had 179 aircraft, the Royal Naval Air Service, 93.

Russia had 224 military airplanes in August 1914, plus 12 airships and 46 captive balloons.

The Italian Royal Army Air Service was made up of 13 squadrons possessing six to nine aircraft each, while the Italian Navy had 15 seaplanes and flying boats.

The U.S., of course, had no intention of becoming involved in Europe's war, and the fact that America rated *fifth* in airpower (behind Japan) at the end of 1914 alarmed none of the nation's leaders, civilian or military. The U.S. Army Signal Corps had between 20 and 24 flyable airplanes at that time; the Navy listed 12 seaplanes and flying boats in active inventory.

Airpower was not a decisive factor in WWI. For all practical purposes, the airplane was but six years old in 1914, and although four years of war accelerated advances in airframe and engine design, the airplane of 1919 still represented a prophecy.

During the early months of WWI, airplanes on both sides were used exclusively for reconnaissance. None was armed, although crews soon began carrying pistols and rifles—and even sacks of bricks—should an enemy aircraft be encountered at close range (the bricks were intended for propellers).

The first armor was cast-iron stove lids from field kitchens, which were placed beneath seat cushions after a Royal Flying Corps (RFC) pilot was wounded in that area by a rifle bullet fired by a German soldier on the ground. The pilots developed the belief during this period that they were reasonably safe from ground fire above 2500 feet.

The First Aircraft Armament

Aircraft armament was inevitable. Having established that aerial reconnaissance was highly useful to an army, each side recognized the desirability of denying access to its airspace to the other, along with the need to ensure the success of its own forays over enemy territory.

At first, the Germans sought to improve their ground-based antiaircraft defenses rather than arm their airplanes, while the British and French looked for ways to mount automatic weapons on their aircraft. Early in 1915, the British and French had flexible machine guns in the noses of several pusher-type airplanes such as the Vickers Gunbus, Farman, and Voisin. France's Rolan Garros was the first to mount a forward-firing, line-of-sight machine gun on a single-seater to presage the advent of the true fighter airplane. The weapon was an 1885 Hotchkiss machine gun and it was not synchronized to fire between the revolving propeller blades, that problem being circumvented by the attachment of triangular-shaped steel deflectors on the back of each blade so the one-in-seven projectiles that struck a blade was deflected away.

Garros' idea was adopted by few of his squadron mates, possibly because he was soon forced down in enemy territory and some suspected that he had shot away his own propeller.

That was not the case, and the Germans viewed Garros' installation with interest, then asked Tony Fokker to design something better. Fokker did so within 48 hours.

Fokker reasoned that the engine should interrupt the firing of the weapon when a propeller blade was in line with the gun's muzzle. A cam attached to the front end of the crankshaft activated a system of levers that accomplished this, and Fokker E-1 Eindeckers, fitted with Fokker gun installations, were in service by August 1915. Later, Fokker engineers perfected a superior system in which the interrupter worked from the engine's camshaft, and a flexible cable replaced the system of levers.

The best and most widely employed synchronizer employed by the Allies later in the war was the hydraulically-activated system invented by Rumanian George Constantinesco early in 1917. Generally referred to as the "CC gear," it was used by Britain well into the '30s.

Meanwhile, the British Vickers company devised a gun synchronizer known as the Vickers-Challenger. This went into action in April 1916, installed in the Sopwith "One and one-half Strutter." Other English gun sychronizing systems appeared at about the same time, all similar to the Fokker mechanism. (It is true that Franz Schneider in Germany patented such a device in 1913, and even sued Fokker for patent infringement, but the Schneider system proved unworkable.)

Other WWI "Firsts"

☐ The British began using wireless telegraphy in their observation planes for the control of artillery fire in March 1915.

☐ The first airplane shot down by hostile fire from another aircraft was a German two-place Aviatak destroyed by a jury-rigged Hotchkiss fired by a Corporal named Quennault riding in a French Voisin pusher flown by Sgt/Pilot Franz; 4 October 1914.

☐ The first effective bombing raid was carried out by two British Navy pilots flying Sopwith Tabloids. They placed several small bombs on a Zeppelin shed at Dusseldorf on 2 October 1914.

☐ The first air unit devoted exclusively to bombing, and the first to fly in formation, was the German "Carrier Pigeon" *staffel* based at Ostend in October, 1914.

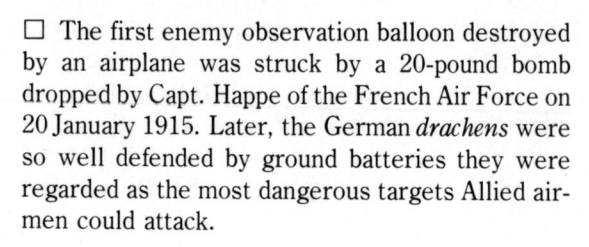

☐ The first enemy observation balloon destroyed by an airplane was struck by a 20-pound bomb dropped by Capt. Happe of the French Air Force on 20 January 1915. Later, the German *drachens* were so well defended by ground batteries they were regarded as the most dangerous targets Allied airmen could attack.

☐ The world's first strategic bombing force was

the German rigid airships that began raiding British and French cities 19 January 1915. Commonly called "Zeppelins," some were, and some were built by Shutte-Lanz. The Zeppelin attacks tapered off late in 1916 as home defense fighter planes reached them. The Zeppelins dropped more than 5000 bombs and killed 557 people.

☐ The first Zeppelin brought down by a defending airplane fell to the guns of RFC pilot Lt. W. Leef-Robinson on the night of 2 September 1916. The last Zeppelin raid was 17 May 1917.

U.S. Airpower in WWI

While airpower did not ultimately prove to be the decisive factor in the outcome of WWI, it should be noted that, in theory, it may have drastically altered that conflict at the very beginning. Within two weeks after their arrival in France, the four British squadrons on the Continent, containing less than 40 airplanes (mostly, the 80-mph BE2c), were caught up in the historic Allied retreat from Mons. Between 24 August and 12 September, reconnoitering over rapidly advancing and vastly superior German forces, these few pilots and their observers saved the British army from encircling movements that could have meant certain annihilation. Military historians believe that, had the British army been defeated at that time, WWI would have ended in victory for the Germans within a month.

In any case, the lesson to be taken from that near-disaster was not lost on military planners on either side or in the U.S., and though it was a valid lesson at the time, it resulted in the dominance of observation-type airplanes in the Air Forces of the Western nations until the eve of WWII. Ground commanders did all the planning during those years, and controlled aircraft procurement.

That was proper during WWI. True, immediately after abandoning the Zeppelin raids, the Germans resumed strategic bombing, mostly against British targets, with fleets of twin-engined bombers (and were planning mammoth four-engine machines, as were the British and Italians), but these aircraft could neither carry sufficient tonnage nor deliver it accurately enough nor far enough to alter the thinking of the ground generals, especially those in the U.S.

The fighter airplane concept evolved quickly because the ground commanders saw the need to both deny enemy aircraft access to the airspace above their armies, and to protect their own reconnaissance airplanes in enemy airspace. What that implied, of course, was control of the air, but planning for *that* went no further in WWI than efforts to produce better fighter airplanes.

From early in 1915 into the first months of 1918, the war seesawed back and forth over a few miles of eastern France while ten million men died in the mud and the conflict became a stalemate, both sides near exhaustion.

However, the Central Powers (principally Germany, with allies Austria-Hungary, Bulgaria, and Turkey) may well have defeated the British, French, and Italian Allies during the summer of 1918 had the U.S. not intervened. The Italians had been soundly beaten at Caporetto and were more burden than help to the beleaguered British and French. The German armies on the Eastern Front had been freed to fight in the west when the Bolshevik Revolution took Russia out of the war, and it seemed clear that Germany had both the will and the means to mount final, massive attacks—which she did, and which may well have proven successful but for the arrival of the Americans in force and in time for those crucial battles.

The series of five German drives began 21 March 1918 and continued until mid-July. The Germans had 82 divisions on the Western Front and by 30 May had driven to within 40 miles of Paris. By that time, however, American ground troops were in action to stop the Germans in three key sectors, and then spearheaded Allied drives, beginning on 18 July, that ended in victory on 11 November 1918.

After the U.S. declaration of war on 6 April 1917, it took one year to put 100,000 American ground troops in France. Then, during May and June 1918, 10,000 per day were arriving, and a total of 29 U.S. combat divisions was in the field when the end came five months later. But building an Air Force from scratch was another matter.

The U.S. Air Service (actually, still the Aviation Section of the Signal Corps, but the term "Air Service" was commonly used from early in 1917; it officially became the Air Service in 1920) was able to put one squadron of pursuits and one observation in action by April 1918. In the meantime, many Air Service pilots and support people served with British, French, and Italian air units, and most were not reclaimed for duty under their own flag until July.

The trickle of pilots and trained support personnel coming from hastily constructed facilities in America reached France without equipment. Another 1800 with only ground school training, scheduled for flight training in France, were idle for months. In October 1917, construction began at Issoudun, France, on a huge flight training facility (characterized by Gen. Pershing as the "worst mudhole in France"), with the American student pilots doing the constructing. Eventually, 776 pursuit pilots were graduated from Issoudun, trained in 15-, 18-, and 23-meter Nieuports purchased from France.

The ground schools in the U.S. were quickly established at eight major universities which, by the end of the war, had passed 18,000 students. Of those, 14,835 entered primary flight training while 8688 of that number received commissions and were rated as Reserve Military Aviators at one of the 27 air training fields in the U.S. or the 16 fields in France, Britain, and Italy.

The Air Service's 94th Pursuit Squadron was the first to enter combat, and Lts. Douglas Campbell and Alan Winslow shot down two German airplanes on 14 April 1918 to score the first American air victories. (Americans flying with the French and British had scored earlier, including 12 of the 38 American volunteers who made up the *Escdrille Lafayette,* a French squadron formed 20 April 1916, manned almost exclusively by Americans. This became the 103rd Pursuit Squadron, U.S. Air Service, on 18 February 1918.)

At the end of the war there were 45 American combat squadrons in action with a strength of 740 airplanes, approximately 800 pilots, and 500 observers. Air Service losses were 289 airplanes and 48 observation balloons, compared to confirmed claims of 781 enemy aircraft and 73 balloons shot down.

There were 83 American aces (five or more air victories confirmed) in the U.S. Air Service. Six Americans who served with the French throughout the war became aces, and 18 Americans who fought exclusively with the British were aces, including "forgotten" ace Lt. Col. (in WWII) William C. Lambert, who accounted for 19½ enemy airplanes (the "½" a shared victory) and two balloons. His record did not come to light until 1968, uncovered by Royal Frey of the U.S. Air Force Museum, whereupon Britain's Ministry of Defense provided Frey with a transcript of Lambert's official record, but offered no explanation as to how the record had remained undiscovered all those years. Lambert flew with RAF No. 24 squadron (The RFC became the Royal Air Force in 1918). He died in his home state of Ohio in 1982 at age 87. Had he achieved the same number of victories with an American unit, he would have been the second ranking U.S. Air Service ace in WWI, ahead of the legendary Frank Luke, Jr., who scored 21.

Luke scored his 21 victories—16 of which were the most dangerous targets he could have chosen, German observation balloons, protected by massed batteries of defensive guns—in a 17-day period, and nine of those days he did not fly!

Following the loss of two wingmen, Luke twice disobeyed orders and left his airfield. His commanding officer then ordered Luke's arrest. Luke ignored that order as well, and took off alone to attack three enemy *drachens* tethered along the Meuse River near Dun. It was near sunset on a Sunday evening, 29 September 1918.

After downing the three balloons and two enemy fighters, Luke was forced to land behind enemy lines with a dead engine. He was killed by German infantrymen when, called upon to surrender, he answered them with his pistol.

Lt. Frank Luke, Jr., is the only American airman to be awarded the Congressional Medal of Honor for a deed done while under military arrest.

The top U.S. Air Service ace in WWI was Edward V. Rickenbacker, with 26 confirmed victories. Capt. Rickenbacker was a well-known race car driver before the war (and a very successful president of Eastern Airlines after WWII).

The combat record of the U.S. Air Service in WWI was indeed a proud one, particularly in view of its limited size and crippling lack of air commanders. Most squadrons were commanded by field grade officers who had attained their rank in the cavalry. But many adjusted, and a few would champion the cause of Army aviation after the war.

Billy Mitchell

The most important air leader to emerge in the U.S. Air Service was Col. William "Billy" Mitchell, who had been a foot soldier for 18 years before learning to fly. Mitchell had enlisted in the Army in 1898 at the outbreak of the Spanish-American War, and rose through the ranks to become an officer in the Signal Corps. By 1916, his record had earned him an assignment to Signal Corps Headquarters in Washington and promotion in rank to major. Mitchell learned to fly at a civilian school at his own expense and, early in 1917, convinced his superiors that he should be sent to Europe to study Allied air tactics. He was there when the U.S. declared war on Germany.

He flew as an observer with the French, and went to England where he was received by Maj. Gen. Sir Hugh "Boom" Trenchard. Trenchard would merge the Royal Naval Air Service and Royal Flying Corps to form the Royal Air Force (RAF) in 1918, and it was Trenchard's concepts of the proper role of airpower in military operations that greatly influenced Mitchell's thinking. (It has been said that Mitchell took part of his doctrine from Italy's Gen. Douhet, but while some of Douhet's theories were similar, neither Trenchard nor Mitchell believed in indiscriminate bombing of civilian populations as did Douhet.) Briefly stated, Trenchard and Mitchell believed that a nation's Air Force should be an independent service commanded by air officers, that a strategic air arm should be committed to the destruction of an enemy's critical industries, that tactical aircraft should be sent against the enemy's lines of supply and communications and, in short, that the airplane should always be employed as an *offensive* weapon.

Mitchell had one opportunity to establish an important part of this basic air philosophy—the proper use of a tactical Air Force—after Air Service Commander Maj. Gen. Mason Patrick (a classmate of Pershing's at West Point) recommended Mitchell for the post of Chief of Air Service, First Army, on 27 July 1918. Mitchell then had six weeks to plan for the great Allied attack on St. Mihiel.

Support from the other Allies was total, and Mitchell had at his disposal 1476 aircraft, 20 balloons, and 30,000 airmen and support personnel from British, French, and Italian Air Forces, including nearly 600 American airplanes and crews. On the ground, 400,000 men awaited the signal to advance.

The drive began on 12 September and the battle was won on the first day, the ground forces moving swiftly beneath U.S. Air Service planes while the British from one end of the 30-mile front and the French and Italians from the other attacked far behind enemy lines to immobilize enemy transport and destroy communications. Denied both help and an avenue of retreat, German resistance collapsed, and the attacking American Army took thousands of prisoners.

General Pershing was impressed, and Mitchell was promoted to the temporary rank of Brig. General a month later.

It later appeared, however, that Pershing never understood that a key factor in the St. Mihiel victory was the way Mitchell employed tactical air units behind the German lines. Pershing would dominate U.S. military planning for ten years after WWI, and his opinion of the Air Service during those years seemed to be based on his experience with the First Aero Squadron during the fruitless 1916 expedition into Mexico.

Navy and Marine Air in WWI

At the time of America's call to arms in April 1917, one Naval Air Station was in operation and 48 aviators and student aviators were available, along with 54 aircraft. When the war ended 19 months later, the U.S. Navy had 27 bases in France, England, Ireland, and Italy, one in the Azores, two in Canada, one in the Canal Zone, and twelve in the

U.S. Spread among these bases were 6716 officers and 30,693 men in Navy units, and 282 officers, 2180 men in Marine Corps units, with 2156 aircraft, 15 dirigibles, and 215 observation balloons. Of these, 18,000 officers and men and 570 aircraft had been sent abroad. There was but one Navy ace, Lt. David S. Ingalls, who scored five air victories flying a Sopwith Camel with RAF No. 213 Squadron.

Navy and Marine air units had little opportunity for air combat, and were not so equipped, because they were primarily assigned to sea patrol duties in Curtiss flying boats. In that role they were officially credited with attacking 25 German U-boats and damaging 12.

U.S. Navy pilots began flying patrols from an RAF station at Killingholme, England, in February 1918; Ens. John F. McNamara carried out the first attack on a German U-boat by an American pilot on 25 March.

The First Marine Aviation Force (its official title) reached France 30 July 1918 for operations as the Day Wing, Northern Bombing Group, USN. Its three squadrons were designated 7, 8, and 9, completing Navy squadrons 1 through 6 of the Northern Bombing Group. This Marine unit had been formed 15 April 1918 at the Miami NAS.

The first raid in force by the Navy's Northern Bombing Group was made by eight planes of Marine Day Squadron 9, which dropped 17 bombs totaling 2218 pounds on the German-held railroad junction at Thielt, Belgium, 14 October 1918. For extraordinary heroism on this and an earlier flight in engaging enemy aircraft at great odds, 2nd. Lt. Ralph Talbot, and his observer, Gunnery Sgt. Robert G. Robinson, were later awarded the Medal of Honor.

The Navy fostered development of a "flying bomb" during WWI, and on 17 October 1918 tested an N-9 training plane converted to an "automatic flying machine" with the forerunner of the automatic pilot—a gyroscope and clock device. The drone flew a prescribed course, launched from Copiague, Long Island, but failed to land at its preset range of 14,500 yards. It was last seen over the Bay Shore Air Station at an altitude of 4000 feet, flying eastward.

WWI Aircraft Production and Procurement

At the end of WWI there were 196 American-built combat airplanes in action with the U.S. Air Service in France. The U.S. Navy and Marines appear to have possessed, at that time, approximately 280 American-built aircraft in service on their overseas stations. Nineteen months after the U.S. entered the war, the committment to "darken the skies over Germany" with American-built warplanes had put no more than 800 American-built machines (counting combat and operational losses) in action against the enemy. Most of the Air Service airplanes in combat were bought from France, England, and Italy.

Actually, 1440 American-built deHavilland DH-4 observation planes were shipped overseas during 1918 from a total of 4846 built by three manufacturers (3106 by Dayton-Wright), but none were fit for service and those that reached the front were rebuilt by personnel of the Air Service depot at Romorantin, France.

Altogether, the U.S. Air Service accepted 11,754 American and Canadian-built airplanes between 6 April 1917 and 11 November 1918. Prior to war's end, the Air Service purchased 4881 French, 258 British, and 19 Italian airplanes.

Therefore, the Air Service received a total of 16,912 airplanes during the war; the Navy and Marines, 2156, for a grand total of 19,068. Contracts for 61,000 machines were canceled due to several reasons, principally the cessation of hostilities.

However, except for 6000 Spads, most of the cancellations involved U.S. manufacturers, and had the war continued into 1919, the Air Service could have put 10,000 airplanes in combat with trained crews.

The production turnaround came in May 1919, following a Senate investigation that prompted President Wilson to fire "swivel-chair Colonels" Howard Coffin and Edward Deeds (who had selected the Liberty engine for mass production, and therefore the DH-4), and reorganize the Aircraft Production Board.

Both Deeds and Coffin would again figure

prominently in aviation matters in years to come. Deeds and his son would make a great deal of money from their investment in Pratt & Whitney. Coffin would sit on President Coolidge's Morrow Board in 1925, the recommendations of which publicly justified the court-martial of Billy Mitchell and denigrated the value of airpower to the nation's security.

Another of the WWI "Detroit Gang" was Harold E. Talbott, Jr., associated with Deeds and Charles Kettering (DELCO) in the Dayton Metal Products Company, which received large wartime contracts from both the U.S. and British governments. Talbott would be appointed Secretary of the Air Force by President Eisenhower in 1952.

There are those who contribute, and those who exploit. Somtimes, it's hard to tell the difference.

Aviation Section Personnel and Direct Appropriations 1912-1919

1912	51	$ 125,000
1913	114	100,000
1914	122	175,000
1915	208	200,000
1916	311	801,000
1917	1,286	18,681,666
1918	195,023	735,000,000
1919	25,603	952,304,758

Military Aircraft Production* Calendar Years 1912-1919

1912	16
1913	14
1914	15
1915	26
1916	142
1917	2,013
1918	13,991
1919	682

*U.S. Department of Commerce; Army and Navy

U.S. Signal Corps Airplanes 1909 through 1916

Aircraft	Year Delivered	Number Delivered	HP	Engine
Wright B	1909	1	30	Wright
Curtiss D	1911	1	50	Curtiss
Wright B	1911	2	40	Wright
Burgess F	1911	1	45	Sturtevant
Curtiss E	1911	3	60	Curtiss
Wright C	1911-12	7	50	Wright
Burgess H	1912	7	75	Renault V-8
Curtiss Flying Boat	1912-13	3	80	Curtiss O
Burgess Scout	1913	2	60	Sturtevant
Wright Scout	1913	2	50	Wright
Curtiss Scout	1913	2	80	Curtiss O
Curtiss J	1914	2	90	Curtiss OXX
Martin TT	1914	17	90	Curtiss OX-2
Curtiss N	1914	1	90	Curtiss OX
Burgess-Dunne Tailless	1915	1	125	Salmson B
Wright Tin Cow	1915	1	90	Daimler 6
Curtiss JN-2	1915	10	90	Curtiss OX

Aircraft	Year Delivered	Number Delivered	HP	Engine
Martin S-Hydro	1915	6	125	Hall-Scott
Curtiss JN (N-8)	1916	4	90	Curtiss OX-2
Curtiss R-2	1916	12	160	Curtiss VX
Standard (Sloan) H-2	1916	3	125	Hall-Scott
Curtiss JN-4	1916	93	90	Curtiss OX-2
Standard H-3	1916	9	125	Hall-Scott
Curtiss Twin JN	1916	8	100	Curtiss OXX-2
Martin R	1916	2	125	Hall-Scott
Sturtevant Trainer	1916	7	135	Sturtevant
Lawson-Willard-Fowler	1916	20	135	Sturtevant

American-Built U.S. Air Service Airplanes 1917-1918

Aircraft	Year First Delivered	Number Delivered	HP	Engine
Lawson-Willard-Fowler	1916	3	135	Sturtevant
Standard SJ-1	1917	23	90	Hall-Scott
Thomas D-5	1917	2	135	Thomas
Sturtevant S-4	1917	4	150	Sturtevant
Curtiss R-4	1917	53	200	Curtiss V-2
Aeromarine M-1	1917	6	90	Hall-Scott
Burgess Model U	1917	6	90	Curtiss-OX-2
Curtiss S-3 Triplane	1917	4	100	Curtiss OXX-2
Curtiss N-9 Seaplane	1917	14	100	Curtiss OXX-3
Gallaudet Hydro	1917	4	125	Hall-Scott
Curtiss L-2	1917	4	100	Curtiss OXX-2
Curtiss R-3	1917	18	200	Curtiss V-2
Wright-Martin R	1917	9	150	Hall-Scott
Wright-Martin Seaplane	1917	3	150	Hall-Scott
Curtiss JN-4	1917	603	90	Curtiss OX-5
Curtiss JN-4B	1917	5	90	Curtiss OX-2
Curtiss JN-4C	1917	2	90	Curtiss OXX-3
Curtiss JN-4A	1917	1	90	Curtiss OX-5
Lowe-Willard-Fowler	1917	112	135	Sturtevant
Standard Seaplane	1917	1	125	Hall-Scott
Standard SJ-1	1917	750	90	Curtiss OX-5
Pigeon-Fraser Pursuit	1917	2	100	Gnome 9
Burgess Seaplane	1917	1	150	Sturtevant
Curtiss JN-4D	1917	1,404	90	Curtiss OX-5
Heinrich Pursuit	1917	4	100	Gnome 9
Thomas Morse S4B	1917	100	95	Gnome B-9
Boeing EA	1917	2	90	Curtiss OXX-3
Orenco A	1917	2	105	Duesenberg

Aircraft	Year First Delivered	Number Delivered	HP	Engine
Standard JR-1	1917	6	150	Hispano A
Dayton Wright SJ-1	1917	400	150	Hispano A
Wright Martin SJ-1	1917	51	125	Hall-Scott
Standard M	1918	2	80	Le Rhone C9
Schaefer & Sons Trainer	1918	1	100	Gnome 9
Breese Penguin	1918	301	28	Lawrance
U.S. Aircraft JN-4D	1918	50	90	Curtiss OX-5
Dayton-Wright DH-4	1918	3,106	400	Liberty
Fowler Corp. JN-4D	1918	50	90	Curtiss OX-5
Howell & Lesser JN-4D	1918	75	90	Curtiss OX-5
Liberty Iron Works JN-4D	1918	100	90	Curtiss OX-5
St. Louis Acft. JN-4D	1918	450	90	Curtiss OX-5
Springfield Co. JN-4D	1918	585	90	Curtiss OX-5
Curtiss JN-4D2	1918	1	90	Curtiss OX-5
Liberty Iron Works JN-4D2	1918	100	90	Curtiss OX-5
Curtiss JN-4H	1918	402	150	Hispano A
Curtiss JN-4HB Bomb Trainer	1918	100	150	Hispano A
Curtiss JN-4HG Gunnery tr.	1918	427	150	Hispano A
Curtiss JN-6HB (4 Ailerons)	1918	154	150	Hispano A
Curtiss JN-6HG-1	1918	560	150	Hispano A
Curtiss JN-6HG-2	1918	90	150	Hispano A
Curtiss JN-6HO Observation	1918	106	150	Hispano A
Curtiss JN-6HP Pursuit tr.	1918	125	150	Hispano A
Curtiss R-4L	1918	6	360	Liberty
Curtiss R-6 Seaplane	1918	10	200	Curtiss V2
Curtiss (rebuild) SE-5a	1918	57	180	Hispano E
Curtiss Bristol Fighter	1918	27	400	Liberty
Fisher Body DH-4	1918	1,600	400	Liberty
Standard DH-4	1918	140	400	Liberty
McCook Field USAC-1 Exp.	1918	1	400	Liberty
McCook Field Bristol Exp.	1918	1	300	Hispano H
McCook Field USB2 Exp.	1918	1	290	Liberty V-8
McCook Field USD-9 Exp.	1918	2	400	Liberty
Dayton-Wright USD-9 (DH-9)	1918	2	400	Liberty
McCook Field USD-9A	1918	5	400	Liberty
Dayton-Wright USD-9A	1918	4	400	Liberty
Lewis and Vought VE-7	1918	14	150	Hispano A
Ordnance Orenco C	1918	6	80	Le Rhone C9
Packard LUSAC-11 (Le Pere)	1918	30	400	Liberty
Thomas Morse MB-1	1918	2	400	Liberty
Standard E-1	1918	30	100	Gnome B9
Standard JR-1B	1918	6	175	Hall-Scott
Fisher Body SJ-1	1918	400	125	Hall-Scott
Thomas Morse S-4C	1918	50	100	Gnome B9

Aircraft	Year First Delivered	Number Delivered	HP	Engine
Thomas Morse S-4C	1918	447	80	Le Rhone C9
Martin GMB Bomber	1918	10	400	Liberty (2)
Motor Products SX-6	1918	1	150	Hispano A
Packard LUSAGH-11 (Le Pere)	1918	1	425	Liberty
Standard Caproni Bomber	1918	2	350	Liberty (3)
Fisher Body Caproni	1918	3	350	Liberty (3)
Standard E-1	1918	98	80	Le Rhone C9
Standard (Handley Page)	1918	107	350	Liberty (2)
Thomas Morse MB-2	1918	2	400	Liberty
Wright-Martin (Loening) M8	1918	2	300	Hispano H

U.S. Air Service Airplanes Purchased from France During WWI

Aircraft	First Delivery	Number Obtained	HP	Engine
Nieuport 21	1917	181	80	Le Rhone C9
Nieuport 80	1917	147	80	Le Rhone C9
Nieuport 81	1917	173	80	Le Rhone C9
Nieuport 83	1917	244	80	Le Rhone C9
Caudron G.3 Trainer	1917	126	80	Le Rhone C9
Avion Renault 1	1917	22	200	Renault
Caudron G.4	1917	10	80	Le Rhone C9
Morane 21 (Penguin)	1917	138	50	Anzani
Nieuport 17	1917	75	120	Le Rhone J9
Nieuport 24	1917	121	130	Le Rhone J9b
Nieuport 24 (Square Wing)	1917	140	130	Le Rhone J9b
Nieuport 27	1917	287	120	Le Rhone J9
Spad VII	1917	189	180	Hispano
Avion Renault 2	1918	120	190	Renault
Breguet 14 Observation	1918	47	300	Renault
Breguet 14	1918	100	240	Fiat
Caudron G.3	1918	66	80	Le Rhone C9
Farman 40	1918	30	130	Renault
Farman 50	1918	2	250	Lorraine
Nieuport 21	1918	17	120	Le Rhone J9
Nieuport 23	1918	47	80	Le Rhone C9
Nieuport 28	1918	297	170	Gnome 9N
Salmson 2	1918	705	270	Salmson CU9-Z
Sopwith 1	1918	236	130	Le Rhone J9b
Sopwith 1	1918	148	130	Clerget 9bc
Sopwith 1	1918	130	135	Clerget 9ba
Spad XI	1918	35	235	Hispano
Spad XIII	1918	893	235	Hispano

Aircraft	First Delivery	Number Obtained	HP	Engine
Voisin 8	1918	8	220	Peugeot
Voisin 10	1918	2	300	Renault
Breguet 14	1918	90	285	Fiat
Breguet 14	1918	139	310	Renault
Caudron R.11	1918	2	215	Hispano
Morane 30	1918	51	120	Le Rhone
Nieuport 17 Experimental	1918	1	130	Clerget
Nieuport 23	1918	3	120	Le Rhone J9
Spad XII (Cannon)	1918	1	235	Hispano
Spad XVI	1918	6	250	Lorraine

Purchases from Britain

Aircraft	First Delivery	Number Obtained	HP	Engine
Sopwith F-1 Camel	1918	143	130	Clerget
A.V. Roe Avro 504-K	1918	52	110	Le Rhone
Aircraft Mfg. Co. DH-9	1918	2	del. minus engine	
Royal Acft. Factory BE2E	1918	12	90	R.A.F.
R.A.F. SE-5a	1918	38	210	Hispano
Sopwith FE2B	1918	30	160	Beardmore
Sopwith Dolphin	1918	5	210	Hispano

Purchases from Italy

Aircraft	First Delivery	Number Obtained	HP	Engine
S.I.A. (Fiat)	1918	19	260	Fiat

WWI Navy and Marine Airplanes

Designation	First Order (fiscal year)	First Delivery (D) or First Flight (F)	Number Accepted	Manufacturer	Models Accepted
DH-4	1918	5/24/18 (D)	332	Dayton-Wright	DH-4, -4B
F-5	1918	7/15/18 (F)	30	Canadian Aeroplanes	F-5
			60	Curtiss	F-5
			137	NAF	F-5

Designation	First Order (fiscal year)	First Delivery (D) or First Flight (F)	Number Accepted	Manufacturer	Models Accepted
H-12	1917	3/17 (D)	20	Curtiss	H-12
H-16	1918	2/1/18 (D)	124	Curtiss	H-16
			150	NAF	H-16
HS	1918	10/21/17 (F) (with Lib. engine)	678	Curtiss	HS-1, -2, -3
			250	LWF	HS-2
			80	Standard	HS-2
			60	Gallaudet	HS-2
			25	Boeing	HS-2
			2	Loughead	HS-2
NC	1918	10/4/18 (F)	4	Curtiss	NC-1 thru -4
			6	NAF	NC-5 thru -10
R	1916	11/16 est. (D)	200	Curtiss	R-3, -5, -6, -9

The Curtiss Triad, the U.S. Navy's first airplane, was delivered 1 July 1911, although the Navy officially considers 8 May 1911 as the birthdate of Naval air, the date the A-1 Triad was ordered. (USN)

The first Naval air operations were carried out by a civilian pilot with a civilian airplane. On 18 January 1911, Curtiss pilot Eugene Ely landed and took off from a specially built platform on the bow of the battleship *Pennsylvania* in San Francisco Bay. Ely had made a similar take off from the cruiser *Birmingham* at Hampton roads the previous November in this 50-hp Curtiss machine. (USN)

The first aircraft launched by catapult from a ship was this Curtiss AB-2 flying boat flown by Lt. Cmdr. H.C. Mustin from the stern of the *North Carolina* in Pensacola Bay, 5 November 1915. (USN)

The Aviation Section of the Signal Corps procured twelve Curtiss R-2s in 1916; these were powered by the Curtiss VX engine of 160 hp. Maximum speed was 85 mph. The red star on the rudder was an early, although unofficial, attempt at a national insignia for Army airplanes. (USAF)

The Curtiss S-3 scout, pictured at Langley Field, 1917, was the 322nd airplane delivered to the Signal Corps. Its engine was the 100-hp Curtiss OXX-2. This photo was taken after 17 April 1917, when the red-white-blue rudder stripes were officially adopted and serial number moved from fuselage to rudder. Four S-3s were built; top speed was 112 mph. (Marty Copp)

Remains of a Curtiss S-3. Aircraft development sometimes seemed to be the result of an apparently endless series of post mortems. We won our wings at high cost; each small advance claimed both life and treasure. (National Archives)

In 1915 French pilot Roland Garros mounted an 1885 Hotchkiss machine gun on his Morane-Saulnier "fighter" for line-of-sight aiming. His propeller was protected from those rounds that did not pass between the rotating blades by wedge-shaped steel deflectors on the backsides of each blade.

The air war over Europe had been in progress for more than three years when the first American air units arrived there in 1917, and all the air fighting, then and later, was in aircraft of foreign design. Pictured is a French Morane-Saulnier Type L observation plane powered with a 110-hp LeRhone rotary engine. Top speed was 97 mph. (USAF)

The French Voisin 10 had a 300-hp Renault engine and saw service throughout the war. It could reach 6500 feet in 15 minutes and its configuration allowed the gunner-observer a good forward field of fire. (Joe Durham)

German mechanics service an Albatros fighter and observation plane, 1916. (Joe Durham)

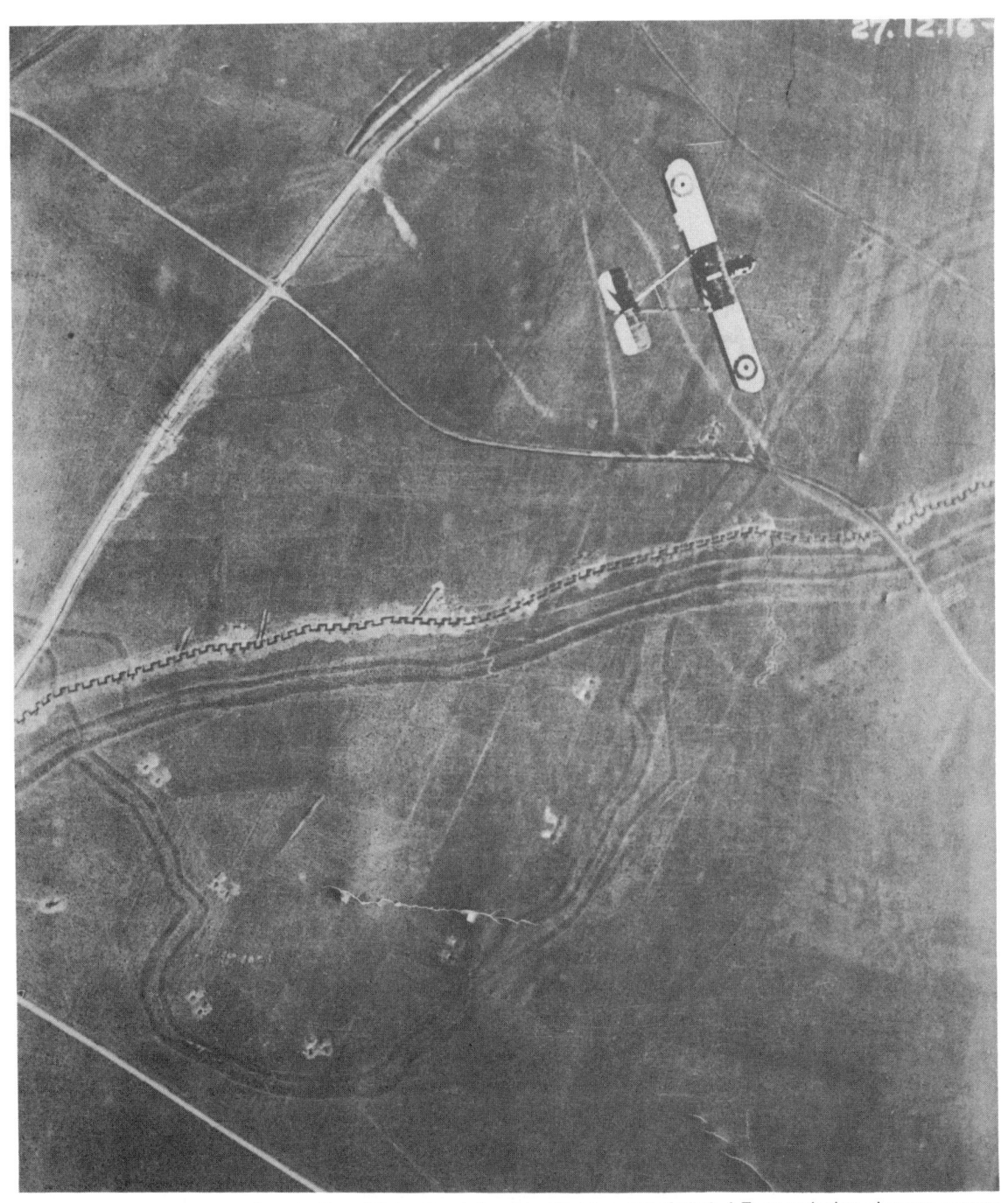

A British F.E.2b over the trenches, the Royal Flying Corp's toughtest fighting ship of 1916. A Farman design, observer-gunner rode in nose with a Lewis machine gun on a flexible mount. Pusher engine was a 120-hp Beardmore. Richthofen was shot down by one and never fully recovered; Max Immelmann was shot down and killed by a Corporal Waller, the gunner of F.E.2b #4272. (The Germans have never recognized Waller's victory, claiming instead that Immelmann's Eindecker suffered structural failure.) (USAF)

The world's first strategic bombers were Germany's WWI rigid airships built by Luftschiffbau-Zeppelin and Shutte-Lanz. Almost 100 of these air giants were built. The first Zeppelin raid against England occurred 19 January 1915, and London was bombed 31 May. The Zeppelins dropped 5800 bombs and killed 557 people but their effectiveness rapidly diminished as British pilots learned to find them in the night sky. (Imperial War Museum)

A standard German bomber of 1917-18 was the Freidrichshafen, which shared that duty with the Gotha. Engines were 300-hp Daimlers (two) mounted as pushers. Span was 78 feet; bomb load, 1200 pounds. The crew of four had five defensive guns. (USAF)

The Fokker Dr-I Dreidecker (three-winger) entered service with the German Air Force in late August 1917, and was gradually phased out during the spring of 1918; about 320 were built and equipped the Richthofen and Voss Geschwaders. On 21 April 1918 von Richthofen died in one, apparently shot down by Capt. Roy Brown of RAF 209 Sqdn. (five Autralian gunners on the ground also claimed the red triplane, but the victory was officially given to Brown). The DR-I was normally equipped with a 110-hp Oberursel rotary engine and it had a maximum speed of 115 mph; weighing but 893 pounds empty, it was highly maneuverable. (USAF)

The best German fighter of WWI was the Fokker D-VII designed by Reinhold Platz. Powered by a 180-hp Mercedes or 185-hp B.M.W., the D-VII had a top speed of 116 mph. It entered service in May 1918, and by late summer about 800 were at the front. (Burrell Tibbs)

247

The Nieuport Model 11 was the standard French fighter during 1915. Weighing but 760 pounds empty, and fitted with the 80-hp Gnome rotary engine, it had a top speed of 97 mph and was highly maneuverable. It was also known as the "Bebe," and the "13 meter" Nieuport. (Burrell Tibbs)

The Spad S.13 entered service during the fall of 1917 and was the principal French fighter during 1918. Sixteen American squadrons flew the Spad. It was faster—at 130 mph—than the Fokker D-VII, but less maneuverable. Engine was the 200-hp Hispano-Suiza. (Joe Durham)

The German Gotha bombers partially replaced the vulnerable rigid airships as bomb carriers to London by mid-1917, at first making daylight raids, and then switching to night bombing as British home fighter defenses stiffened. The Gotha G-V had a span of 77 feet, a maximum speed of 87 mph with two 260-hp Mercedes pushers, and a normal bomb load of 1100 pounds. (USAF)

The British Sopwith F.1 Camel reached the front in July 1917 and from that time until war's end was the victor in more aerial combats than any other fighter of the war, its pilots destroying a total of 1294 enemy airplanes. Fitted with the 130-hp Clerget rotary engine, its top speed was 115 mph. (USAF)

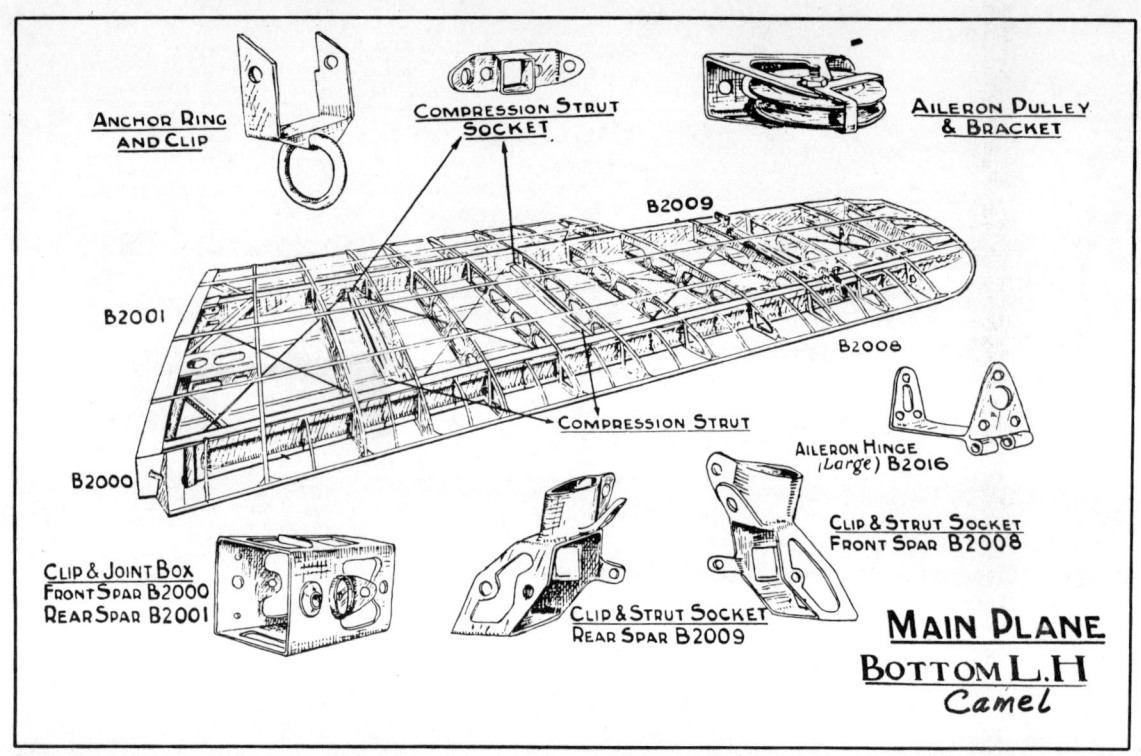

Revealed here is the importance of the numerous metal fittings that held everything together in wood-framed aircraft, in this case, a Sopwith Camel.

Curtiss JN-4 trainer, OX-5-powered, was the Army's standard pilot trainer during WWI. Top speed was alleged to be 80 mph. (Earl C. Reed)

DH-4s used for pilot transition training at Ft. Sill's Henry Post Field, Oklahoma, 1918. The Air Service officially recognized 21 different versions of the DH-4. Top speed was 124 mph. (Joe Reed)

Curtiss HS-1L flying boat at San Diego, 1918, used for pilot training, as were the Curtiss N-9 floatplanes (flying). The N-9 was a Jenny with extra wingspan and 100-150 hp. (Lloyd Stearman)

251

Curtiss F Boat was the principal flying boat trainer used by the Navy in WWI. A 1914 design, powered with the Curtiss OX engine of 90 hp. it had a top speed of about 70 mph. Approximately 150 were built. (USN)

Second/Lt. Frank Luke, Jr. scored 18 victories in 17 days, and nine of those days he did not fly. Downed behind enemy lines with a crippled airplane, Luke fought to the end with his pistol. (USAF)

German observation balloon begins to burn following attack by Spad believed to be that of Frank Luke. (Archiv Krueger)

Army balloners' primary mission was artillery fire correction, and their basic vehicle from 1917 until the mid-'30s was the C-3 "Sausage" balloon, anchored to a winch truck by a steel cable with telephone line attached. (Ft. Sill Field Artillery Museum)

Army observation balloon winch truck was an "all terrain" vehicle of cab-over-engine configuration. (M/Sgt. Harvey Nelson)

The Sopwith Snipe, a heavier and more powerful follow-on to the Camel, appeared too late to prove itself in WWI. Its 230-hp Bently engine gave it a top speed of 118 mph at 10,000 feet. (Joe Durham)

The French AR-1 Avion Renault was original equipment of the U.S. 12th Aero Squadron in France, an observation unit. The AR-1s were replaced by Salmsons in mid-1918. (USAF)

Capt. E.V. Rickenbacker (L) was America's leading ace at the end of WWI with 26 confirmed victories. Capt. Douglas Campbell shot down six enemy planes before he was wounded in June 1918. (USAF)

The 12-cylinder 400-hp Liberty engine. Officially, the "U.S. Standard Aircraft Engine," more than 13,000 were built during WWI for the DH-4. Production continued until 22,000 had been built in a massive boondoggle. With Army warehouses bulging with surplus Liberties, the Air Service was forced to use them throughout the '20s. The last one was retired at Kelly Field in 1935. (USAF)

Workmanship on the American-built DH-4s was poor. Those sent to France were rebuilt before going to the front. Remanufactured DH-4s remained in service in the U.S. throughout the '20s. (USAF)

The Thomas-Morse SC-4 Scout resembled the Sopwith Camel. It was an American design that appeared too late for combat, but offered no performance advantage over most 1917 foreign designs.

View of the Curtiss Buffalo plant, wing assembly section, on 11 November 1918, the day WWI ended, and everyone left to celebrate.

Chapter 11

The Tumultuous '20s

Despite the fact that both Army and Navy aviation existed on short rations during the 1920s, significant advances were made—and so were future air commanders.

Both air arms were groping uncertainly for their proper roles in the nations's defense. Air advocates in both services knew that theirs was a weapon of tremendous potential, but a number of factors would cloud that issue until our very survival was threatened.

During the '20s, our wings were still relatively frail; there was no discernible military threat to America or the Western Hemisphere, and President Harding was elected with the slogan "Back to Normalcy" which, to the average American, meant a form of hemispheric isolationism. We wanted "no more foreign entanglements" that could trap us into a war not of our making. We would enforce the Monroe Doctrine with our Navy and Marines, but we did not need a large Army—the Atlantic and Pacific Oceans were our great natural defenses against any conceivable major aggressor—and the need for military airplanes seemed limited indeed.

If that seemingly unassailable posture was not enough to keep the Army and Navy air arms at the end of the line when the services' bare-bones appropriations were divvied up, there were other debilitating factors, including interservice and intraservice prejudices. Those would not immediately surface following WWI and were manifested only by what could be called the benign neglect of military air by the Army and Navy brass, until the old horse soldiers on the Army's General Staff and the battleship admirals on the Navy's General Board were eventually pushed into a hardening of their positions by the air advocates, the most vocal of which was Gen. Billy Mitchell.

Mitchell was appointed Assistant Chief of the Air Service in 1919 (actually, the Aviation Section did not officially become the Air Service until 1920, but it had been called nothing else since early 1917), a position he had certainly earned as a result of his initiative and leadership in France. He was acceptable to Gen. Pershing because he was regular Army, although it was highly unlikely that he would ever go any further because he was not a West Pointer. The fact that he had come up through the ranks to the temporary rank of Brigadier General (one star) was a minor miracle in itself. So Mitchell had reached the apex of his career, and no one knew that better than he.

Chief of Air Service Gen. Charles T. Menoher was Mitchell's immediate superior. Menoher had commanded the American Rainbow Division in France and was not a flier.

Meanwhile, the Navy, while firmly in the hands of the saltiest of old sea dogs, nevertheless contained its share of air advocates. As early as 1913, Capt. Washington I. Chambers recommended a separate bureau of air, as did Capt. (later Rear Adm.) Mark L. Bristol three years later. Bristol also proposed the construction of two aircraft carriers at that time. Chambers was marked for retirement, and Bristol was "allowed" to return to sea from his desk job in Washington. Rear Adm. William S. Sims, who had commanded the U.S. Fleet in the Atlantic during WWI, asked for aircraft carriers again in 1919 and would, along with Rear Adm. Bradley Fiske, later support Billy Mitchell's contention thet airplanes could sink capital ships.

The Navy General Board would be forced to ask Congress to approve a Navy Bureau of Aeronautics when it appeared that Billy Mitchell's preachments were being listened to on Capitol Hill (between 1919 and 1925 there were 20 separate congressional hearings on the "aviation question"). In March 1921, Capt. William A. Moffett was named Director of Naval Aviation pending such legislation, which sailed through the Congress on 12 July, and Moffett, promoted to Rear Admiral, became Chief of the Bureau of Aeronautics (BuAer) when it began functioning on 1 September 1921.

Moffett believed only in Naval airpower. He was against an independent Air Force and, in fact, saw no need for any military airplanes except those operated by the Navy and Marines. He detested Mitchell, characterizing Mitchell as being "of unsound mind and suffering from delusions of grandeur." Moffett did a great deal for Naval aviation during the '20s, his only aberration being his strong belief in rigid airships for long-range sea patrol.

First Atlantic Flight

There was, of course, a number of Naval officers below flag rank who buttressed the cause of Naval air with deeds rather than words. The ink was scarcely dry on the Armistice agreement that ended WWI before Cmdr. John H. Towers, Naval Aviator No. 3, turned up in Washington to visit with Assistant Secretary of the Navy Franklin D. Roosevelt. Towers mentioned the Curtiss four-engine flying boats under construction (designed to fly the Atlantic to avoid the German U-boat menace), suggesting that an Atlantic flight by navy airplanes would be good public relations and a demonstration of the Navy's progress in aeronautics.

Roosevelt agreed, and cleared the project with Secretary of the Navy Josephus Daniels.

Had Towers submitted his proposal to his Navy superiors it is unlikely that it would have been approved, but his audacity was rewarded because it resulted in a U.S. Navy airplane being the first to fly the Atlantic.

At about 1800 hours 16 May 1919, three Navy flying boats took off from Trepassey Bay, Newfoundland, bound for the Azores. These were the flagship, NC-1, commanded by Cmdr. John Towers; NC-3, commanded by Lt. Cmdr. P.N.L. Bellinger, and NC-4 under the command of Lt. Cmdr. Albert C. Read. The big air boats had spans of 126 feet, were powered with four Liberty engines, and cruised at about 75 mph.

Lt. Cmdr. Read and his crew of five reached Horta in the Azores 18 hours later, and on 20 May flew to Ponta Delgada. From there, on the 26th, Read took the NC-4 on to Lisbon, and then completed the flight to Plymouth, England, on 30 May.

Towers and Bellinger both landed at sea in dense fog. Bellinger and his crew were picked up by the Greek freighter *Ionia*, and the NC-3 sank following unsuccessful attempts to tow her. Cmdr. Towers and his men sailed and taxied the NC-1 some 250 miles to Ponta Delgada after heavy seas damaged their aircraft's lower wings to the extent that further flight was not possible.

Just two weeks later, on 14 June 1919, RAF Capt. John Alcock and Lt. Arthur Whitten-Brown flew a twin-engine Vickers Vimy bomber from Newfoundland to Ireland—1890 miles—in 16 hours and 28 minutes to claim a cash prize of $50,000 offered by a British newspaper publisher for the first nonstop trans-Atlantic flight.

And two weeks after *that*, on 2 July 1919, the British rigid airship R.34 left Scotland bound for New York. commanded by RAF Maj. G.H. Scott and carrying a crew of 30, the R.34 arrived in New York on 6 July after flying 3600 miles in 108 hours. Three days later, Maj. Scott took R.34 back to Britain, thus completing the first aerial round trip across the Atlantic.

The Lindbergh flight would follow eight years later, and it would capture the fancy of the world because Lindbergh flew alone, in a single-engine airplane, and symbolically linked two of the world's greatest cities.

By the time of the Lindbergh flight to Paris in 1927, at least 68 people had crossed the Atlantic by air, including four Air Service pilots who flew around the world in 1924, and the 25-man crew that delivered the German-built Zeppelin *Los Angeles* to the U.S. Navy that same year.

Army and Navy Racers

Beginning in 1920, the Army and Navy competed in the annual Pulitzer Trophy air race with a series of specially built racing planes, dominated by a few Curtiss-designed machines fitted with souped-up Curtiss D-12 engines. These competitions gained a world speed record for the U.S. and provided some new concepts in fighter airplane configurations. Army Thomas Morse entries in 1922 were all-metal. That same year saw several low-wing designs with retractable landing gears compete, including the Verville-Sperry R-3, and the Booth and Thurston *Bee Line Special*, the Vervilles flown by Air Service pilots, the Booth and Thurston a Navy entry. Neither service would have fighter airplanes embodying all of these features, including cantilever wing construction, until the Seversky P-35 and Curtiss P-36 were acquired 15 years later. Lacking money to buy new airplanes in significant numbers, the services meanwhile regarded the races as a proving ground for new ideas, as well as a useful recruiting attraction. Enlistments lagged alarmingly in the immediate postwar years.

While the low-wing retractables were the progenitors of America's future fighters, they were too far ahead of their time, and it was the tiny Curtiss biplanes, with D-12 engines souped-up to as much as 525 hp, that won most of the races. A thin racing wing (C-27 airfoil) was developed by Curtiss in their own wind tunnel, and most of the wing surfaces served as radiators for the engine's water cooling system, the surfaces being covered with thin brass sheets formed with passages through which the water circulated.

The Pulitzer races were open to anyone, although all were won by military machines for the six years the event was held, 1920-1925 inclusive. After that, Army and Navy airplanes continued to win the top races until 1930, when Walter Beech's low-wing Travel Air "Mystery Ship" easily outsped Army and Navy Hawk fighters in an unlimited event at the National Air Races. During the '30s, civilian racing planes—most of which were essentially homebuilts—were faster than anything possessed by the U.S. military. Roscoe Turner won the 1938 and 1939 Thompson Trophy closed-course events with average speeds above 280 mph flying his Laird-Turner Special. The fastest U.S. fighters in service at that time were the Seversky P-35 and Curtiss P-36, with speeds of 281 mph at 10,000 feet, and 302 mph at 14,000 feet, respectively.

Throughout the '20s the services brought new airplanes in small batches, all of which, whatever the mission, were variations of a single, basic concept: fabric-covered open-cockpit biplanes. The exceptions were a few Ford and Fokker trimotor transports. The first concession to modern design was the 225-mph Boeing P-26 of 1933, a low-wing, all-metal Air Corps pursuit, although it retained externally-braced wings, an open cockpit, and fixed landing gear. It could have been designed at least five years earlier with the same engine, and many observers suspected that the Army wrote specifications for an airplane so configured because of the embarrassing loss to Beech's Mystery Ship at the 1930 National Air Races.

The Curtiss biplane racers of the early '20s began with the rather conservative CR-1 of 1921, which had drag-producing external Lamblin radiators and 405 hp. It was Navy-owned, but flown to victory in the 155-mile Pulitzer at an average speed of 176 mph by Curtiss test pilot Bert Acosta. The following year the Navy had two Curtiss racers, and the Army two, all four featuring wing-skin radiators; these craft claimed the first four places in the Pulitzer, and Army R-6 taking first place (with a speed of 205.8 mph) flown by Lt. Russell Maughan. Four days later, Gen. Billy Mitchell flew the black-and-yellow R-6 over a three-kilometer course four times at an average speed of 224.4 mph and a new world's speed record.

In 1923, the Navy took first and second in the Pulitzer with two new Curtiss R2C-1s flown by Lts. Alford Williams and H.J. Brow at 243.7 mph and 241.8 mph, respectively. The Army's pair of Curtiss racers did no better than 216.5 mph. Later that year, Lt. Al Williams raised the world's speed record to 266.6 mph in his R2C-1.

There were but four entries in the 1924 Pulitzer, all of them Army: the two R-6 racers, a modified Curtiss PW-8 Pursuit, and a Verville-Sperry R-3. The R-3 took first place at 215 mph with Lt. H.H. Mills at the controls. One of the R-6s disintegrated during its diving start when its wooden prop failed; its pilot, Capt. Burt Skeel, was killed.

The last Pulitzer race was flown in 1925, and the two services each entered an identical Curtiss R3C-1 racer, essentially the same as the R2C-1s except for a new Curtiss V-12 engine of 620 hp, and Army Air Service Lt. Cyrus Bettis won at 249 mph. Two weeks later, Lt. Jimmy Doolittle won the International Schneider Cup race for seaplanes in Bettis' R3C-1 fitted with floats. Doolittle's speed was 232.6 mph, a new world's speed record for seaplanes. The R3C-1 on floats was officially the R3C-2. Later, Doolittle upped the record mark to 245.7 mph in the same airplane.

Gen. Billy Mitchell's court-martial would come at the end of 1925, and there would be no more records established by U.S. military aircraft, and virtually no advance in design, until the mid-'30s. The old generals and admirals, firmly in control of the U.S. military between the two world wars, effectively muzzled the air advocates after Mitchell was disposed of, and only "tame" officers served as chiefs of the Air Corps.

Metal Propellers

Capt. Skeel's death as the result of a wooden propeller failure in the 1924 Pulitzer race convinced the services that they should switch to metal propellers. Metal props has become available in limited numbers in 1923 when both the Pulitzer and Schneider races were won by machines so equipped.

These propellers were developed by Dr. Sylvanus Albert Reed, a retired engineer who had been investigating acoustics with spinning metal paddles when it occurred to him that his experiments could contribute useful data on airplane propellers.

Dr. Reed went to Curtiss late in 1920, where Mike Thurston worked with him in designing metal propeller blades that could be turned at tip speeds well beyond those possible with wooden laminated propellers. The first Curtiss-Reed propellers were made from a forged aluminum alloy plate, cut and shaped by machining and then twisted to the desired pitch. When Reed and Thurston were satisfied with what they had, Reed took a propeller to McCook Field where, during 1922, tests proved so successful that the thin-blade Curtiss-Reed props were fitted to the Curtiss racers in 1923. The decision not to use them in 1924 appears to have been because McCook Field had a new wooden propeller made of birch it wanted to prove in the races. That cost the life of Capt. Skeel.

McCook Field

McCook Field, the Air Service research and

experimental test center on the outskirts of Dayton, Ohio, was established in 1917 and during the early '20s was easily the most significant place in American aviation. Most of the new aerial hardware was either tested or originated there. Although the Navy had its own aircraft factory, it possessed no aeronatical research facility comparable to the Army's McCook Field.

At McCook, a lot was accomplished with relatively little money. Personnel was a mixture of Air Service officers, many of whom had been in aviation before WWI, and some low-paid or non-paid but talented civilians. The McCook Field wind tunnel, for example, was operated by Don Berlin, a recent engineering graduate from Purdue who would later design the P-36/P-40 as chief engineer at Curtiss-Wright. Charles N. Monteith, later Boeing's top engineer, was chief of the Airplane branch, and in the Structures Branch, Joe Newell and Alfred Niles wrote the industry's first structural design handbook. In the Powerplant Section was Sam D. Heron, designer of the air-cooled engine cylinders that made the Whirlwinds and Wasps possible; Dr. Sanford Moss, who was actually on the payroll of General Electric, developed turbosuperchargers at McCook, a quest that began there in 1918. The phenomenon of aileron flutter was identified and explained by Col. Carl Green there in 1922, and in the Propeller Test Section, experiments with controllable and reversible-pitch props were in progress by 1919.

Head of the Flight Test Section was Maj. Rudolph "Shorty" Schroeder, who entered aviation in 1912 as a mechanic for early bird pilot Otto Brodie; two of Schroeder's pilots were Lt. Jimmy Doolittle and Capt. George Kenney. An early Curtiss engineman was George E.A. Hallett, who headed the Powerplant Section. Also during the early '20s, several manually-operated free-fall parachutes were perfected and tested at McCook, including that of Floyd Smith—and just in time to save the life of Lt. Harold R. Harris, chief of the Engineering Division, who bailed out of an experimental Loening fighter when it shed its wings during a mock dogfight with Lt. John Macready.

In 1927, following passage of the 1926 Air Corps Act (more fallout from the Mitchell trial), McCook moved to Wright Field about eight miles away, site of today's Wright-Patterson AFB.

1921 Bombing Tests

The Navy's successful Atlantic flight in May 1919 undoubtedly influenced the vote on the Naval Appropriations Act for fiscal 1920, which was passed 11 July 1919 authorizing acquisition of an aircraft carrier, two seaplane tenders, and two Zeppelin-type rigid airships.

Gen. Billy Mitchell and his superior, Chief of Air Service Gen. Menoher, certainly took note of such Congressional munificence at a time of drastically reduced military spending. Therefore, Mitchell organized a mass transcontinental flight of Air Service DH-4s, which he called "an evaluation and reliability test," with 30 machines leaving New York bound for San Francisco as 30 others left San Francisco headed for New York. A few were forced down with engine trouble, but the transcontinental air route had been blazed, and the news coverage was satisfying.

The following June, four DH-4s, commanded by Capt. St. Clair Street, left Mitchel Field on Long Island for Nome, Alaska. They returned in October, having covered 4345 miles—much of it over uncharted wilderness—in 112 flying hours.

But those demonstrations were merely prelude to the famous airplanes vs. battleship bombing test in the fall of 1921.

When the Navy announced in the spring of 1921 that it would conduct aerial bombing tests against some German warships taken as prizes at the end of the late war, Mitchell challenged the Navy to allow the Air Service to participate in the experiments.

The news media, aware of Mitchell's views on airpower, took up the "dare" and the Navy could hardly refuse—although not without a tad of smugness, because the Navy had itself conducted aerial bombing tests the previous November against the old battleship *Indiana* that resulted in only superficial damage to the obsolete dreadnought.

The tests began 21 June 1921 when Navy flying boats sank the German submarine U-177; on the

29th, Navy airplanes dropped some dummy bombs on the old battleship *Iowa* to no apparent purpose. Then, on 13 July, Mitchell's bombers were allowed to attack the German destroyed G-102 with bombs not to exceed 300 pounds in weight. The destroyer went to the bottom in 19 minutes, sent there by 16 Martin bombers, each carrying six bombs of the prescribed size.

Five days later, under complicated rule, Air Service bombers were allowed a chance at the cruiser *Frankfurt*. The 300-pounders were not effective, but when the Martin crews planted 14 600-pound bombs on or very close to the *Frankfurt*, she went down in 35 minutes.

On 20 July came the crucial test—the German battleship *Ostfriesland*. Again, under conditions set down by the Navy, Navy and Marine aircraft were to have first chance at the big battlewagon, employing bombs up to 1000 pounds in size. After that (since the navy had every reason to believe that the ship would still be afloat), Mitchell's bombers could try, with the same limitations.

The Naval aviators did not damage the battleship. Mitchell's crews dropped five 600 pound bombs, scoring two direct hits from an altitude of 1500 feet, but the *Ostfriesland* did not go down.

The next morning, Mitchell led a formation of eight Martin bombers, each carrying two 1000-pound bombs, off the Virginia Capes from Langley Field. The *Ostfriesland* was anchored 75 miles out to sea, with the Navy transport *Henderson* nearby carrying reporters and various dignitaries to observe the action. Five of the 1000-pound bombs were released, three scoring direct hits, but the dreadnaught remained afloat.

The Navy, however, wasn't alone in hedging its bets on the outcome of the tests. Mitchell had arranged for the fabrication of some 2000-pound bombs. He justified the blockbusters by taking the position that in war, neither side would enjoy the privilege of specifying the size of the bombs to be used. That afternoon six of the 2000-pounders placed in the water close beside the *Ostfriesland* opened her seams below the waterline and she sank just 20 minutes after the first bomb was released.

Whether or not the tests proved anything depended upon one's prejudices. They changed no minds among those sitting on the Navy's General board, and Mitchell's boss, Gen. Menoher, asked Secretary of War Weeks to replace Mitchell as Assistant Chief of the Air Service—a proposal that was so thoroughly denounced in the press that Mitchell remained in his job.

The Navy's First Aircraft Carrier

Congressional approval for the Navy's first aircraft carrier in 1919 was certainly influenced by Adm. W.S. Sims' support of the proposal. Adm. Sims had championed the cause of the battleship at the turn of the century, and he commanded the U.S. Atlantic Fleet during WWI. At the time of the 1921 bombing tests, Adm. Sims was serving as President of the Naval War College, and prior to the sinking of the *Ostfriesland* he said that a fleet superior in aircraft carriers "would sweep an enemy fleet clean of its airplanes and proceed to bomb the battleships . . . " It was all a question, he said, of "whether the airplane carrier . . . is not the capital ship of the future."

The aircraft carrier that was paid for with Navy money may not have been quite what its proponents had in mind. It was a converted collier—which is to say, a coal barge with a flight deck erected on a superstructure. The two seaplane tenders turned out to be a pair of war surplus merchant ships.

Since the new carrier would need airplanes, Capt. Thomas Craven, Director of Naval Aviation, asked for 108 small, specially designed fighters, to be powered with the Lawrance J 200-hp radial engine (the engine that would be developed into the Wright J-5 Whirlwind with Navy prodding), assuming that some would be assigned to shore bases as well. That figure was cut to 75, and then to 10, as Adm. Benson, Chief of Naval Operations*, can-

---

*The office of Chief of Naval Operations was established 2 March 1915 and the scope of CNO's authority was challenged for years thereafter by the Navy bureau chiefs. Not until President Truman issued Executive Order 9635 on 29 September 1945 were the CNO's authority and duties clearly defined.

celed the carrier project altogether late in 1920.

Capt. Craven went directly to Secretary of Navy Daniels and got the carrier program reinstated. That appears to have cost Craven his job. He was replaced by Capt. William Moffett who, as earlier related, was in that office when the Bureau of Aeronautics was established in 1921, and Moffett became its first chief, with a promotion to the rank of Rear Admiral.

Eventually, by 1925, a total of 43 of the little fighters was built for service on the new carrier, the *Langley*. These airplanes, originally designated the TS-1s, were later modified as the F4Cs. Thirty-two were built at the Naval Aircraft Factory; Curtiss built eleven.

The *Langley*, known throughout the Navy as the "Covered Wagon," was placed in commission at Norfolk, Virginia, under the command of her Executive Officer, Cmdr. Kenneth Whiting, on 20 March 1922.

It should be noted that Japan commissioned her first carrier, the 7470-ton *Hosho*, in October 1922. The British, having experimented with divided flight decks and partial decks built atop several different hulls during the war, placed in commission the world's first *true* aircraft carrier, *HMS Argus*, in August 1918. *HMS Eagle* and *Hermes* followed in 1920 and 1923 respectively.

The first takeoff from the deck of the *Langley* was 17 October 1922. Lt. Cmdr. V.C. Griffin was the pilot, flying a Vought VE-7. Nine days later, Lt. Cmdr. Godfrey deC. Chevalier, flying an Aeromarine trainer, made the first landing aboard the *Langley* while underway off Cape Henry.

On 1 July 1922 the U.S. Congress authorized conversion of the unfinished battle cruisers *Lexington* and *Saratoga* to aircraft carriers. Both would be commissioned late in 1927; both would be lost to enemy action in WWII. The *Langley* would become a seaplane tender in 1936.

The First World Flight

After Secretary of War John Weeks refused to remove Gen. Mitchell from his post as assistant chief of the Air Service as Mitchell's boss, Maj. Gen. Menoher, had requested during the ship-bombing controversy, Menoher himself resigned, and was replaced by Maj. Gen. Mason M. Patrick, under whom Mitchell had served during the late war. Patrick generally shared Mitchell's views regarding American airpower, but was much more tactful about it.

Meanwhile, Mitchell continued to foster headline-gathering flights. There wasn't much else the Air Service could do with its limited resources. During fiscal 1923 it received slightly over $13 million, and Gen. Patrick agreed with his deputy that since such budgets provided little more than the Air Service's basic housekeeping expenses, they could not buy new airplanes in meaningful numbers; therefore they would invest in the future, building good public relations and aiding aircraft development through air racing and pioneering flights. In 1923, Air Service pilots Lts. John Macready and Oakley Kelly accomplished the first nonstop transcontinental flight in a Fokker T-2 monoplane (Macready had set a world's altitude record of 34,563 feet the year before), and in August 1923, Lts. Lowell H. Smith and J.P. Richter remained airborne over Rockwell Field for 37 hours, refueling 16 times through a 40-foot hose lowered by another plane flying above. But the Air Service project that seemed almost unvelievable, in view of the equipment available for it, was announced by Gen. Patrick in the fall of 1923: Four U.S. airplanes would fly around the world! Just 20 years after the Wright's first flights, the Air Service intended to circle the Earth on wood and cloth wings.

Success would depend as much on the courage and resourcefulness of the fliers as on the adequacy of their machines. There would be few landing fields, no navigational aids, no radio, and no weather reports. The art of instrument flying was still in the future, and the fliers themselves would have to perform all maintenance on their airplanes, including half a dozen complete engine changes in each of the four aircraft. They dared not trust the Liberty engine—the only engine of sufficient power available to them—beyond 50 hours of operation.

A small airframe builder in Santa Monica, California, Donald Douglas, was awarded a contract to build five World Cruisers. Four would attempt the epic journey, with the fifth held in reserve. The World Cruisers would be completed by mid-March 1924. They would be identical, each with a span of 50 feet, fuel capacity of 465 gallons, gross weight of 8200 pounds, would have a service ceiling of 800 feet, and cruise at 90 mph.

Meanwhile, a great deal of planning was necessary. Gasoline had to be cached along the route, extra floats and wheels and, at some places engines delivered to appropriate stops along the way. The Navy offered to help by stationing ships at preselected positions across the North Pacific and North Atlantic.

Equipped with floats, the big biplanes took off from Sand Point Field near Seattle shortly after sunup 6 April 1924. The four planes were the *Seattle*, crewed by Maj. Martin and Sgt. Harvey; the *Chicago*, manned by Lts. Lowell Smith and Leslie Arnold; the *New Orleans*, with Lts. Eric Nelson and Jack Harding, and the *Boston*, carrying Lts. Leigh Wade and Henry Ogden.

The World Cruisers followed the Inside Passage to Alaska through spring blizzards, losing the *Seattle* when Maj. Martin, blinded by snow, flew into the ground. He and Harvey, neither seriously injured, walked to civilization, and the remaining airplanes continued across the North Pacific with Lt. Lowell Smith in command.

Bad weather forced an unscheduled landing in Russian waters off the Komandorski Islands, but by the time the Russians decided that such a landing could not be allowed, the Cruisers had refueled from a U.S. Bureau of Fisheries boat and were ready to continue to Paramushiru, Japan, and the completion of the first flight across the Pacific.

New engines were installed for the second time in Tokyo (the first engine change was in Alaska), and then the Cruisers flew to Shanghai and down the South China coast to Haiphong and Saigon in French Indochina. The *Chicago's* engine blew up over the Gulf of Tonkin, but Smith and Arnold hired a native chief to tow them with a fleet of war sampans to a village dock where a new Liberty, delivered by Navy destroyer, was installed. The *Chicago* was able to catch up with its sisters at Saigon, where the flight turned northwestward for Bangkok, Rangoon, and Calcutta, flying most of the way in poor weather.

In Calcutta, the three Cruisers switched from floats to wheels again and then made good time across India (despite a newspaperman stowaway in the *Boston's* tool compartment, sand storms, and 120-degree heat) until, just short of Karachi, the *New Orleans'* engine swallowed a valve.

Again, all three planes received new engines, and took off for Baghdad on 7 July. Then they crossed the Arabian Desert, picked their way through passes in the Tarus Mountains, and landed in Constantinopole on 11 July. Five days later they were in London.

On floats once again, the three Cruisers took to the air from Houten Bay in the Orkneys on 2 August. They would stop at Iceland and Greenland for fuel, although no airplane had ever before flown to either place. The Navy had placed fuel caches there for them, and units of the Atlantic Fleet were positioned along their route in case of trouble. That proved a fortunate circumstance for Wade and Ogden.

About halfway to Iceland, the *Boston* was forced to land in the North Atlantic with a broken oil line. Smith and Arnold saw her go down and circled to land in a rescue attempt, but Wade and Ogden waved the *Chicago* away, believing the sea to be too rough. Smith then headed for the nearest Navy ship, the *Billingsby*, 100 miles away. A note dropped on the destroyer's deck sent her to the stricken *Boston*.

The *Billingsby* picked up Wade and Ogden late that afternoon, but the *Boston* sank following an unsuccessful attempt to take her in tow.

Smith and Arnold continued to Iceland where they joined Nelson and Harding. Then the *Chicago* and *New Orleans* set out for Frediriksdal on the southern tip of Greenland, 850 miles away. Much of that flight was a nightmare; fog forced them to the deck, where visibility of less than 200 feet demanded a succession of last-second evasive maneuvers to avoid icebergs. They landed in the harbor at Frediriksdal after eleven hours in the air—wet, cold, and exhausted, but they managed a brief

265

celebration. Just 560 miles across Davis Strait lay North America.

They were to have other troubles: fuel pump failure on the *Chicago* a hundred miles off Labrador requiring frantic manual pumping of gasoline, near-collision with a steamship in fog off Cape Charles, and a forced landing at Casco Bay. But rejoined by Wade and Ogden flying the backup Cruiser, the fliers touched down in Boston Harbor to a tumultuous homecoming.

Replacing their floats with wheels once again, they continued across the U.S. to their starting point at Seattle, arriving there 28 September. The big Douglas biplanes had flown 26,345 miles in 177 days, and had been in the air 363 hours and 7 minutes for an average speed of 72½ mph.

Those figures were not really important. The important thing was that the world had been circumnavigated by airplane. The important thing was the promise—or threat—implicit in that simple fact.

The Mitchell Trial

Even before the world flight was planned, Chief of Air Service Gen. Patrick had asked the War Department to appoint a panel of high-ranking Army officers to investigate the condition of the Air Service and recommend a realistic policy. This board, under the chairmanship of Gen. William Lassiter, reported the facts and proposed a long-term program that would maintain Air Service personnel strength at 4000 officers, 2500 students, 25,000 enlisted men, and 2500 aircraft (total Air Service personnel at the end of 1923 was 9441). These recommendations were approved by Secretary of War Weeks and forwarded to the Joint Army and Navy Board (comparable to today's Joint Chiefs), and that was the end of it.

In the meantime, Patrick had sent Mitchell on two fact-finding missions overseas. Mitchell went first to Europe, where he found that the Germans were combining "resourcefulness and ability" to create superior aircraft despite the restrictive conditions of the Versaille Treaty and economic chaos.

German aviation could "put France out of business," his report stated, "provided that they have a chance to build."

Mitchell's second foreign tour, in 1924, took him to the Far East, and he returned from there genuinely alarmed. According to Patrick, Mitchell's classified report was likely to be "of extreme value some ten or fifteen years hence."

It certainly *could* have been, but it wasn't, because it was ignored. Mitchell predicted the attack on Pearl Harbor with uncanny foresight, even pinning down the day of the week on which it would occur, and he missed the actual time of day that it did occur by only 25 minutes!

Mitchell's report also told of the confused command situation in Hawaii, stating that the relations between the top Army and Navy commanders were such that they refused to attend the same social functions. "Each claimed to be the senior officer," Mitchell reported, and the Commanding General told Mitchell that in case of war, he would take over Honolulu from the Admiral—by force, if necessary. Sadly, that situation would not change until after the actual attack on Pearl Harbor 16 years later. Mitchell's report concluded that "If our warships there were to be found bottled up in a surprise attack from the air and our airplanes destroyed on the ground, nothing but a miracle would help us to hold our Far East possessions."

In 1925, Mitchell's tour of duty as Assistant Chief of Air Service ran out and he was not reappointed to that post. He was allowed to revert to his permanent rank of Colonel and, just in case he didn't get the message, was sent to Ft. Sam Houston as an Air Service liaison officer in the Eighth Corps Area. Mitchell had testified before the House Select Committee of Inquiry into Operations of the U.S. Air Service—commonly called the Lampert-Perkins Committee—and his testimony had brought a reprimand from Secretary of War Weeks, who said that Mitchell had made statements that seriously reflected upon the War Department's "efficiency and management"—which, of course, Mitchell had been doing for years.

Adms. Sims and Fiske joined Gen. Patrick in support of Mitchell, but President Coolidge apparently regarded Mitchell's statements as an indictment of the Coolidge Administration as well as the high commands of the Army and Navy. A series of articles in the *Saturday Evening Post* magazine by Mitchell further angered the President, and some newspapers speculated that Mitchell was flirting with a court-martial.

No one knew that better than Mitchell. His crusade had failed, and his military career was ended. Characteristically, he chose to go down with all guns firing.

Before daylight on the morning of 3 September 1925, the Navy's rigid dirigible *Shenandoah* was torn apart in a violent line squall over Byesville, Ohio. The control car and aft hull of the giant airship fell directly to the ground, taking the lives of 14, including that of the airship commander, Lt. Cmdr. Zachary Lansdowne. The forward section of the hull, with seven men aboard, free-ballooned for an hour to land all hands safely twelve miles from the crash site.

The *Shenandoah* had just started out on a tour of county fairs as a public relations gesture, although the threat of thunderstorms—common over the Midwest in late summer—had argued against the tour. The *Shenandoah* had been in service two years. (The Navy's other airship, the *Los Angeles*, built in Germany, had joined the fleet the previous October.)

On the day the *Shenandoah* was lost, a Navy PN-9 flying boat, attempting a flight from San Francisco to Hawaii, had been unreported for three days. These two navy losses sent news reporters to Mitchell in San Antonio for his reaction. Mitchell did not disappoint them.

He waited 48 hours, and then on 5 September handed the newsmen a lengthy prepared statement that made bold headlines across the nation. Mitchell charged that the two accidents were the direct result "of incompetency, criminal negligence, and almost treasonable administration of the national defense by the War and Navy Departments." All aviation policies were, he said, "dictated by the non-flying officers of the Army and Navy who know practically nothing about it." He accused the brass of forcing junior officers to give false testimony before Congressional committees. He attacked the War Department for retaining in service the "Flying Coffins" (DH-4s), and derided the Navy for sending the PN-9 across the Pacific with insufficient fuel to reach Hawaii.

He concluded by saying, "As a patriotic American citizen, I can stand by no longer and see these disgusting performances by the Navy and War Departments, at the expense of the lives of our people, and the delusion of the American public."

It was heavy stuff indeed, and it could not be left unanswered. While the service chiefs and the President were regrouping, Mitchell issued another statement, calling for an investigation of the "War and Navy Departments and their conduct in this disgraceful administration of aviation." The investigators, he said, should be representative Americans instead of members of the Army and Navy bureaucracies.

The President did appoint a commission, although no one ever accused its members of being representative Americans. As its chairman, Coolidge selected an old college chum, Dwight Morrow, a partner in the J.P. Morgan banking empire (and Charles Lindbergh's future father-in-law). Among the eight other members were Rear Adm. Frank Fletcher (during WWII, Vice Adm. Frank Fletcher would lose the Fleet Carrier *Lexington* in the Battle of the Coral Sea, and then lose the *Yorktown* at Midway); retired Gen. James Harbord, who was on record as being against "Mitchellism;" Connecticut Senator Hiram Bingham (who would co-sponsor the 1926 Air Commerce Act, also known as the Bingham-Parker-Merritt Bill), along with Representatives James Parker and Carl Vinson; Federal Judge Arthur Denison and—of course—Howard E. Coffin.

The Morrow Board, as this panel was known, convened on 21 September 1925 and was ready to issue its recommendations by 25 October, which was good timing for its primary purpose: official rejection of Mitchellism. This report, however, was not immediately made public. It was to be introduced by the prosecution in the court-martial of Billy Mitchell. The gist of the board's findings was summed up thus: "The next war may well start in the air, but in all probability it will wind up, as the last one did, in the mud."

Five days after the Morrow Board gave its report to the President, Coolidge himself ordered Mitchell's trial by a military court. Mitchell was charged, under the 96th Article of War, with "conduct prejudicial to good order and military discipline."

The proceedings lasted from 28 October to 17 December, although there was never any doubt as to the verdict once the six senior Army officers sitting in judgment ruled that *truth was not a defense*. Mitchell was not charged with lying, they pointed out. He had sassed his superiors and publicly questioned their competence. He had corroded a lot of brass, and the *truth* contained in his charges was not at issue.

Nevertheless, when Mitchell's turn came to speak, he spent seven weeks presenting a detailed indictment of his accusers. His purpose was to make the facts a matter of public record. Witnesses who appeared in support of his charges were Lt. Leigh Wade and Majs. H.H. "Hap" Arnold, Carl A. Spaatz, Gerald C. Brandt, and Lewis Brereton. They paid for their testimony. All were banished to minor posts in the boondocks after the trial.

The court voted five to one for Mitchell's conviction. The dissenting vote was cast by Gen. Douglas MacArthur.

Mitchell's punishment was set at suspension from rank, command, and duty, with forfeiture of all pay and allowances for five years. The new Secretary of War, Dwight Davis, recommended that Mitchell receive half pay during his suspension, and President Coolidge agreed, but then had a few things to say about Mitchell's "defiance toward his military superiors."

As expected, Mitchell responded by resigning his commission. He died in 1936.

The immediate result of the Mitchell trial was a second Morrow Board, which made recommendations that were enacted into law as the Air Corps Act of 1926. The net results of that were a change in name for the Army's air arm from Air Service to Army Air Corps, several hundred new airplanes delivered over a five-year period, and establishment of a new primary flight training facility (Randolph Field, added to the Army's air training complex at San Antonio, Texas).

The new airplanes were an inprovement over the old DH-4s and Martin bombers chiefly because of their new air-cooled radial engines. Airframes were still fabric-covered open-cockpit biplanes, whether bomber, pursuit, or observation type. A new class—the attack plane—especially reflected the ground commanders' concept of airpower, and all combat units in the Air Corps remained under the command of the various corps area commanders stateside, and the department commanders overseas.

Advancement in rank for air officers was agonizingly slow. There was, literally, no place for them at the top, and it seems remarkable today that so many very able men remained in the Air Corps. Some did not. Jimmy Doolittle, still a 1st Lieutenant after 13 years of outstanding service, left the Air Corps in 1930 to accept a much better paying position with the Shell Oil Company.

In 1921, Air Service personnel represented 5.1 percent of total regular Army manpower. In 1930, 9.8 percent of the regular Army personnel was in the Air Corps. During those years, the Army appropiation averaged about $300 million per year. While the Air Service received $35 million in 1921, that was down to $13.4 million in 1925, then rose to $34.9 million in 1930.

Viewed in this way, it would seem that the Air Service and Air Corps was receiving its share of the Army appropriations. But that was never an issue. Actually, the Army ground forces were as poorly equipped as the air arm during these years. But a study of the tables at the end of this chapter will reveal that the Air Service money was cut dramatically 1922-1925 inclusive, and those figures provide convincing support for Mitchell's charges.

The *Sara* and *Lady Lex* Join the Fleet

For 15 years—from 1921 to 1936—the American people, their representatives in Congress, and their Presidents thought that the United States could and should avoid future wars with other major powers. They believed the nation could achieve this goal by maintaining a minimum of defensive military strength, avoiding entangling commitments with Old World nations, and yet using American "good offices" to promote international peace and the limitation of armaments. The U.S. took the initiative in 1921 in calling a conference in Washington to consider an arms agreement.

Generally referred to as the "Washington Conference" this meeting, with representatives of nine nations in attendance, lasted until 6 February 1922. In the end it produced the Washington Treaty, under which the U.S., Britain, Japan, France, and Italy agreed to limit the tonnages of their capital ships to ratios of 5—5—3—1.7—1.7 respectively. The same ratios for aircraft carrier tonnage set overall limits at 135,000—135,000—81,000 tons for the U.S., Britain, and Japan. The treaty also limited any new carrier to 27,000 tons, with a provision that, if the total carrier tonnage was not exceeded, these nations could build two carriers of not more than 33,000 tons each, or obtain them by converting existing or partially built ships that would otherwise be scrapped by this treaty.

Since the sole function of an aircraft carrier is to take airplanes to sea, it would appear that the U.S., at the time of the Washington Treaty, accepted the probability that the airplane was a viable weapon against warships. However, America continued to plan its Naval defense strategy around the battleship.

Japan renounced the Washington Treaty in 1934, but the U.S. made no effort to build carriers at a matching rate until the Naval Expansion Act was passed in May 1938.

The U.S. Navy took advantage of the provision in the Washington Treaty that allowed conversion to carrier configuration of a ship that would otherwise be scrapped in accordance with the agreement. Two 36,000-ton cruisers under construction thus became the *USS Saratoga*, the first carrier and fifth ship of the Navy to bear the name, and the *USS Lexington*, first carrier and fourth ship so named. Both were officially listed at 33,000 tons, and joined the fleet in November and December (respectively) 1927, Capt. H.E. Yarnell commanding the *Saratoga*, and Capt. A.W. Marshall commanding the *Lexington*.

Thirteen months later, the two new carriers participated in fleet exercises for the first time, attached to opposing forces. The most notable event of the problem was the employment of the *Saratoga* by the attacking Black Fleet to achieve the theoretical destruction of the Panama Canal. The *Sara* was detached from its main force and, with an escorting cruiser, sent on a wide southward sweep before turning north to approach within striking distance of the Canal.

On the morning of 26 January, while it was still dark, she launched a strike group of 69 aircraft that arrived over the target and completed the "destruction" of the Miraflores and Pedro Miguel Locks without opposition.

This demonstration made a profound impression on Naval tacticians, and in maneuvers the following year a tactical unit, built around the aircraft carrier, appeared in force organization for the first time.

Although the Navy was relatively better funded between World Wars One and Two because it was seen as America's first line of defense, the Great Depression of the early '30s (that caused military spending cuts across the board), Adm. Moffett's strong belief in rigid dirigibles (which diverted significant Naval air funds to a lost cause), and with repeated attempts by the U.S. to reaffirm arms limitations agreements (particularly with regard to capital ships) all conspired to delay acquisition of additional U.S. aircraft carriers. The keel of the *USS Ranger* was laid in 1931 and it was commissioned in June 1934, but the 14,500-ton *Ranger* accommodated but 35 airplanes and was comparatively slow (30 kts). The Attack Carriers *Yorktown* and *Enterprise*, each with air group complements of 80 aircraft, would join the fleet in 1937 and 1938, respectively. The *Wasp* would follow in 1940 and the *Hornet* in 1941 to give the U.S. Navy a total of seven carriers at the time of the attack on Pearl Harbor.

The Japanese had nine carriers at that time and, as demonstrated, fully understood how to most effectively employ them.

U.S. Air Service Personnel and Direct Appropriations 1920-1930

1920......	9,050	$28,123,503
1921......	11,649	35,124,300
1922......	9,642	25,648,333
1923......	9,441	13,060,000
1924......	10,547	12,626,000
1925......	9,670	13,476,619
1926......	9,674	15,911,191
1927......	10,078	15,256,694
1928......	10,549	21,117,494
1929......	12,131	28,911,431
1930......	13,531	34,910,059

Military Aircraft Production* Calendar Years 1920-1930

1920......	256		1926......	532
1921......	389		1927......	621
1922......	226		1928......	1,219
1923......	687		1929......	677
1924......	317		1930......	747
1925......	447			

*Total Army and Navy

U.S. Navy Combat Aircraft 1919-1930

Manufacturer	Designation	Contract	Procured
	Attack		
Martin	MT & MBT	1919	9
Naval Acft Factory	PT series	1921	33
Douglas	DT series	1921	41
Naval Acft Factory	DT series		6
Lowe-Williard-Fowler	DT series		20
Dayton-Wright	DT series		11
Curtiss	CS/SC	1922	8
Martin	CS	1922	75
Martin	T3M	1925	124
Martin	T4M	1927	103
Martin	BM	1928	34
Great Lakes	TG-1	1929	50
	Fighter		
Vought	VE-7	1920	60
Naval Acft Factory	VE-7		69
Thomas Morse	MB-3	1921	11
Curtiss	TS-1	1921	34

Manufacturer	Designation	Contract	Procured
	Fighter		
Naval Acft Factory	TS-2/3	1921	9
Boeing	FB-1/5	1925	43
Curtiss	F6C series	1925	75
Boeing	F2B-1	1926	33
Boeing	F3B-1	1927	74
Curtiss	F7C-1	1927	18
Curtiss	F8C series	1928	124
Boeing	F4B series	1928	188
	Patrol		
Douglas	P2D-1/T2D-1	1925	30
Douglas	PD-1	1927	25
Hall	PH-1	1927	10
Martin	P3M	1928	9
Consolidated	XPY-1	1928	1
Keystone	PK-1	1929	18
Martin	PM-1/2	1929	55
	Observation		
Loening	M-8	1919	17
Naval Acft Factory	M-80/81	1919	36
Vought	VE-9	1922	21
Vought	UO series	1922	163
Martin	MO-1	1922	36
Boeing	O2B-1	1924	30
Loening	OL series	1924	84
Keystone	OL-9	1924	26
Vought	O2U series	1926	291
Curtiss	OC-2 (F8C-1)	1927	27
Berliner-Joyce	OJ-2	1929	40
Vought	O3U series (SU series)	1930	330

U.S. Air Service/Air Corps Aircraft 1919-1930*

Model	Manufacturer	Cal Yr 1st Del	Number Procured	Engine/hp	
		Pursuit			
PW-1	McCook Field	1920	2	Packard	350
PN-1	Curtiss	1920	2	Liberty	220**
PW-1A	McCook Field	1921	(1)	Packard	350
PA-1	Loening	1921	2	Wright	350

*Figures in parentheses indicate modification of previously listed aircraft
**Liberty six-cylinder version

271

Model	Manufacturer	Cal Yr 1st Del	Number Procured	Engine/hp
PG-1	Aeromarine	1921	3	Wright 330
PW-2	Loening	1921	2	Hisso 320
PW-2A	Loening	1921	4	Hisso 320
PW-3	Orenco	1921	3	Hisso 320
PW-4	Gallaudet	1921	1	Packard 350
TP-1	McCook Field	1922	2	Liberty 400
PW-5	Fokker	1922	11	Wright 330
PW-7	Fokker	1922	3	Curtiss 440
XPW-8	Curtiss	1923	3	Curtiss 440
PW-8	Curtiss	1924	25	Curtiss 460
XPW-8A	Curtiss	1924	(1)	Curtiss 460
XPW-8B	Curtiss	1924	(1)	Curtiss 440
XPW-9	Boeing	1923	3	Curtiss 440
PW-9	Boeing	1924	30	Curtiss 440
PW-9A	Boeing	1926	25	Curtiss 440
PW-9B	Boeing	1926	(1)	Curtiss 440
PW-9C	Boeing	1926	39	Curtiss 440
PW-9D	Boeing	1927	16	Curtiss 440
P-1	Curtiss	1925	10	Curtiss 435
P-1A	Curtiss	1926	25	Curtiss 435
P-1B	Curtiss	1927	25	Curtiss 435
P-1C	Curtiss	1929	33	Curtiss 435
P-1D	Curtiss	1929	(24)	Curtiss 435
P-1E	Curtiss	1929	(4)	Curtiss 435
P-1F	Curtiss	1929	(24)	Curtiss 435
P-2	Curtiss	1925	5	Curtiss 500
XP-3	Curtiss	1926	(1)	Curtiss 400
XP-3A	Curtiss	1927	(1)	P&W 500
P-3A	Curtiss	1928	5	P&W 450
XP-4	Boeing	1926	(1)	Packard 510
XP-5	Curtiss	1927	1	Curtiss 435
P-5	Curtiss	1927	4	Curtiss 435
XP-6	Curtiss	1927	(1)	Curtiss 600
P-6	Curtiss	1929	18	Curtiss 600
XP-6A	Curtiss	1927	(1)	Curtiss 600
P-6A	Curtiss	1930	(8)	Curtiss 600
XP-6B	Curtiss	1929	(1)	Curtiss 600
P-11	Curtiss	1930	2	Curtiss 600
XP-21	Curtiss	1930	(1)	P&W 450
XP-22	Curtiss	1930	(1)	Curtiss 600
YP-20	Curtiss	1930	(1)	Wright 525
XP-7	Boeing	1928	(1)	Curtiss 600
XP-8	Boeing	1928	1	Packard 600
XP-9	Boeing	1928	1	Curtiss 600

Model	Manufacturer	Cal Yr 1st Del	Number Procured	Engine/hp	
Pursuit					
XP-10	Curtiss	1928	1	Curtiss	600
P-12	Boeing	1929	9	P&W	450
XP-12A	Boeing	1929	1	P&W	525
P-12B	Boeing	1929	90	P&W	525
XP-13	Thomas-Morse	1929	1	Curtiss	600
XP-16	Berliner-Joyce	1929	1	Curtiss	600
XP-17	Curtiss	1930	(1)	Wright	550
Attack					
GAX	McCook Field	1919	2	Liberty	400
GA-1	Boeing	1920	10	Liberty	435
GA-2	Boeing	1921	2	McCook	750
XA-2	Douglas	1926	(1)	Liberty	420
A-3	Curtiss	1927	76	Curtiss	435
A-3B	Curtiss	1930	78	Curtiss	435
XA-4	Curtiss	1928	(1)	P&W	410
XA-7	Fokker	1930	(1)	Curtiss	600
XA-8	Curtiss	1930	1	Curtiss	600
Observation					
IL-1	Orenco	1919	2	Liberty	400
CO-1	McCook Field	1921	2	Liberty	400
CO-1	Gallaudet	1922	1	Liberty	420
CO-2	McCook Field	1921	2	Liberty	390
XCO-4	Fokker	1922	3	Liberty	420
CO-4A	Fokker	1922	5	Liberty	435
XCO-5	McCook Field	1923	(1)	Liberty	420
XCO-6	McCook Field	1923	2	Liberty	420
XCO-7	Boeing	1924	1	Liberty	435
XCO-7A	Boeing	1924	1	Liberty	435
XCO-7B	Boeing	1924	1	Liberty	420
XCO-8	Atlantic	1924	1	Liberty	435
XO-1	Curtiss	1924	1	Liberty	420
O-1	Curtiss	1925	10	Curtiss	435
O-1A	Curtiss	1925	2	Liberty	420
O-1B	Curtiss	1927	25	Curtiss	435
O-1E	Curtiss	1929	37	Curtiss	435
O-1F	Curtiss	1929	(1)	Curtiss	435
XO-2	Douglas	1924	1	Liberty	420
O-2	Douglas	1925	46	Liberty	435
O-2A	Douglas	1925	18	Liberty	435
XO-2B	Douglas	1925	6	Liberty	435

Model	Manufacturer	Cal Yr 1st Del	Number Procured	Engine/hp	
Observation					
O-2C	Douglas	1926	36	Liberty	435
O-2D	Douglas	1926	2	Liberty	435
O-2E	Douglas	1926	1	Liberty	435
O-2H	Douglas	1927	73	Liberty	435
O-2K	Douglas	1929	59	Liberty	435
O-5	Douglas	1924	5	Liberty	435
XO-6	Thomas-Morse	1925	2	Liberty	435
O-6	Thomas-Morse	1925	4	Liberty	435
O-7	Douglas	1925	3	Packard	504
O-8	Douglas	1925	1	Curtiss	400
O-9	Douglas	1925	1	Packard	500
XO-10	Loening	1927	(1)	Wright	480
XO-11	Curtiss	1927	(1)	Liberty	420
O-11	Curtiss	1927	66	Liberty	435
O-11A	Curtiss	1927	(1)	Liberty	435
XO-12	Curtiss	1928	(1)	P&W	410
XO-13	Curtiss	1927	(1)	Curtiss	600
XO-13A	Curtiss	1927	(1)	Curtiss	600
O-13B	Curtiss	1928	(1)	Curtiss	600
YO-13C	Curtiss	1929	3	Curtiss	600
YO-13D	Curtiss	1930	(1)	Curtiss	600
XO-14	Douglas	1928	1	Wright	225
XO-15	Keystone	1928	1	Wright	225
XO-16	Curtiss	1928	1	Curtiss	600
XO-17	Consolidated	1928	(1)	Wright	225
O-17	Consolidated	1928	29	Wright	225
XO-17A	Consolidated	1928	1	Wright	225
XO-18	Curtiss	1928	(1)	Curtiss	600
XO-19	Thomas-Morse	1928	1	P&W	450
O-19	Thomas-Morse	1928	4	P&W	500
O-19A	Thomas-Morse	1929	1	P&W	500
O-19B	Thomas-Morse	1930	70	P&W	450
YO-20	Thomas-Morse	1929	(1)	P&W	525
XO-21	Thomas-Morse	1929	(1)	Curtiss	600
XO-21A	Thomas-Morse	1930	(1)	Wright	525
YO-22	Douglas	1929	3	P&W	500
YO-23	Thomas-Morse	1929	1	Curtiss	600
O-25	Douglas	1929	(1)	Curtiss	600
O-25A	Douglas	1930	53	Curtiss	600
O-25B	Douglas	1930	(3)	Curtiss	600
Y10-26	Curtiss	1929	1	Curtiss	600
XO-27	Fokker	1929	2	Curtiss	600
O-28	Vought	1929	1	P&W	450
O-29	Douglas	1929	(1)	Wright	525

Model	Manufacturer	Cal Yr 1st Del	Number Procured	Engine/hp	
Observation					
XO-31	Douglas	1930	2	Curtiss	600
O-32	Douglas	1930	(1)	P&W	450
O-32A	Douglas	1930	30	P&W	450
Y10-33	Thomas-Morse	1930	(1)	Curtiss	600
YO-34	Douglas	1930	(1)	Curtiss	600
XO-35	Douglas	1930	1	Curtiss	600
XO-36	Douglas	1930	1	Curtiss	600
O-38	Douglas	1930	46	P&W	525
Amphibian					
XCOA-1	Loening	1924	1	Liberty	420
COA-1	Loening	1925	10	Liberty	435
XOA-1A	Loening	1925	(1)	Wright	480
OA-1A	Loening	1926	15	Liberty	435
OA-1B	Loening	1927	9	Liberty	435
OA-1C	Loening	1928	10	Liberty	435
XOA-2	Loening	1926	1	Wright	480
OA-2	Loening	1929	8	Wright	480
Light Bombardment					
XLB-1	Huff-Daland	1923	1	Packard	800
LB-1	Huff-Daland	1926	10	Packard	800
XLB-2	Atlantic	1926	1	2 P&W	525
XLB-3A	Keystone	1927	1	2 P&W	525
XLB-5	Huff-Daland	1926	1	2 Liberties	420
LB-5	Huff-Daland	1927	10	2 Liberties	420
LB-5A	Keystone	1928	25	2 Liberties	420
XLB-6	Keystone	1928	(1)	2 Wrights	525
LB-6	Keystone	1929	17	2 Wrights	525
LB-7	Keystone	1929	18	2 P&W	525
LB-8	Keystone	1929	(1)	2 P&W	550
LB-9	Keystone	1930	(1)	2 Wright	575
LB-10	Keystone	1930	(1)	2 Wright	525
LB-10A	Keystone	1930	63	2 P&W	525
LB-11	Keystone	1930	(1)	2 Wright	525
LB-11A	Keystone	1930	(1)	2 Wright	525
LB-12	Keystone	1930	1	2 P&W	575
LB-13	Keystone	1930	7	2 P&W	525
LB-14	Keystone	1930	3	2 P&W	575

Model	Manufacturer	Cal Yr 1st Del	Number Procured	Engine/hp	
Bombardment					
DB-1	Gallaudet	1921	1	McCook	750
DB-1B	Gallaudet	1922	1	McCook	700
NBS-1 (MB-2)	Martin	1921	(20)	2 Liberty	420
NBS-1	Curtiss	1921	50	2 Liberty	420
NBS-1	LWF	1921	35	2 Liberty	420
NBS-1	Aeromarine	1922	25	2 Liberty	420
XNBS-3	Elias	1922	1	2 Liberty	420
XNBS-4	Curtiss	1922	2	2 Liberty	435
XNBL-1	Barling-Witteman	1923	1	6 Liberty	420
XB-1B	Keystone	1927	1	2 Curtiss	600
XB-2	Curtiss	1926	1	2 Curtiss	600
B-2	Curtiss	1928	12	2 Curtiss	600
B-3A (LB-10A)	Keystone	1930	(36)	2 P&W	525
Y1B-4 (LB-13)	Keystone	1930	(5)	2 P&W	575
B-5A (LB-10A)	Keystone	1930	(27)	2 Wright	525
Transport					
T-1 (GMB)	Martin	1919	1	2 Liberty	400
T-2	Fokker	1922	2	1 Liberty	420
XT-3	LWF	1923	1	1 Liberty	400
XA-1	Cox-Klemin	1923	2	1 Liberty	420
C-1	Douglas	1925	9	1 liberty	420
C-1C	Douglas	1926	17	1 Liberty	420
C-2 (Fokker)	Atlantic	1926	3	3 Wright	225
C-2A	Atlantic	1928	8	3 Wright	225
C-2B (Fokker)	General	1929	(1)	3 Wright	300
C-3	Ford-Stout	1928	1	3 Wright	225
C-3A	Ford-Stout	1929	7	3 Wright	235
C-4	Ford-Stout	1929	1	3 P&W	450
C-5	General	1929	1	3 P&W	450
C-6	Sikorsky	1929	1	2 P&W	450
C-6A	Sikorsky	1930	10	2 P&W	450
C-7	General	1929	(5)	3 Wright	300
C-7A	General	1930	(6)	3 Wright	300
XC-8	Fairchild	1929	1	1 P&W	410
C-9	Ford-Stout	1929	(7)	3 Wright	300
XC-10	Curtiss	1929	1	1 Warner	110
Miscellaneous					
Dart	Driggs	1926	1	Wright-Morehouse	26

Model	Manufacturer	Cal Yr 1st Del	Number Procured	Engine/hp	

Miscellaneous

Model	Manufacturer	Cal Yr 1st Del	Number Procured	Engine/hp	
M-1	Dayton-Wright	1920	1	dePalma	50
M-1	Lawrance-Sperry	1921	6	Lawrance	64
M-1A	Lawrance-Sperry	1922	20	Lawrance	64
MAT	Lawrance-Sperry	1922	6	Lawrance	64
DT-2	Douglas	1924	2	Liberty	420
World Cruisers	Douglas	1923	5	Liberty	420
Eagle	Curtiss	1920	3	3 Liberty	400
JL-6	Junkers-Larsen	1920	2	BMW	245
Owl	LWF	1920	1	3 Liberty	400
S-1	Loening	1922	8	Liberty	420
XPS-1	Dayton-Wright	1922	3	Lawrance	200
YF-1	Fairchild	1930	8	P&W	450

Racers

Model	Manufacturer	Cal Yr 1st Del	Number Procured	Engine/hp	
R-1 (VCP-1)	McCook-Verville	1920	(1)	Packard	660
R-2 (MB-6)	Thomas-Morse	1921	(1)	Hispano	400
R-3	Verville-Sperry	1922	3	Wright	400
R-4	Loening	1922	2	Packard	600
R-5	Thomas-Morse	1922	2	Packard	600
R-6	Curtiss	1922	2	Curtiss	460
R-8	Curtiss-Navy	1924	1	Curtiss	500
R3C-1	Curtiss	1925	1	Curtiss	620

WWI Models Purchased and/or Modified Postwar

Model	Manufacturer	Cal Yr 1st Del	Number Procured	Engine/hp	
DH-4M-2	Atlantic	1924	135	Liberty	420
DH-4M	Boeing	1923	53	Liberty	420
DH-4M-1	Boeing	1923	97	Liberty	420
MB-3A	Boeing	1921	200	W. Hispano	300
Orenco D	Curtiss	1920	50	W. Hispano	300
XB-1A	Dayton-Wright	1920	40	W. Hispano	330
SE-5E	Eberhardt	1922	50	W. Hispano	180
VCP-1A	McCook	1920	(1)	W. Hispano	300
VCP-2	McCook-Verville	1921	2	Packard	350
VE-9	Lewis & Vought	1922	27	W. Hispano	180
NBS-1 (MB-2)	Martin	1920	20	2 Liberty	420
Orenco D-2	Ordnance	1920	3	W. Hispano	300
MB-6	Thomas-Morse	1921	3	W. Hispano	400
MB-7	Thomas-Morse	1921	1	W. Hispano	400
S-6	Thomas-Morse	1920	1	Le Rhone	80

Model	Manufacturer	Cal Yr 1st Del	Number Procured	Engine/hp	
		Trainers			
TA-1	Elias	1920	3	Lawrance	140
TA-2	Huff-Daland	1920	3	A.B.C.	170
TA-3	Dayton-Wright	1921	13	Le Rhone	80
TA-4	McCook	canceled			
TA-5	Dayton-Wright	1923	1	Lawrance	220
TA-6	Huff-Daland	1923	1	Lawrance	220
TW-1	McCook	1920	2	Liberty 6	230
TW-2	Cox-Klemin	1921	3	W. Hispano	180
TW-3	Dayton-Wright	1922	2	W. Hispano	150
TW-3	Consolidated	1923	20	W. Hispano	180
TW-4	Fokker	1922	1	Curtiss	90
TW-5	Huff-Daland	1923	5	W. Hispano	180
TW-5C	Huff-Daland	1923	1	W. Hispano	190
PT-1	Consolidated	1925	221	W. Hispano	180
XPT-2,3	Consolidated	1927	(2)	Wright	225
PT-3	Consolidated	1928	130	Wright	225
PT-3A	Consolidated	1929	120	Wright	225
XPT-4	Consolidated	canceled			
XPT-5	Consolidated	1929	(1)	Curtiss	175
XPT-6	Fleet	1930	1	Kinner	100
YPT-6	Fleet	1930	10	Kinner	100
PT-6A	Fleet	1930	5	Kinner	100
XPT-7	Mohawk	1930	1	Kinner	100
XPT-8, 8A	Consolidated	1930	(2)	Packard	220
BT-1 (O-2K)	Douglas	1930	(40)	Liberty	420
BT-2 (O-32)	Douglas	1930	(31)	P&W	450
AT-1	Huff-Daland	1925	10	W. Hispano	180
AT-3 (PW-9)	Boeing	1926	(1)	W. Hispano	180
XAT-4 (P-1A)	Curtiss	1927	(1)	W. Hispano	180
AT-4	Curtiss	1927	35	W. Hispano	180
AT-5	Curtiss	1927	5	Wright	220
AT-5A	Curtiss	1928	31	Wright	220

Brig. Gen. William Mitchell with his personal Thomas-Morse MB-3 pursuit plane, 1920. The MB-3 was a 1918 design, 54 of which were ordered in 1919 and delivered to the Air Service in 1920. The Navy bought 11 in 1921. The Air Service procured an additional 200 in 1921-1922, these latter craft being built by Boeing. (USAF)

Air Service Sgt./Pilot Joe R. Reed stands unhappily by his Curtiss JN6H-1—Hisso-powered Jenny—following a crash landing near Ft. Sill's Henry Post Field, Oklahoma, in 1920. (Joe Reed)

The Vought VE-7 was purchased by both the Navy and Air Service late in 1918 and 1919. Fitted with both the 150 and 180-hp Hispano-Suiza engines built by Wright-Martin, its top speed was 106 and 114 mph, respectively. (USN)

The Navy's Curtiss NC-4, first airplane to fly the Atlantic 16-30 May 1919. A trio of NCs ("Nancies") attempted the flight; two landed on the ocean in fog, one crew was rescued by a passing ship, one crew taxied their damaged NC to the Azores, and the NC-4, with Lt. Cmdr. Albert Reed in command of a crew of five, flew to London with stops for fuel in the Azores and Portugal. The big flying boats were powered with four Liberty engines and cruised at 75 mph. (USN)

Air Service Loening R-4 with a 600-hp Packard engine was intended for the 1922 Pulitzer Trophy Race but encountered aileron flutter problems in test, the first time such a problem had been observed. The full cantilever low wing was an advanced concept at the time. Maximum speed attained was 174 mph. (USAF)

The Verville-Sperry R-3 was another 1922 Air Service racer and research aircraft. Powered with a Wright Hisso of 400 hp, it featured a rectractable landing gear and full cantilever wing. The objects beneath the fuselage are Lamblin radiators for the engine coolant. Maximum speed was 191 mph. America's young aeronautical engineers did not lack advanced ideas; their problem was lack of money for research. (USAF)

The ubiquitous deHavilland DH-4. Three manufacturers built 4846 DH-4s during WWI; obsolescent when it was built, this aircraft nevertheless served the Air Service, Navy, Marines, and the U.S. Air Mail Service until the late '20s, receiving countless modifications. (Joe Reed)

The Curtiss 18-T Wasp was a two-place fighter built for the Navy at the end of WWI. Powered by a 400-hp Curtiss K-12 engine (from which the D-12 evolved), it established both speed and altitude records, but was unsuited to Navy needs and never fully developed. Top speed was 163 mph. Two were built for the Navy, one for Bolivia. The Air Service tested a biplane version. (USN)

The Navy's Curtiss R2C-1 (L) and Wright F2W racers at St. Louis for the 1923 Pulitzer race. The 500-hp R2C-1s placed first and second at 243.7 mph and 241.8 mph, while the 700-hp F2W averaged 240.3 mph for the 155-mile event. Later that year, Navy Lt. Al Williams established a new speed mark of 266.6 mph in one of the R2C-1s. (Hudek Collection)

The GMB Martin bomber of 1918. Ten were built; Lawrence Bell was factory manager of the Martin plant at Baltimore, and Donald Douglas was the designer. Both would build airplanes under their own names later. The GMB, fitted with two Liberties, served into the '20s and at least three were modified to carry the mail. (USAF)

The tenth GMB Martin bomber was converted into a transport airplane, the T-1, in 1919. (USAF)

The Curtiss PN-1 night fighter of 1921 employed Fokker D-VII wing strut arrangement, no bracing wires, and semi-cantilever wings. Engine was a six-cylinder version of the Liberty. Two were built. Identifier on rudder was the McCook Field number. (USAF)

The Army Air Service test and research center during the first half of the '20s was McCook Field, on the edge of Dayton, Ohio. (Merle Olmsted)

Loening PW-2A powered with a Wright-Hispano engine of 300 hp was an attempt at a monoplane pursuit; Lamblin radiators beneath nose. Top speed was 136 mph. Four were built in 1922. (USAF)

Inventors Leslie Irving and Floyd Smith perfected their free-fall parachutes at the Air Service's McCook Field in the early '20s. Test jumper above, leaving a DH-4, was a brave but anonymous enlisted man; most tests were performed with dummies. The first aviator to be saved in an emergency jump was Air Service Lt. Harold R. Harris, who bailed out over Dayton when an experimental Loening fighter shed its wings 20 October 1922. (USAF)

Army Air Service pilot Lt. Joe R. Reed in standard winter flying gear, mid-'20s. (Joe Reed)

An important factor in supercharger development (besides money) was a reliable controllable-pitch propeller. Although this need was recognized by the end of WWI, and had progressed by 1924 to installations such as the one above on a McCook Field DH-4, not until 1934 did a service-ready controllable-pitch prop appear.

Curtiss PW-8s of the 1st Pursuit Group (PG) during winter maneuvers at Oscoda, Michigan, in February 1925. Martin bombers are in left background with a pair of deHavilland DH-4s in center background. (Merle Olmsted)

The Curtiss PW-8 of 1924 was the Air Service's first significant post-WWI pursuit design, not counting ten 1922 PW-5 Fokkers that did not see squadron service. D-12 powered, the PW-8's maximum speed was 161 mph. Dark areas on top wing are wing-skin radiators. (Hudek Collection)

Curtiss wing-skin radiators, made of thin brass sheets, added almost nothing to total drag but required a lot of maintenance and were too vulnerable to gunfire to be practical on fighter aircraft. Note filler and expansion headers on top wing with overflow tubes extended beyond trailing edge. Trailing edge is made of shaped brass tubing and is part of the cooling system. (Air Force Museum)

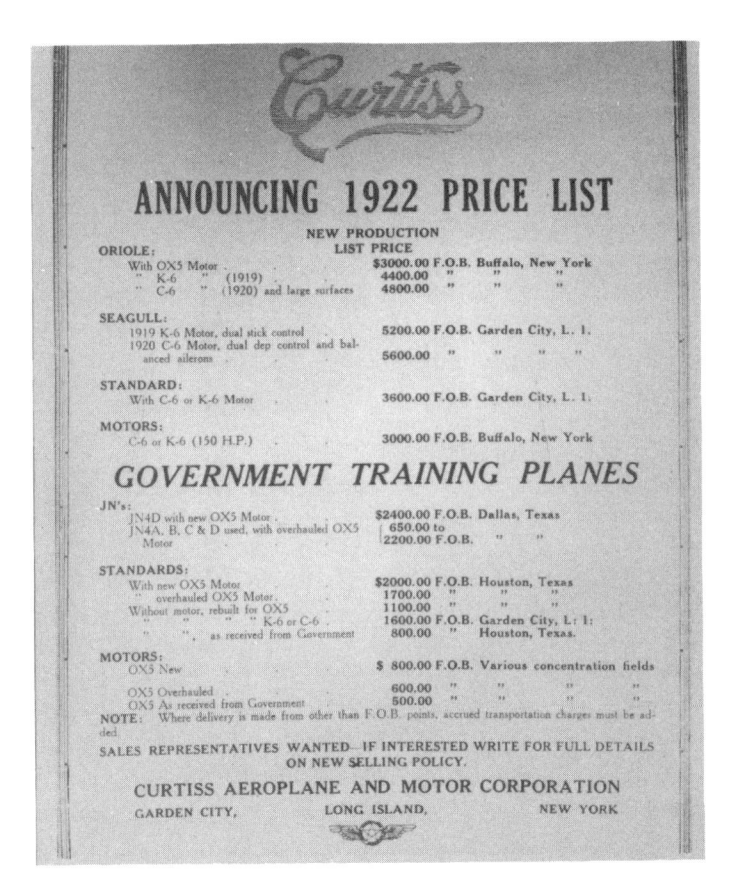

Curtiss advertisement that appeared in the 3 April 1922 issue of *Aviation* magazine.

The Douglas DT-2, first production Douglas airplane, 1922. Marking identifies this one as being aircraft Number 10 of the Navy's Torpedo Squadron One. Douglas produced 24, and the Navy contracted for 55 more from other manufacturers. (USN)

The 1921 Martin MB-2, redesignated NBS-1 (Night Bombardment, Short Distance), was built by Curtiss, L.W.F., and Aeromarine, in addition to Martin. Two 420-hp Liberties gave this craft a maximum speed of 98 mph. (USAF)

The monstrous Witteman-Lewis XNBL-1 of 1923, designed by E.D. Barling, had a span of 120 feet, gross weight of 42,500 pounds, and was powered by six 420-hp Liberties. Top speed was 95 mph. Only one was built. (Donovan Berlin)

The Air Service/Air Corps primary trainer from 1925 until the first Stearmans were delivered in 1936 was the Consolidated PT-1 through PT-3A series. The PT-1s had 180-hp Hissos completely exposed as shown. The PT-3As had 225-hp Wright Whirlwinds. (USAF)

Cockpit detail is revealed in this construction photo of the Douglas World Cruisers. Instrument panel contained but three flight instruments. The 1924 World Cruiser airframes were based on the DT-2 Navy torpedo plane. (McDonnell-Douglas)

Two of the four World Cruisers are shown on the morning of 17 March 1924 at Santa Monica's Clover Field, just after rollout from the nearby Douglas plant. (McDonnell-Douglas)

The *Chicago* on floats over Japan. The World Cruisers switched to wheels in India; returned to floats for the westward flight across the Atlantic, and then back to the wheeled undercarriage for triumphal crossing of the U.S. to their starting point at Seattle. (McDonnell-Douglas)

The *Boston* sinks in the North Atlantic after flying three-fourths of the way around the world. Wade and Ogden were rescued, and the *Chicago* and *New Orleans* flew on to complete their epic journey. (McDonnell-Douglas)

The Air Service procured 113 Boeing PW-9 pursuits between 1923 and 1928. The Navy bought it as the FB series during the same period. The Army versions had Curtiss D-12 engines rated at 440 hp; beginning with the FB-2, the Navy switched to the 510-hp Packard V-12. Maximum speeds were 155 to 160 mph for all. (USAF)

Deliveries of the famed Curtiss Hawk series began with the P-1s to the Air Service in 1925. The P-1 through P-1F were D-12 powered and had maximum speeds in the 155-160 mph range. Altogether, the Air Service/Air Corps received 247 biplane Hawks, P-1 through XP-23. Pictured is a P-1B, November 1926. (Francis Dean)

The Navy also purchased a number of biplane Hawks. This is one of five F6C-1s, which was practically identical to the Army's P-1. This one is shown in the 1928 markings of Fighting Squadron Two, plane number three, a squadron made up of Naval aviators possessing the rank of chief petty officer. (USN)

Cockpit of the Curtiss P-1, s/n 25-410, delivered to McCook Field 22 July 1925. (USAF)

The AT-5A Hawk was fitted with the nine-cylinder 220-hp Wright Whirlwind engine, and was expected to serve as an advanced trainer. However, all were later given D-12 engines of 435 hp and redesignated P-1Es. (Merle Olmsted)

A P-1D Hawk of the 43rd School Sqd., Kelly Field, Texas. The bumblebee insignia, outlined in yellow, carries the legend, "Cielito Lindo, Mexico," ("Beautiful Sky"), and the aircraft weights, fuel, oil capacities, and rigging data are stenciled beneath the cockpit. Wings were chrome yellow; struts, fuselage, and fin, olive drab; horizontal tail surfaces, chrome yellow. Maj. (later General) Clarence Tinker is in the cockpit. He disappeared flying a B-17 in a raid on Wake Island during WWII, and today's Tinker AFB is named for him. (Merle Olmsted)

A Curtiss Hawk F6C-4 of VF-9M, the Marines' famed Fighting Nine which, during this period, formed the "Rojo Diablos" (Red Devils) aerobatic demonstration team to counter threats to abolish Marine aviation. (Rowland Gill)

Curtiss test pilot Temple Joyce in cockpit of the second production F7C-1, 26 November 1928. The Seahawks were obtained for the Marines and reflected the Marines' "no frills" philosophy in the sturdy P-6 type landing gear and elimination of prop spinners. Fighting Nine received the first F7C-1s. (NASM)

A Navy F6C-4 pilot is caught in an exuberant moment over Haines Point, Washington, D.C. (NASM)

The Curtiss XP-6B was a reworked P-1C with a Conqueror engine of 600 hp and a 250-gal. fuel capacity for an attempted record flight from New York to Nome, Alaska, and return. Capt. Ross G. Hoyt made the attempt in July 1929, but was forced down on the return trip with contaminated gasoline. (NASM)

The Curtiss Hawk II was the export version of the F11C-2. This one, taking off from New York Harbor 22 March 1933, was sold to Colombia for $21,250, according to C-W billing records. (NASM)

Seventy-six Curtiss A-3 attack planes were delivered to the Air Corps in 1927 and 1928, followed by 78 more A-3Bs in 1930. Engine was the 435-hp D-12. This same basic airframe first appeared as the Curtiss 0-1 series in 1925. As the 0-11, 66 were purchased for National Guard units and fitted with Liberty engines in 1927. Maximum speed was 155-160 mph for all. (USAF)

A total of twelve P-6D Hawks resulted when ten P-6As, plus two P-11s, were given turbosuperchargers and three-bladed props. This P-6D belonged to the 37th Attack Sqdn., 8th PG, at Langley Field. The turbosupercharger had been under development at McCook Field since 1918 under the direction of General Electric's Dr. Sanford Moss. Lack of funds, and proper metals, made for slow progress. (Merle Olmsted)

Curtiss B-2 Condor of 1927 was powered with 600-hp Conquerors. Rear gunners were in extended rear of each engine nacelle. Bomb capacity was 2200 pounds, range 780 miles, and maximum speed 125 mph. Only twelve were built. (USAF)

The Navy's ZR.3 *Los Angeles* does a headstand while moored to its mast. Americans never seemed to get the hang of rigid airship operation. (USN)

The first Vought Corsairs were Navy scouts and scout-bombers. Pictured is an O2U, 291 of which were built 1926-1930 inclusive. This one was attached to the cruiser *Raleigh* and launched by catapult. All the biplane Corsairs were Wasp-powered (R-1340s), because Fred Rentschler at P&W, Chance Vought, and Bill Boeing merged in 1928 to form United Aircraft, and had been good friends before that. (Francis Dean)

The Corsairs assigned to battleships and cruisers were retrieved by crane after landing in the water alongside the mother ships. Wheeled versions served on both the *Lexington* and *Saratoga* until 1932. Maximum speed was in the 140 mph range. (Francis Dean)

The SU2 and O3U series Corsairs were delivered during the early '30s and served both the Navy and Marines. These SU2s were photographed in 1935 and had acquired only detailed improvements over the early Corsairs. Last of the line were SBUs purchased in 1936, top speed of which was near 200 mph. (USN)

The Boeing P-12 series—F4B in the Navy—was very successful for the Boeing Company, more than 550 being built for the Navy, Air Corps, and export. The P-12B pictured, delivered to the Air Corps in 1929, had a maximum speed of 165 mph at 5000 feet; engine was the P&W Wasp rated at 450 hp. (USAF)

The Air Corps' Boeing P-12K and Navy F4B-4 had a top speed of 189 mph at 6000 feet with the 550-hp R-1340 Wasp engine. Army deliveries began in 1931, Navy in 1932. (USAF)

The Navy procured 52 Curtiss N2C-1 and -2 Fledgling trainers in 1927-1928, powered with the Curtiss R-600 Challenger of 180 hp and the Wright J-5 Whirlwind of 200 hp. The Challenger was a twin-row, six-cylinder radial, and it gave the gentle Fledgling a cruising speed of 60 mph. (Hudek Collection)

The Air Corps took delivery of eight OA-2 Keystones in 1929, the last of 54 similar machines designed by Grover Loening beginning in 1924. Most were powered with reworked Liberty engines, which were mounted inverted to raise the propeller shaft as high as possible for such an application. (USAF)

The Curtiss XP-10 of 1928 was an all-new design. Conqueror-powered, it reverted to wing-skin radiators for engine cooling. Top speed was 173 mph. Only one was built. (USAF)

James Harold Doolittle with a P-6 Hawk demonstrator, 14 February 1931. Although an outstanding pilot and aeronautical engineering graduate of MIT, Doolittle was still a Lieutenant after twelve years of service, and therefore left the Air Corps in 1930. He would return when needed, ten years later. (Bill Sweet)

Chapter 12

The 1930s: Time of Decision

The Army Air Corps moved out of the biplane era during the early 1930s, its last combat biplane purchase being eight Douglas O-38Fs in 1934. By that time it had the Boeing P-26 all-metal low-wing pursuit, the Martin B-10 monoplane bomber on order, and was asking for proposals on a 300-mph pursuit. The Navy, sensitive about high landing speeds for carrier operations, made the decision to go to monoplane configuration for its combat airplanes by mid-1936, when it began tests of the Brewster F2A and Grumman F4F fighter prototypes, while concurrently beginning development of a monoplane dive bomber design at the Naval Aircraft Factory. However, the Navy was still buying biplane dive bombers (the Curtiss SBC-4) as late as August 1939, and its first monoplane fighter, the F2A Buffalo, did not enter service until December of that year.

While the attack on Pearl Harbor came as a surprise to America and her leaders, the threat of a world war had been building for years, and evidence that the President and the Congress were increasingly (if vaguely) aware of it is revealed by the fact that from 1935 onward, military appropriations steadily increased. Between 1935 and 1939, the Air Corps saw its money more than double.

During the first part of that decade, America was preoccupied with the devastating effects of the Great Depression, but across the oceans, the clouds of war began to form in 1931 when the Japanese seized Manchuria and then defied the diplomatic efforts of the League of Nations and the United States to pry them loose. In 1933, Japan quit the League, and a year later announced that it would no longer be bound by the naval limitation treaty after it expired in 1936. Japan was poor in natural resources, possessing no oil or other raw materials essential to an industrial society. The Japanese leaders saw military conquest as the only means of attaining economic independence and what they perceived to be Japan's rightful status in the world.

Japan's aggressions in the Far East created uneasiness in Washington and London (and some alarm in the Netherlands, because the oil, tin, and rubber essential to Japanese industry—and its military—were to be found in abundance in the Netherlands East Indies), but in America we knew little about the Japanese and understood them even less. Unfortunately, they would misjudge us as badly as we did them.

Meanwhile, in Europe, Adolf Hilter came to power in Germany in 1933, and by 1936 Nazi Germany had denounced the Treaty of Versailles, began rearming, and occupied the demilitarized Rhineland.

Hilter's rise to power was practically guaranteed by the terms of the 1919 Versailles Treaty that the victorious Allies imposed upon Germany. Had it not been Hitler, it would have been someone similar who would offer hope to the German people when there seemed no hope. The Versailles document, primarily authored by France (the U.S., to its credit, refused to sign it and made a separate peace agreement with Germany in 1921), effectively doomed Germany to what amounted to economic servitude, stripping away many of its possessions, and called for huge amounts to be paid in damages. France repeatedly sent troops into Germany attempting to collect; by 1923 German money was worthless and there were food riots in the streets of Berlin. If the world was surprised at Hitler's apparent appeal to the German people (actually, he received only one-third of the popular vote and was a compromise candidate), the true surprise should have been that so much time elapsed before a strong leader emerged there.

Hitler's partner in dictatorship, Italy's Benito Mussolini, began his external aggressions by attacking Ethiopia in 1935. A revolution in Spain the following year produced a third dictatorship and an extended civil war that became a proving ground for WWII. The neutrality acts passed by the U.S. Congress between 1935 and 1937 reflected the determination of the American people to stay out of overseas conflicts, and no quick changes in U.S. military policy occurred. Two great oceans insulated us from the ambitions of foreign aggressors, and the U.S. Navy guarded our shores.

The Navy's Airships

Chief of the Navy's Bureau of Aeronautics Rear Adm. William Moffett (he wasn't the kind of man one called "Bill") believed that Zeppelin-type airships should operate with the fleet to provide a degree of long-range reconnaissance unmatched by any other means, and Moffett vigorously championed the cause of the great airships throughout the '20s and early '30s.

The Navy operated four great airships. The ZR-1, *Shenandoah,* a copy of the WWI German Zeppelin L-49, was built at the Naval Aircraft Factory, Philadelphia, and assembled at the nearby Lakehurst Naval Air Station, New Jersey. It first flew in September 1923. The ZR-3, *Los Angeles*, was built by Germany's Zeppelin Works and flown to the U.S. by a German crew commanded by Dr. Hugo Eckener for delivery to the Navy at Lakehurst 15 October 1924. The ZRS-4, *Akron,* built in America by the Goodyear Zeppelin Company, was commissioned 27 October 1931; the ZRS-5, *Macon,* which entered Navy service 23 June 1933, was also built by Goodyear. The ZR-2 was never delivered. Built in England as the R.38, a copy of the WWI German L-33, the R.38 crashed into the Humber River during tests, taking the lives of 16 U.S. Navy men along with 28 of its British crew.

The lifting gas in the Navy's airships was nonflammable helium, and America had the only known helium deposit in recoverable quantity in the world (discovered near Amarillo, Texas, in 1918) until the Soviets announced a find after WWII.

Most Americans alive today were born after the day of the huge airships, and perhaps associate them with photos of the burning *Hindenburg,* which fell at Lakehurst in 1937. The *Hindenburg,* as all German-operated airships, was filled with highly-flammable hydrogen gas, because the U.S. Congress refused to allow export of helium to Nazi Germany.

The end of the *Shenandoah* was described in the previous chapter. The *Los Angeles,* too small for the mission required of the Navy's airships, nevertheless proved useful for many years, particularly as a training machine and a test vehicle for the launch and retrieval of scout planes from airships. It served well until decommissioned as an economy measure in June 1932. It was finally scrapped in 1939.

The *Akron* crashed 4 April 1933 in a storm off Barnegat Light, New Jersey, when it was driven by vertical wind gusts near the surface and its stern struck the water. Only three of its crew survived; among the 73 lost was Adm. Moffett.

The *Macon*, sister to the *Akron*, survived but 20 months in service. A structural weakness was discovered in a stern former that anchored the leading edge of the upper vertical tail. Inexplicably, it was not repaired before the Macon was dispatched for participation in fleet maneuvers in the Pacific. On 12 February 1925, the *Macon*, too, was caught in a storm and crashed at sea near Big Sur, California. Only two off the 81 men aboard were lost, but that disaster ended the Navy's rigid airship program.

The *Akron* and *Macon* were 785 feet in length, 133 feet in diameter, contained 6.5 million cubic feet of helium, and were powered with eight 650-hp German-built Maybach engines, which gave these airships a speed of 72 kts and a 6500-mile range. The engines were mounted inside the airship's hulls, with swiveling propellers mounted on outriggers.

An interesting feature of the *Akron* and *Macon* was their take-along fighter airplanes. Each airship could carry five Curtiss F9C-2 fighters inside; these were lowered on a "trapeze" in flight and launched then later recovered by means of "sky hooks" atop the Sparrowhawks' center sections. The little fighters multiplied the effectiveness of the mother ship by scouting up to 100 miles in any direction from their airborne aircraft carrier and, if necessary, would protect the airship from air attack. When operating at sea, the Sparrowhawks' wheeled landing gears were usually removed unless the exercise called for operation from the *Lexington* or *Saratoga* as well as the dirigible. Eight Sparrowhawks were built and delivered to the Navy between March 1931 and January 1933.

No Sparrowhawks were lost with the *Akron*. The prototype and two others were scrapped after the *Macon* went down; four were destroyed with the *Macon*, and the remaining example is today on display in the National Air and Space Museum.

The trapeze, and the operation of airplanes from it, were proven in tests conducted with the *Los Angeles* employing Consolidated N2Y trainers and Vought UO-1s, beginning in 1929. In 1934, a pair of WACO XJ1-Ws (similar to the civilian RNF) were equipped with sky hooks and served as general utility and air-to-ground taxis from the *Macon*.

Navy Blimps

The Navy's interest in lighter-than-air vehicles dates back to 1915, when it ordered a non-rigid dirigible from the Connecticut Aircraft Company of New Haven, CT. It was a primitive craft and soon abandoned. The Navy then built its own 175-foot non-rigid, the hydrogen-filled DN-1, which performed no better. Undiscouraged, and after much study, the Navy produced a sound design designated the Type B and constructed 16 for use during WWI, with Goodyear the main contractor. Their primary mission was submarine patrol off the U.S. coasts.

It has often been reported that the term "blimp" originated with those Navy B-class non-rigids. Such craft have no internal structure and are limp when deflated, and it was said that the Navy's description, "B, limp, airship," resulted in blimp. Another version has it that the term comes from the sound that results when one thumps the inflated envelope.

The C-class blimps naturally followed the Bs, but after WWI the Navy neglected the non-rigids in favor of the large rigid airships. However, Goodyear kept the concept alive with half a dozen commercial blimps of its own (and still does). After the *Macon* disaster, the Navy acquired no more large rigid airships and turned once again to blimps. At the time of the attack on Pearl Harbor, the Navy had four Type K patrol blimps and five smaller training types.

During WWII, the Navy operated 14 fleet blimp squadrons from some 50 bases, patrolling U.S. coasts and accompanying ship convoys. At the peak of operations in 1944, there were 1500 officer-pilots and 3000 enlisted crewmen, plus about 8000 ground support personnel, assigned to the blimp squadrons, which operated 135 K-type patrol blimps. In 1944, Blimp Squadron 14 ferried its little airships across the Atlantic via the Azores.

The L-8 Mystery

On a Sunday morning in August 1942, a little before midday, the citizens of Daly City, California, were shocked to see U.S. Navy blimp L-8 settle into the street before a row of apartment buildings. The L-Class blimp, smaller than the Ks, had a deep crease across her back, revealing the loss of some of her lifting gas.

Police and fireman were quickly on the scene and, finding no one in the control car, and being unfamiliar with such vehicles, slashed open the envelope to look inside for the L-8's crew. As the blimp's hull collapsed, it became clear that there *was* no crew. The L-8 was a ghost ship.

It actually was. It had been taken on patrol at 0600 hours that morning by two experienced officers, Lt. j.g. Ernest D. Cody and Ensign Charles E. Adams, both survivors of the *Macon* crash. They had reported to Moffett Field by radio at 0750 hours when investigating an oil slick on the water east of the Farallon Islands. They were never heard from again.

OS2U search planes from Alameda went to look for the L-8 after repeated attempts to contact it by radio proved futile. Several of these pilots caught brief glimpses of the L-8 flying in and out of clouds, and at 1020 hours the captain of a Pan Am Clipper reported sighting a blimp above the clouds some miles out to sea.

The L-8 came down ashore a little after 1100 hours. Her engine ignitions were in the "On" position and there was fuel in her tanks, although both engines had stopped. Her control car door was open, but nothing inside had been disturbed. No hint as to the fate of her crew was ever discovered.

U.S. Navy Blimp L-8's envelope was repaired and she was returned to service with her secret.

U.S. Army LTA

The U.S. Army's use of lighter-than-air vehicles may be traced back to the Civil War, when Prof. Thaddeus Lowe volunteered himself and his hydrogen balloon for service in the Union Army. For two years, wind and weather permitting, Lowe ascended in his balloon *Intrepid*, tethered in forward positions behind the Union lines, to direct artillery fire and report on Confederate troop movements. A second balloon, the *Union*, often swayed in the breeze above the Capital grounds.

The Army's first maneuverable lighter-than-air craft was a small, non-rigid dirigible purchased from Capt. Thomas S. Baldwin in 1908, which was being flown at Ft. Myer, Virginia, concurrent with Orville Wright's demonstration there of the Army's first airplane.

The Army operated captive observation balloons (called "kite balloons" in the Navy, and otherwise known to the troops as "sausage balloons" and "rubber elephants") in WWI, throughout the '20s, and into the '30s. Until 1936, all, including the free balloons used for training, were hydrogen-filled due to the higher cost of helium. The Army's peacetime balloon stations were Scott Field, Illinois, and Henry Post Field at Ft. Sill, Oklahoma.

In the mid-'30s, the Army turned to helium and began acquiring some motorized observation balloons. Some of these were convertible; they flew to their observation sites, where the control gondola and its attached engines were traded for the conventional wicker basket and the "blimp" turned balloon then ascended to be held in place by a steel cable attached to a winch truck to serve the artillery.

The Army transferred its lighter-than-air vehicles to the Navy in mid-1937.

Earlier, while the *Shenandoah* was under construction by the Navy, the Army purchased a semi-rigid, 1,240,000-cubic foot dirigible from Italy, the *Roma*, which was dismantled and sent to the U.S. by ship, arriving at Langley Field, Virginia, in August 1921. Filled with hydrogen, the *Roma* first flew in America 15 November 1921, and lasted only until 21 February 1922. Her nose collapsed in flight, apparently due to rotten fabric. Out of control, the *Roma* crashed and burned at Norfolk, Virginia, taking 34 lives; eleven survived.

Toward War

The U.S. Army Air Corps procured 1133 combat airplanes 1935-1940 inclusive. Only 91 were bombers and 247 were fighters. Of the remaining 795 machines, 329 were observation planes and 466 were attack aircraft. These figures alone reveal how the Army General Staff—headed by one Malin Craig—expected to employ the Air Corps in case of war. The observation planes, mostly North American O-47s, would direct artillery fire; the attack planes, mostly Northrop A-17s, would of course work with the infantry in a ground-support role. The General Staff was prepared to fight the next war exactly as it had the last one.

Navy planners did little better. True, the Navy commissioned or started construction of five new aircraft carriers during the '30s (most of that time limited by treaty in capital ship tonnage), but during the late '30s the Navy also planned 30 new capital ships, which President Roosevelt expected to fund, at least in part, with public relief money. When that plan was declared unconstitutional, Roosevelt got Congressional approval anyway, although that program was substantially modified after the Japanese demonstrated the striking power of an aircraft carrier task force at Pearl Harbor.

It is equally true that the President called for more aircraft production as early as September 1938—while England's Prime Minister Chamberlain was in Munich trading Czechoslovakia's Sudetenland to Hitler for "peace in our time." But Roosevelt's dramatic statement did not result in any meaningful orders to U.S. planemakers for modern combat airplanes. Neither did the U.S. Congress make any money available to the American aircraft industry for plant expansion until passage of the Lend-Lease Act in March 1941. America's aircraft industry *did* greatly expand between 1938 and 1941, but it did so with private money, and because of orders for airplanes received from Britain and France.

The German annexation of Austria in March 1938, followed by the Munich crisis in September of that year, should have been enough to alert the world to the imminence of another great world conflict. It had already begun in the Far East with Japan's invasion of China in 1937. After Germany seized Czechoslovakia in March 1939, Great Britain and France decided that they must fight rather than yield anything more to Hitler.

In August 1939, Germany made a deal with Russia that provided for a partition of Poland and a Soviet free hand in Finland and the northern Baltic states. Then, on 1 September 1939, Germany invaded Poland. When France and Great Britain responded by declaring war on Germany, they certainly knew that they would prevail only with a great deal of aid from the U.S.

The American people wanted to stay out of the war, and the President immediately proclaimed U.S. neutrality, but it was a neutrality that allowed delivery of essential war material to Britain and France—an ambiguity endorsed by most Americans, and one that allowed the U.S. another 27 months of an uneasy peace in which to strengthen her military forces and expand industrial capacity. That wasn't time enough for the awful trial we ultimately faced, and the Congress, reflecting the mood of the people, continued to act as if America could somehow avoid direct involvement in the war while cautiously supporting the President's proclamation that we serve as the "Arsenal of Democracy."

In the end, America had no choice but to fight. In fact, any choice we may have had was gone long before we recognized the danger. Indeed, there may never have been a time when war with Japan was avoidable. If we mark Japan's actions from the time she defeated Russia in 1905, her determination to rule supreme in Southeast Asia was apparent long before Billy Mitchell returned from there with his warning in 1924.

The time to take Germany into the family of peaceful western nations was at the end of WWI. We blew that opportunity.

When the U.S. at last had to fight, it came as no surprise to most Americans. The only surprise was the attack on Pearl Harbor. One can imagine what the volatile Billy Mitchell would have said about *that*. (While much of the Mitchell doctrine was soundly based, we should note that his belief that all U.S. airpower be incorporated into a single service would have been a disaster if implemented. The Navy's need for its own aircraft, commanded by Naval air officers, cannot now be questioned, and the Army has retained its organic aviation arm—conceived in 1942—for its unique mission. The British integrated their naval air into the RAF much to their regret.)

When Germany invaded Poland to start WWII in Europe, President Roosevelt proclaimed a limited national emergency, authorized increases in regular Army and National Guard enlisted strengths to 227,000 and 235,000, respectively, and asked Congress to end the prohibition on munitions sales to nations at war embodied in the Neutrality Act of 1937. Less than a year later, in mid-1940, Hitler was in possession of Denmark, Norway, the Low Countries, and France; the British stood alone against the Nazi juggernaut in the west, both pathetic and magnificent in their terrible peril.

Hitler invaded Russia 22 June 1940, and his air offensive against England began on 12 August. Arrayed against three *Luftflotten* (Air Fleets) of Reichsmarschall Hermann Goering's Luftwaffe with 2618 combat aircraft was Sir Hugh Dowding's 46 squadrons of the RAF Fighter Command with 620 Hurricanes and Spitfires. This critical aerial showdown, known as the Battle of Britain, was decided within a month when Goering, perhaps influenced by high losses, gave up the struggle for control of the air above England and, on 7 September, turned to nighttime bombing attacks on British cities. "The few to whom so many owed so much" in turn owed a lot to Goering's bad decision because, as Prime Minister Churchill later admitted, RAF losses were averaging 120 per week, and Fighter Command could not have long survived at that rate.

A total of 146,777 civilians were killed or seriously injured by the nighttime bombing (and later V-1 and V-2 attacks) in Britain. Later in the war, German and Japanese cities were heavily bombed, but there is no evidence that civilian losses affected the outcome of WWII. The RAF obliterated Dresden, killing 135,000 German civilians, and American B-29s killed more than 83,000 Japanese in a single night raid on Tokyo. The war went on.

In April 1941, President Roosevelt authorized an active Naval patrol of the western half of the Atlantic Ocean. In May, the U.S. took over responsibility for the operation of military air routes across the North Atlantic via Greenland and across the South Atlantic by way of Brazil, while American troops went to the Azores to prevent the Germans from establishing bases there.

On 6 August, the 33rd Squadron of the 8th Pursuit Group arrived in Iceland. The 33rd flew its 30 P-40Cs and three PT-17s the last 100 miles into Reykjavik from the deck of the *Wasp* to join 4000 men of the 6th Marines and 1st Provisional Marine Brigade to forestall the possibility of German seizure of that strategically located island. The 33rd had orders to accept as "conclusive evidence of hostile intent" the sighting of any German aircraft or ship within 50 miles of shore, and was to immediately attack. Navy Patrol Squadrons 73 and 74, flying PBY-5As, began operation from Iceland at the same time.

Later in the month, President Roosevelt and Prime Minister Churchill met in Newfoundland to draft the Atlantic Charter, which defined the general terms of a just peace for the world. By October the U.S. Navy was fully engaged in convoy escort duties in the North Atlantic, which meant that it was joining the British and Canadians in the shooting war against German submarines. However, the American people were not told that the Navy was so engaged, and when the U.S. Navy Destroyer *Kearny* was torpedoed while joining British warships in an attack on a U-boat, the President was indignant. "We have wished to avoid shooting," he told reporters, "but the shooting was started, and history has recorded who fired the first shot . . . All that will matter is who fired the last shot!"

U.S. Army Corps Personnel and Direct Appropriations 1931-1940

1931	14,780	$ 38,945,968
1932	15,028	31,850,982
1933	15,099	25,673,236
1934	15,861	31,037,769
1935	16,247	27,917,702
1936	17,233	45,600,444
1937	19,147	59,619,694
1938	21,089	58,851,266
1939	23,455	71,099,532
1940	51,165	186,562,847

Military Aricraft Production* Calendar Years 1931-1940

1931	812		1936	1,141
1932	593		1937	949
1933	466		1938	1,800
1934	437		1939	2,195
1935	459		1940	6,028

*Total Army and Navy, includes exports

U.S. Navy Combat Aircraft 1931-1940

Manufacturer	Designation	Contract	Procured
	Attack		
Vought	SBU	1932	126
Grumman	SF	1932	35
Great Lakes	BG-1	1932	61
Curtiss	SBC	1932	258
Curtiss	BF2C	1932	28
Douglas	TBD	1934	130
Vought	SB2U	1934	170
Naval Acft Factory	SBN	1934	30
Douglas (1934-1944)	SBD	1934	5,321
Curtiss (1939-1945)	SB2C	1939	5,516
Consolidated	TBY	1940	180
Vought	XTBU	1940	1
Grumman (1940-1945)	TBF	1940	2,290
	Fighter		
Curtiss	F9C	1931	8
Grumman	FF	1931	28
Grumman	F2F	1932	56
Grumman	F3F	1934	164
Boeing (from spares)	F4B-4	1934	1
Grumman (1936-1945)	F4F	1936	1,978
Brewster	F2A	1936	503
Vought (1938-1953)	F4U	1938	7,829

Manufacturer	Designation	Contract	Procured
Attack			
Goodyear (1940-1944)	FG-1	1940	4,006
Brewster	F3A-1	1940	735
Patrol			
Consolidated	P2Y	1931	47
Consolidated (into 1945)	PBY	1933	2,387
Boeing	PB2B-1,2		290
Naval Acft Factory	PBN-1		155
Vickers	OA-10B for Commonwealth use		230
Consolidated	PB2Y	1936	176
Rohr	PB2Y-3		41
Observation			
Curtiss	SOC	1933	259
Naval Acft Factory	SON-1	1934	44
Vought (into 1942)	OS2U	1937	1,218
Naval Acft Factory	OS2N-1		300
Curtiss (into 1944)	SO3C	1938	794
Naval Acft Factory	SON		included in SOC total

U.S. Air Corps Aircraft 1931-1940*

Model	Manufacturer	Cal Yr 1st Del	Number Procured	Engine/hp	
Pursuit					
XP-6D	Curtiss	1931	(1)	Curtiss	600
P-6D	Curtiss	1931	(12)	Curtiss	600
P-6E	Curtiss	1932	46	Curtiss	600
XP-6F	Curtiss	1932	(1)	Curtiss	600
P-6F	Curtiss	1933	(2)	Curtiss	600
XP-6G, H	Curtiss	1934	(2)	Curtiss	600
P-12C	Boeing	1931	96	P&W	525
P-12D	Boeing	1931	35	P&W	525
P-12E	Boeing	1931	110	P&W	525
P-12F	Boeing	1932	25	P&W	600
XP-12G,H,J	Boeing	1932	(3)	P&W	575
YP-12K	Boeing	1933	(7)	P&W	525
XP-12L	Boeing	1933	(1)	P&W	525
Y1P-16	Berliner-Joyce	1931	25	Curtiss	600
P-16	Berliner-Joyce	1932	(25)	Curtiss	600
YP-20	Curtiss	1931	(1)	Wright	650
XP-21	Curtiss	1931	(1)	P&W	300
XP-22	Curtiss	1931	(1)	Curtiss	600
Y1P-22	Curtiss	1932	(46)	Curtiss	600
XP-23, YP/23	Curtiss	1932	(1)	Curtiss	600

*Figures in parenthesis indicate modification or redesignation of previously listed aircraft

Model	Manufacturer	Cal Yr 1st Del	Number Procured	Engine/hp	
YP-24	Detroit	1932	1	Curtiss	600
Y1P-25	Consolidated	1932	2	Curtiss	600
Y1P-26	Boeing	1932	3	P&W	550
P-26A	Boeing	1933	111	P&W	600
P-26B	Boeing	1933	25	P&W	600
P-26C	Boeing	1934	(23)	P&W	600
Y1P-27, 28	Consolidated	canceled			
YP-29	Boeing	1934	1	P&W	575
YP-29A	Boeing	1934	1	P&W	575
YP-29B	Boeing	1934	1	P&W	575
P-30	Consolidated	1933	4	Curtiss	675
P-30A	Consolidated	1935	50	Curtiss	700
XP-31	Curtiss	1933	1	Curtiss	600
YP-32	Boeing	canceled			
P-33	Consolidated	canceled			
XP-34	Wedell-Williams	canceled			
P-35	Seversky	1936	76	P&W	950
Y1P-36	Curtiss	1937	3	P&W	1050
P-36	Curtiss	1938	(3)	P&W	1050
P-36A	Curtiss	1938	210	P&W	1050
P-36B	Curtiss	1939	(1)	P&W	1100
P-36C	Curtiss	1939	(31)	P&W	1200
XP-36D, E, F	Curtiss	1940	(3)	P&W	1050
XP-37	Curtiss	1937	1	Allison	1150
YP-37	Curtiss	1938	13	Allison	1150
XP-38	Lockheed	1939	1	2 Allison	1150
YP-38	Lockheed	1940	13	2 Allison	1150
P-38	Lockheed	1940	30	2 Allison	1150
XP-39	Bell	1938	1	Allison	1150
YP-39	Bell	1940	13	Allison	1090
XP-39A, B	Bell	1940	(2)	Allison	1150
XP-40	Curtiss	1939	(1)	Allison	1160
P-40	Curtiss	1939	200	Allison	1040
P-40A	designation not used				
XP-41	Seversky	1940	(1)	P&W	1200
XP-42	Curtiss	1940	(1)	P&W	1050
YP-43	Republic	1939	13	P&W	1200
P-43	Republic	1940	54	P&W	1200
P-43A (into '41)	Republic	1940	80	P&W	1200
P-44-1, -2	Republic	not built			
P-45 (P-39C)	Bell	1940	80	Allison	1150
XP-46	Curtiss	1940	2	Allison	1150
XP-47, -47A	Republic	1940	canceled		

Model	Manufacturer	Cal Yr 1st Del	Number Procured	Engine/hp	
XP-47B	Republic	1940	1	P&W	2000
XP-48	Douglas	1940	not built		
XP-49	Lockheed	1940	1	2 Cont.	1350
XP-50 (XF5F-1)	Grumman	1940	crashed before acceptance		

Attack

Model	Manufacturer	Cal Yr 1st Del	Number Procured	Engine/hp	
YA-8	Curtiss	1932	5	Curtiss	600
Y1A-8	Curtiss	1932	7	Curtiss	600
A-8	Curtiss	1933	(12)	Curtiss	600
Y1A-8A	Curtiss	1933	1	Curtiss	600
A-8A	Curtiss	1934	(1)	Curtiss	600
A-8B	Curtiss	1934	canceled		
YA-9, Y1A-9	Detroit	1932	canceled		
YA-10	Curtiss	1933	(1)	P&W	625
Y1A-11	Consolidated	1933	(1)	Curtiss	675
A-11	Consolidated	1933	4	Curtiss	675
XA-11A	Consolidated	1934	(1)	Allison	1000
A-12	Curtiss	1934	46	Wright	600
YA-13	Northrop	1935	1	Wright	710
XA-14	Curtiss	1936	1	2 Wright	735
A-15	Martin	1935	canceled		
XA-16 (YA-13)	Northrop	1935	(1)	2 P&W	800
A-17	Northrop	1936	110	P&W	750
A-17A	Northrop	1937	129	P&W	825
A-17AS	Northrop	1937	2	P&W	600
Y1A-18	Curtiss	1937	13	2 Wright	930
A-18	Curtiss	1939	(13)	2 Wright	930
YA-19	Vultee	1938	7	P&W	1200
XA-19A, B, C	Vultee	1940	(3)	P&W	1200
A-20	Douglas	1940	63	2 Wright	1700
A-20A	Douglas	1940	143	2 Wright	1600
XA-20B	Douglas	1940	(1)	2 Wright	1600
XA-21	Stearman	1940	1	2 P&W	1400
XA-22	Martin	1940	1	2 P&W	1200

Observation

Model	Manufacturer	Cal Yr 1st Del	Number Procured	Engine/hp	
Y1O-35	Douglas	1932	5	2 Curtiss	600
XO-36 (XB-7)	Douglas	1931	1	2 Curtiss	600
O-37	Keystone	1930	canceled		
O-38A	Douglas	1931	(1)	P&W	525
O-38B	Douglas	1931	63	P&W	525
O-38C	Douglas	1932	1	P&W	525
O-38D	Douglas	1933	1	Wright	575

Model	Manufacturer	Cal Yr 1st Del	Number Procured	Engine/hp
O-38E	Douglas	1933	37	P&W 625
O-38F	Douglas	1934	8	P&W 625
O-39	Curtiss	1932	10	Curtiss 600
YO-40	Curtiss	1932	1	Wright 630
YO-40A	Curtiss	1933	1	Wright 690
Y10-40B	Curtiss	1933	4	Wright 690
Y10-41	Thomas-Morse	1932	canceled	
Y10-42	Thomas-Morse	1932	not built	
Y10-43 (Y10-31)	Douglas	1933	(5)	Curtiss 600
O-43A	Douglas	1934	24	Curtiss 675
YO-44	Douglas	1933	(1)	2 Wright 750
XO-45 (YB-10)	Martin	1934	(1)	2 Wright 675
XO-46 (O-43A)	Douglas	1935	(1)	P&W 725
O-46A	Douglas	1936	90	P&W 725
XO-47	N. American	1936	1	Wright 850
O-47A	N. American	1937	164	Wright 975
O-47B	N. American	1939	74	Wright 1060
XO-48	Douglas	1938	canceled	
O-49 (L-1)	Stinson	1940	142	Lycoming 295
O-49A (L-1A)	Stinson	1940	182	Lycoming 295
YO-50	Bellanca	1940	3	Ranger 420
YO-51	Ryan	1940	3	P&W 440

Autogyro

Model	Manufacturer	Cal Yr 1st Del	Number Procured	Engine/hp
YG-1	Kellett	1936	1	Jacobs 225
YG-1A	Kellett	1937	1	Jacobs 225
YG-1B	Kellett	1938	7	Jacobs 225
YG-2	Pitcairn	1936	1	Wright 400

Reconnaissance

Model	Manufacturer	Cal Yr 1st Del	Number Procured	Engine/hp
F-1A	Fairchild	1931	6	P&W 450
F-2	Beech	1940	14	2 P&W 375
XF-3 (A-20)	Douglas	1940	(1)	2 Wright 1700
YF-3 (A-20)	Douglas	1940	(2)	2 Wright 1700

Bombardment

Model	Manufacturer	Cal Yr 1st Del	Number Procured	Engine/hp
B-4A	Keystone	1932	25	2 P&W 575
Y1B-6	Keystone	1931	(5)	2 Wright 575
B-6A	Keystone	1932	39	2 Wright 575
XB-7 (XO-36)	Douglas	1931	(1)	2 Curtiss 600
Y1B-7	Douglas	1933	7	2 Curtiss 600
XB-8 (XO-27)	Fokker	1932	(1)	2 Curtiss 600
YB-9	Boeing	1933	1	2 P&W 525
Y1B-9	Boeing	1933	1	2 Curtiss 600

Model	Manufacturer	Cal Yr 1st Del	Number Procured	Engine/hp	
Y1B-9A	Boeing	1933	5	2 P&W	630
XB-10	Martin	1933	1	2 Wright	675
YB-10	Martin	1933	14	2 Wright	675
YB-10A	Martin	1934	1	2 Wright	675
B-10B	Martin	1934	103	2 Wright	740
YB-11	Douglas	1933	canceled		
YB-12	Martin	1934	7	2 P&W	775
YB-12A	Martin	1934	25	2 P&W	775
YB-13	Martin	1933	canceled		
XB-14	Martin	1934	1	2 P&W	950
XB-15	Boeing	1937	1	4 P&W	1000
XB-16	Martin		not built		
XB-17	Boeing	1936	crashed prior to acceptance		
YB-17	Boeing	1937	13	4 Wright	930
YB-17A	Boeing	1939	1	4 Wright	1000
B-17B	Boeing	1939	39	4 Wright	1000
B-17C	Boeing	1940	38	4 Wright	1200
B-18	Douglas	1937	133	2 Wright	930
B-18A	Douglas	1938	217	2 Wright	1000
B-18B	Douglas	1939	(122)	2 Wright	1000
Y1B-20 (XB-15)	Douglas	1938	(1)	4 P&W	1400
XB-21	N. American	1938	1	2 P&W	1200
B-22	Douglas	1938	canceled		
B-23	Douglas	1939	38	2 Wright	1600
XB-24	Consolidated	1940	1	4 P&W	1200
YB-24	Consolidated	1940	7	4 P&W	1200
B-25	N. American	1940	24	2 Wright	1700
XB-27	Martin	1940	canceled		

Transport

Model	Manufacturer	Cal Yr 1st Del	Number Procured	Engine/hp	
C-4A	Ford	1931	4	3 P&W	450
C-8 (YF-1)	Fairchild	1931	(8)	P&W	450
C-8A	Fairchild	1931	(6)	P&W	450
Y1C-11	Consolidated	1931	1	P&W	575
Y1C-12	Detroit	1931	1	P&W	450
C-13	not used				
Y1C-14	General	1931	18	Wright	525
C-14B, Y1C-14A	General	1932	(2)	Wright	575/525
Y1C-15	General	1932	1	Wright	525
C-15A	General	1932	1	Wright	575
C-16	General	1932	test only		
Y1C-17	Lockheed	1931	1	P&W	600
Y1C-18	Boeing	1931	test only		
YC-19	Northrop	1931	1	P&W	450

Model	Manufacturer	Cal Yr 1st Del	Number Procured	Engine/hp	
Y1C-19	Northrop	1932	2	P&W	450
YC-20	General	1931	test only		
Y1C-21	Douglas	1932	8	2 Wright	350
Y1C-22	Consolidated	1932	3	Wright	575
Y1C-23	Lockheed	1932	1	P&W	450
Y1C-24	American	1932	4	Wright	575
Y1C-25	Lockheed	1933	1	P&W	450
Y1C-26	Douglas	1933	2	2 P&W	300
Y1C-26A	Douglas	1933	8	2 P&W	350
Y1C-27	Bellanca	1933	4	P&W	550
C-27A	Bellanca	1933	10	P&W	650
C-27B	Bellanca	1934	(1)	Wright	675
C-27C	Bellanca	1934	(13)	Wright	750
Y1C-28	Sikorsky	1933	1	P&W	300
C-29	Douglas	1934	2	2 P&W	575
YC-30	Curtiss	1934	2	2 Wright	650
XC-31	Krieder-Reisner	1935	1	Wright	750
XC-32	Douglas	1936	1	2 Wright	750
C-33 (DC-2)	Douglas	1936	18	2 Wright	750
C-34	Douglas	1937	2	2 Wright	750
XC-35	Lockheed	1937	1	2 P&W	550
Y1C-36	Lockheed	1937	3	2 P&W	450
Y1C-37	Lockheed	1937	1	2 P&W	450
C-38	Douglas	1937	(1)	2 Wright	930
C-39	Douglas	1939	35	2 Wright	975
C-40 (Model 12)	Lockheed	1938	3	2 P&W	450
C-40A	Lockheed	1939	10	2 P&W	450
C-40B	Lockheed	1939	1	2 P&W	450
C-41	Douglas	1939	1	2 P&W	1200
C-41A (DC-3)	Douglas	1940	1	2 P&W	1200
C-42	Douglas	1939	1	2 Wright	1200
YC-43	Beech	1939	3	P&W	450
XC-44 (Me-108)	B.F.W.	1939	1	Argus	240
C-45	Beech	1940	11	2 P&W	450

Primary Trainer

Model	Manufacturer	Cal Yr 1st Del	Number Procured	Engine/hp	
PT-6A	Fleet	1931	5	Kinner	100
YPT-9	Stearman	1931	4	Wright	165
YPT-9A, B, C	Stearman	1932	(3)	Cont/Lyc/Kinner	
YPT-10	Verville	1931	4	Wright	165
YPT-10A, B, C, D,	Verville	1932	(4)	Cont/Lyc/Kinner	
Y1PT-11	Consolidated	1931	4	Continental	165
Y1PT-11A, B	Consolidated	1932	(2)	Curtiss/Kinner	
PT-11B	Consolidated	1932	5	Kinner	210

Model	Manufacturer	Cal Yr 1st Del	Number Procured	Engine/hp	
Y1PT-11C, D	Consolidated	1932	(10)	Lycoming	200
PT-11D	Consolidated	1932	21	Lycoming	200
Y1PT-12	Consolidated	1932	10	P&W	300
PT-13	Stearman	1936	26	Lycoming	215
PT-13A	Stearman	1937	92	Lycoming	220
PT-13B	Stearman	1940	255	Lycoming	280
XPT-14	Waco	1939	1	Continental	220
YPT-14	Waco	1940	13	Continental	220
XPT-15	St. Louis	1939	1	Wright	225
YPT-15	St. Louis	1940	13	Wright	225
XPT-16	Ryan	1939	1	Menasco	125
YPT-16	Ryan	1940	15	Menasco	125
PT-16A	Ryan	1940	(14)	Kinner	132
PT-17	Stearman	1940	2,671	Continental	220
PT-18	Stearman	1940	150	Jacobs	225
PT-19	Fairchild	1940	270	Ranger	175

Basic Trainer

Model	Manufacturer	Cal Yr 1st Del	Number Procured	Engine/hp	
BT-2B	Douglas	1931	146	P&W	450
BT-2B (0=32) modified; -2BA thru -2CR			(126)	P&W	450
BT-2C	Douglas	1932	20	P&W	450
YBT-3 (YPT-9)	Stearman	1932	(1)	Wright	300
YBT-4 (O-1E)	Curtiss	1932	(1)	Curtiss	435
YBT-5 (YPT-9)	Stearman	1932	(1)	P&W	300
Y1BT-6 (Y1PT-11)	Consolidated	1932	(1)	Wright	300
Y1BT-7 (Y1PT-12)	Consolidated	1933	(10)	P&W	300
BT-8	Seversky	1936	30	P&W	400
BT-9	N. American	1936	42	Wright	400
BT-9A	N. American	1936	40	Wright	400
BT-9B	N. American	1937	117	Wright	400
BT-9C	N. American	1938	66	Wright	400
BT-9D	N. American	1938	(1)	Wright	400
Y1BT-10 (BT-9C)	N. American	1938	(1)	P&W	600
XBT-11	A.R.C.	1940	canceled		
XBT-12	Fleetwing	1940	1	P&W	450
BT-13	Vultee	1940	300	P&W	450
BT-14	N. American	1940	251	P&W	450

Advanced Trainer

Model	Manufacturer	Cal Yr 1st Del	Number Procured	Engine/hp	
AT-6	N. American	1940	94	P&W	600

Model	Manufacturer	Cal Yr 1st Del	Number Procured	Engine/hp
Basic Combat				
BC-1 (SN-J)	N. American	1938	177	P&W 600
BC-1A	N. American	1040	83	P&W 600
(as AT-6)				
BC-1B	N. American	1940	(1)	P&W 600
BC-11	N. American	1940	(30)	P&W 600
BC-2	N. American	1938	3	P&W 600
BC-3	Vultee	1940	1	P&W 600
Amphibian				
OA-3/OA-4 series,	Douglas, formerly C-26 series		(29)	Wright 350
Y1OA-8	Sikorsky	1937	5	2 P&W 750
OA-9	Grumman	1938	31	2 P&W 450

Douglas O-38 of the California Air National Guard. The O-38s were delivered in 1931, and the emphasis on observation-type aircraft reflected the Army General Staff's concept of airpower. (USAF)

The Curtiss P-6E Hawk was the Air Corps' last biplane pursuit. Top speed was 193 mph at sea level; engine was the Curtiss V-1570 Conqueror of 600 hp, a liquid-cooled V-12. The P-6E pictured belonged to the 33rd PS, 8th PG. (Ray Pritchard)

P-6Es of the 17th PS, 1st PG, Selfridge Field, Michigan. As late as 1 January 1937, 17 P-6Es were still in service at the Air Corps Tactical School, Maxwell Field, Alabama. (USAF)

Lockheed YP-24 (XP-900) of 1931 was that company's first military design. This two-place fighter was powered with a Curtiss Conqueror engine and had a top speed of 214 mph. The prototype was lost due to a landing gear malfunction, and ten more on order for the Air Corps were never built. (USAF)

The Boeing P-26 was the Air Corps' first all-metal, low-wing pursuit aircraft. Unofficially known as the "Peashooter," the P-26s had a top speed of 234 mph at 7500 feet; engine was the R-1340 Wasp rated at 570 hp. One hundred and eleven P-26As were delivered in 1933, followed by 25 P-26Bs in 1934. (USAF)

America's only successful rigid-frame dirigible was the ZR.3 *Los Angeles*, built in Germany by the Zeppelin Company and delivered to the U.S. Navy in October 1924. Filled with nonflammable helium, the *Los Angeles* flew 4320 hours as a training airship before being decommissioned in 1932 as an economy measure. (USN)

One of eight Curtiss F9C-2 Sparrowhawk parasite fighters built for operation from the Navy dirigibles *Akron* and *Macon*. Wright 440-hp R-975 engines gave the Sparrowhawks a maximum speed of 175 mph. This one is piloted by Lt. H.B. Miller, 8 February 1935, just four days before the *Macon* was lost with four F9C-2s aboard. (USN)

The Air Corps' Douglas O-46 observation planes entered service in 1936, and were the last of a series of graceful parasol monoplanes that first appeared in 1933. (USAF)

The Curtiss XP-31 Swift was flown with both a 700-hp Cyclone and the 600-hp geared Conqueror engine. It was accepted by the Air Corps for test with the Conqueror installation on 1 March 1933, but by that time it had already lost out to the new Boeing P-26, an order for which had been placed just a week before. (Francis H. Dean)

The Curtiss SOC-1 Seagull was a 1934 design primarily intended for catapult operation from battleships and Navy cruisers. It filled that role so well that it served throughout WWII. Most were equipped as single-float seaplanes, and power was the 550-hp Wasp. A total of 259 was built. (Rowland P. Gill)

The Curtiss XF11C-1 actually appeared after the F11C-2 Goshawk design was firmed up, apparently to test the experimental twin-row Wright R-1510 engine of 600 hp. The F11C-2, of which 27 were delivered to VF-1B aboard the *Saratoga* in February 1933, was redesignated the BFC-2 in March 1934. VF-1B flew these machines until February, 1938. (USN)

Grumman SF-1 Scout-Fighter was the first of a series of Navy biplane fighters that entered service in 1934. Originally the FF-1 ("Fifi"), the two-placers quickly evolved into the single-place F2F-1, which equipped the *Lexington's* Fighting Two and the *Ranger's* Fighting Four. The F2F series remained the Navy's basic fighter until replaced by the F4F Wildcat. (USN)

In 1933 a production Curtiss Goshawk became the XF11C-3 when it was given a retractable landing gear and reworked aft turtle deck. This was the prototype of the BF2C-1s (and export Hawk IIIs). The Navy's 27 BF2C-1s, however, had metal-framed wings whereas all previous Hawks had wood-framed wings, and it was soon discovered that periodic vibration of the metal structure coincided with the cruising rpm of the Wright R-1820 engine. Therefore, service life of the BF2C-1s was short. (USN)

The Curtiss BF2C-1, with wooden-framed wings, was the Hawk III, 101 of which were sold to China between May 1936 and June 1938 for an average price of $23,000 each less guns. (NASM)

Chinese Hawk IIIs at Lungwa, China, October 24, 1936. These aircraft were assembled in China by the Central Aircraft Manufacturing Corporation, set up by C-W, later owned by C-W sales representative and international wheeler-dealer William D. Pawley. Fixed-gear versions of the P-36 would later be assembled by this company, and Flying Tiger personnel would be recruited by representatives of Central Aircraft Manufacturing Corp., which was by then located in Burma. (Curtiss-Wright)

A Chinese Hawk III, crippled by defensive fire from a Japanese ship, crash-landed at Shanghai 30 September 1937 as the Japanese began their invasion of China. The Chinese Air Force at that time had one squadron of Hawk IIIs, one of Boeing P-26s, and one of Northrop attack aircraft similar to the A-17. Chinese pilots at the time were trained by Italian Air Force personnel sent by Mussolini. Chennault was there in an advisory capacity. (UPI)

The U.S. Army continued to use tethered observation balloons for artillery fire control into the mid-'30s. Balloonists were trained in hydrogen-filled free balloons. Here, the 1st Balloon Squadron at Ft. Sill, Oklahoma, inflates a 35,000 cu/ft free balloon at Henry Post Army Airfield. (M/Sgt. Harvey Nelson)

Army free balloon was "weighed-off" by balancing lift with sand ballast to achieve zero weight just prior to lift off. Release of gas or sand then controlled ascent and descent. America had a supply of helium, but Army didn't use it because of the expense. (M/Sgt. Harvey Nelson)

The Army acquired a few non-rigid helium-filled balloons in 1937 that possessed a four-place, twin-engine gondola, but shortly afterwards all of the Army's lighter-than-air vehicles were transferred to the Navy. (Maj. A.C. Maurer)

The Curtiss P-36 prototype, design 75, first flew in May 1935, fitted with the experimental Wright twin-row radial R-1640. This engine was not developed and it cost Curtiss the fighter competition in 1936. (Donovan Berlin)

The Air Corps selected the Seversky P-35 over the XP-36, ordering 76 of this 285-mph all-metal craft, but hedged its decision by also ordering three YP-36s when Curtiss offered to power them with the new P&W R-1830 Twin Wasp. (USAF)

The Twin Wasp was the proper engine for the P-36. The YP versions performed so well that the Air Corps ordered 210 of them as P-36As, the largest pursuit plane order since WWI. Top speed was 313 mph. (Donovan Berlin)

331

One of the export versions of the P-36 was the Hawk 75H fitted with the 875-hp Wright Cyclone GR-1820-G3. This one was Chennault's personal airplane and it saw some combat during the battle for Shanghai in September 1937. (Donovan Berlin)

A 1939 Air Corps publicity photo of 8th PG officers. It appears that all except the colonel are wearing the new "pinks," a trouser of an odd gray color that the reader may recall (if old enough). Today's Air Force blues were introduced in 1949. L to R: Lt. J.D. Lee, Capt. J.A. Bulger, Maj. M.B. Asp, Lt. Col. W.E. Kepner, Capt. G.H. MacNair, and Lt. W.H. Wise.

The hand-built Curtiss XP-37, painted aluminum to hide scratches, went to the Wright Field in April 1937. Fitted with the early Allison V-1710 and a turbosupercharger mounted beneath the nose, the XP-37 and 13 YP-37s to follow achieved 340 mph at 20,000 feet on those occasions when the superchargers could be induced to function. (Donovan Berlin)

The YP-37s were delivered in 1938, but it would be another four years before America possessed a reliable turbo unit. "It was a noble experiment," Don Berlin said of the P-37 project. (Donovan Berlin)

The Curtiss XP-40 was a P-36 powered with a Allison V-1710-19 of 1090 hp, the coolant radiator located on the aft belly. This craft first flew 14 October 1938. (Donovan Berlin)

The best of the U.S. Army Air Corps—23 September 1940, Langley Field, Virginia. America's aerial might is displayed for visiting dignitaries from Latin America: new P-40s in foreground, a total of four early B-17s, and a passle of Douglas B-18 Bolo bombers, along with a P-36. (USAF)

The first of 13 Lockheed YP-38s delivered to the Air Crops beginning in Spetember 1940. These service test aircraft revealed problems awaiting aircraft engineers as this new generation of fighters encountered compressibility effects in high-speed dives. The XP-38 had been lost 19 months earlier to an unexplained engine failure after spanning the continent in record time. (USAF)

Chief of Air Corps Gen. H.H. Arnold (L), Gen. A.W. Robbins, CO of Wright Field, and Lt. Ben Kelsey, Air Corps Fighter Project Officer, at Wright Field 11 February 1939. Kelsey has just delivered the XP-38 to Wright from March Field, California, in record time, and is ordered by Arnold to continue to Mitchel Field, L.I., New York. Kelsey was not injured when the XP-38 crash-landed there, perhaps because insufficient heat was available from the superchargers to prevent carburetor icing. (Gen. Benjamin Kelsey)

Lightning Mk I (the British gave the name "Lightning" to the P-38) for the RAF had no superchargers, no armor, no counter-rotating props—and no value as a combat airplane. Not surprisingly, the British did not want it. As Lockheed model 322, 143 were built and all but five were taken by the USAAF and used for training. The British order may have been a ploy concocted by Arnold to get a P-38 production line moving since the USAAF had not ordered the P-38 in quantity when the British order was given to Lockheed in April 1940. (USAF)

North American O-47, 239 of which were delivered to the Air Corps 1937-1939 inclusive. Observer had unobstucted view from prone position in belly. These craft were defenseless and of small value once war began. (USAF)

Curtiss O-52 Owl. The Air Corps took delivery of 203 of these anachronisms in 1941. (USAF)

A Curtiss P-36C on the eve of WWII for America. This photo was made after March 1941, when all identifiers were painted over except serial number and squadron aircraft number. This one probably belonged to the 8th PG. (USAF)

The Grumman J2F-1 Duck series was widely used by the Navy as utility aircraft, with first deliveries beginning in 1933 and continuing through WWII. Wright Cyclone-powered, the Ducks were slow but versatile. (USN)

The huge Douglas B-19, like the Boeing B-15, was one of a kind. Construction began in the spring of 1937, and the B-19 first flew 27 June 1941. A research aircraft, it was fitted with Wright R-3350 engines. It was scrapped in 1948. (McDonnell-Douglas)

Douglas SBD Dauntlesses in 1940, the first of a series that remained in service for seven years. The last ones were built in 1944. Six models were contained in the 5396 built. (USN)

The Cessna UC-78, officially the Bobcat (and RCAF Crane), was more commonly called the "Bamboo Bomber" or "Useless-78" by the student pilots who received multi-engine training in it. Cessna originally intended it as a light civil transport, the T-50. Other WWII designations were AT-8 and AT-17. (Cessna)

The Fairchild PT-19 primary trainer of WWII was fitted with the 200-hp Ranger engine of in-line, inverted, air-cooled configuration. The PT-19 above was restored as per original, and photographed at Coatsville, Pennsylvania, in 1974. (Francis H. Dean)

One of the most extensively used basic trainers during WWII was the Vultee BT-13 Valiant, commonly called the "Vibrator" by air cadets. More than 13,000 were built by Convair (now General Dynamics). The BT-13 was powered by the P&W R-985 Wasp Junior rated at 450 hp, and had a maximum speed of 164 mph. (USAF)

The Beechcraft AT-10 multi-engine pilot trainer was fitted with 280-hp Lycomings. The postwar civilian Bonanza's V-tail was tested on an AT-10 during the war. (Beech)

The North American AT-6 (redesignated T-6 in 1948) was the SNJ in Navy dress, an advanced trainer with a 600-hp P&W R-1340 Wasp engine. Maximum speed was slightly over 200 mph. (USAF)

The Beechcraft Model 17 Staggerwing was pressed into war service by both the Navy and Air Force as a general utility aircraft, as the GB-1/2 in the Navy, UC/YC-43 in Army colors. (Beech)

Chapter 13

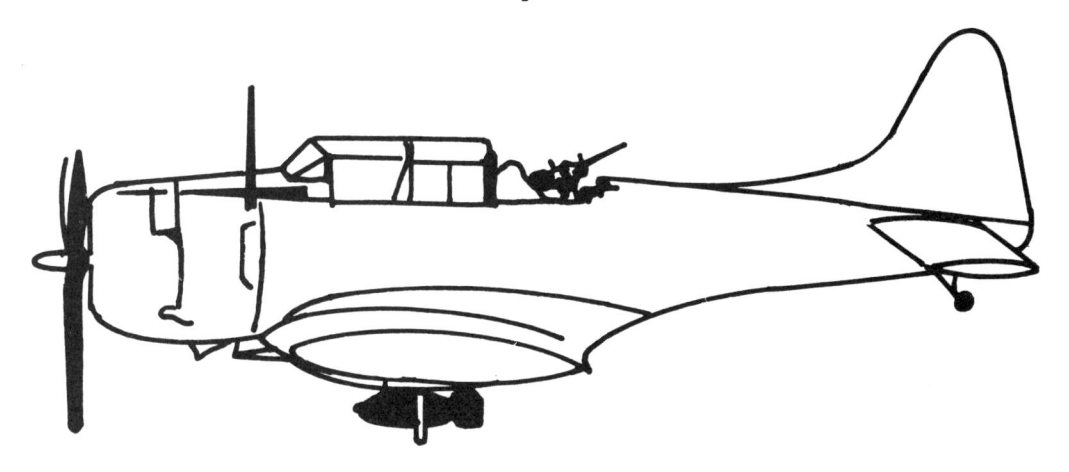

World War II

"The Americans' (air) attacks, which followed a definite system of assault on industrial targets, were by far the most dangerous. It was in fact these attacks which caused the breakdown of the German armaments industry. The attacks on the chemical industry would have sufficed, without the impact of purely military events, to render Germany defenseless." So said Albert Speer, Germany's Reichminister of Armaments and War Production, on 18 July 1945.

"It is my opinion that our loss in the air lost us the war," said Lt. Gen. Masakaza Kawabe, Japan's Commanding General, Air, 14 October 1945.

"Three factors defeated us in the West where I was in command. First, the unheard-of superiority of your air force, which made all movement in daytime impossible. Second, the lack of motor fuel—oil and gas—so that the panzers and even the remaining Luftwaffe were unable to move. Third, the systematic destruction of all railway communications so that it was impossible to bring one single railroad train across the Rhine . . . and robbed us of all mobility," said Marshal von Rundstedt, Commander in Chief of the German Armed Forces in Western Europe, on 14 June 1945.

Similar statements came from other German and Japanese commanders at the end of World War II. This was the completely defeated, frustrated enemy, and implicit in each statement was the hint that many thousands of soliders, on both sides, died needlessly. The invasion of Europe by Allied ground forces may have been as unnecessary as the planned invasion of Japan proved to be. (The atomic bombs did not win the war against Japan, but only shortened it by a month or so at most.)

This was, and has been, the judgment of many air leaders. Gen. LeMay, whose B-29 bombers were systematically destroying Japan's ability to make war in mid-1945, and who was *unaware* that America possessed the A-bomb, recommended against an invasion of Japan, noting that the army assembled there could not long be supplied or fed. LeMay predicted that the Japanese would sue for peace within weeks. As the end neared, there was almost no enemy opposition to the B-29 raids, and LeMay began broadcasting his targets in advance in order to spare enemy civilian lives (U.S. air commanders in Europe did the same).

A negotiated peace with Germany was clearly possible as early as April 1944, according to this scenario. Not a single German general believed that Germany could win the war by then, and the German general staff would have eagerly dumped Hitler and his top henchmen in exchange for peace.

Perhaps these air advocates were (and are) correct. But we never offered to negotiate. Once America was committed to the fight, we never considered any alternative to the total, unconditional surrender of our enemies. At first, that was a natural, largely emotional position. In the end, our military might had grown to such proportions that it had a compelling momentum of its own and, with victory in sight, Allied leaders were loath to settle for less.

In the beginning, however, none had been so confident, and in fact America entered WWII with no air strategy. The Army and Navy bosses had, at the President's request, given Roosevelt a plan—"Rainbow 5"—in mid-1941, which represented their collective view of what it would take to win the war should the U.S. be drawn into it. While that plan called for 239 combat air groups, it also envisioned the mobilization of 215 Army divisions and a grand total of 8.8 million men under arms. Five million of these would be committed to the ground war against Germany. Rainbow 5 did not include plans for a strategic air offensive against specific industrial targets in Germany and Japan. Its call for 20,000 airplanes and five million ground troops for the assault on Germany revealed that our military planners, just five months before the attack on Pearl Harbor, envisioned another war fought in the mud of France, and U.S. airpower was not expected to play a decisive role.

It should be noted, however, that a strategic bombing offensive of the kind that later brought Germany to her knees had never been tried. There was no precedent for it, and the lessons offered up to that time by the war in Europe were false ones. Hitler had no strategic air arm. The Luftwaffe was structured for direct support of the highly mobile German Army and its armored divisions (panzers) in the conquest of Europe. That was *blitzkrieg*, lightning warfare, and it impressed U.S. observers who failed to consider that the Luftwaffe met no significant resistance in the air until England was attacked. Then, when the German bombing raids on Britain failed to appreciably effect the course of the war, that, too, was misinterpreted. The Luftwaffe's twin-engine light bombers carried small bomb loads and, raiding by night, destroyed few military targets. Hitler had never believed that the British would fight after France fell, and when they did, his air force was unsuited to the task.

America is Attacked

President Roosevelt was sufficiently alarmed by the threats posed by Germany and Japan by 1939 that he thrust aside military sensibilities and politics to appoint Gen. George C. Marshall Chief of Staff—the first non-West Pointer to hold the position since 1914. It was time to select the best people for the job at hand.

General Marshall rescued the remaining members of the "Billy Mitchell Crowd" still in minor posts, presided over the buildup of an American Army of 1.5 million men during the ensuing two years, and supported administrative reorganization of the Army Air Corps, which became the U.S. Army Air Forces in July 1941, with Gen. Henry H. "Hap" Arnold in command. By December 67 air combat groups existed in varying stages of organization (including some still on paper). Four groups were in Panama, one in Alaska (a composite group), four in Hawaii, and two in the Philippines at the time of the attack on Pearl Harbor, and, ten hours later, the attack on the Philippines. The USAAF could count 1157 first-line combat aircraft, of which slightly over 100 P-40s were in the Philippines, along with 35 B-17s. There were 99 P-40s and 30 P-36s in Hawaii, along with 12 B-17s. A mixture of obsolescent aircraft (including P-35s and O-52s) was also in the Philippines. The same situation occurred in Hawaii, where the Army was still flying some B-18s as well as the more modern Douglas A-20.

At that time, the Navy had three carriers in the Pacific—the *Lexington, Enterprise,* and *Saratoga.* The Navy had 165 shore-based patrol planes in Hawaii, two patrol squadrons in the Philippines, and a Marine fighter squadron on Wake Island.

Altogether, counting aircraft aboard the *Lexington* and *Enterprise (Saratoga* was at San Diego, just out of overhaul), the U.S. Navy and USAAF in the Pacific could muster a total of about 600 combat aircraft when America was thrust into WWII.

The Japanese attack on Pearl Harbor was a brilliant tactical feat, but a strategic blunder of enormous proportions. It united the American people in a grim determination to fight that was unprecedented in U.S. history. Prior to the attack, most Americans—including the U.S. Congress—would have firmly rejected war; afterwards, even the isolationists were breaking out their flags and demanding vengeance.

From six aircraft carriers, which had approached undetected to a position 200 miles north of Oahu, 352 Japanese airplanes—171 Val dive bombers, 102 Kate attack bombers, and 79 Zero fighters—roared out of the mists at 0755 hours on Sunday morning 7 December 1941 to visit unbelievable destruction on Hawaii's military installations. The prime targets were the Navy's capital ships anchored in a row beside Ford Island; the torpedo-laden Kates struck there, while the Vals attacked the Marine airfield at Ewa, as well as the Army's Wheeler and Hickam Fields and Navy shore installations at Kaneohe, then, along with the Zeros, strafed targets of opportunity.

Official Air Force records for that date are incomplete, but the following can be established regarding air action by U.S. pilots: Lewis Sanders and P. Rasmussen each claimed a Kate shot down, apparently, but not certainly, flying P-36s. Six other pilots engaged the enemy flying P-36s: Lts. Harry Brown, Mike Moore, Bob Rogers, Gordon Sterling, Tony Thacker, and John Daines. Brown shot down a Kate, and shared another with Mike Moore. Army antiaircraft gunners shot down John Daines. The P-36s were flown from Haleiwa, a gunnery airstrip. Lts. Kenneth Taylor and George Welch joined the fight in P-40s, also from Haleiwa, and Welch is credited with downing four of the attackers. At Bellows Field, the P-40s of the 44th Pursuit Squadron (PS) were fueled and armed while under attack, and Lts. George Whitman, Hans Christianson, and Sam Bishop attempted takeoff. Christianson's machine was exploded during its takeoff roll; Whitman was shot down just as his wheels left the runway, and Zeros shot out Bishop's controls at about 400 feet. Bishop went into the ocean, but swam ashore uninjured. There is some evidence that Lt. John Webster took off in a P-40, but we found no official record of his action.

Except for the three unfortunate pilots of the 44th PS (18th Pursuit Group), all the others belonged to the 47th PS. And since, officially, there were only six P-36As at Haleiwa, it would seem that Sanders and Rasmussen flew P-40s. Previous accounts do not include these two pilots as being airborne that day, and our inclusion of them at this date is based solely on the fact that each was credited with shooting down a Kate.

The Japanese attack, which came in two waves, ended at approximately 0945 hours. Official U.S. records list Japanese losses as 49 aircraft and five midget submarines. Japanese records show only 28 aircraft lost.

The departing enemy left behind 2400 American dead, 1500 wounded, four battleships sunk, 14 ships heavily damaged, and 233 U.S. warplanes destroyed. Fortunately, no aircraft carriers were anchored at Pearl. The *Lexington* was about 425 miles south of Midway, toward which she was headed to deliver a Marine scout bombing squadron, and *Enterprise* was about 200 miles west of Pearl Harbor, returning from Wake Island after delivering a Marine fighter squadron there. According to Navy records, *Enterprise*'s Scouting Squadron Six, launched earlier that morning to land at Ewa Airfield, arrived during the attack and "engaged enemy aircraft."

Word of the surprise attack on Pearl Harbor was flashed to the world by Navy radio 34 minutes after the first bombs fell, or at 0829 hours Honolulu time. It was then 0259 hours, 8 December, in the Phillipines, where Adm. Hart's radioman picked up the stunning message. Hart immediately alerted the headquarters of Gen. Douglas MacArthur in Manila, and British, Dutch, and Chinese forces in the area. Adm. Thomas Hart's Asiatic Fleet consisted of three cruisers, 13 old destroyers, six motor torpedo boats, 32 patrol bombers, and 29 submarines.

The Navy signal from Hawaii was read in Washington about 1 p.m. or 1300 hours, and the President asked Congress for a declaration of war against Japan on the following day. Japan's Axis partners, Germany and Italy, then declared war on the U.S. and America was embarked on an all-out two-ocean war.

Meanwhile, about ten hours after the Japanese attacked Pearl Harbor, or 1220 hours Manila time (official records do not agree as to the exact time), the Japanese struck the Army airfields in the Manila area with 108 bombers and 84 Zero-sens. The result was similar to the debacle in Hawaii.

The Far East Air Force (FEAF) officially contained 162 fighter airplanes and 35 heavy bombers, including the 6th Philippine Pursuit Squadron's eleven obsolete Boeing P-26As. That total also included 46 Seversky P-35s, eleven of which were out of commission (the flyable ones were unfit for combat against the Japanese), and 107 Curtiss P-40s, 35 of which were grounded for maintenance or lack of parts. The Japanese destroyed 31 of the 72 flyable P-40s and 16 of the 17 B-17s in their attacks on Clark and Iba Fields.

The next day, Nichols and Del Carmen Fields were hit; the Navy base at Cavite was destroyed on the 10th, and enemy fighters occupied Vigan Field on Luzon's north coast as the Nipponese came ashore in force to begin their march on Manila.

At the end of a week, the 24th Pursuit Group had but 24 flyable P-40s remaining, and that number steadily dwindled as the Americans, flying four and six-plane missions, were daily confronted by enemy formations of up to 30 aircraft.

Although the official Army history claims that serious efforts were made to resupply the 15,000 Americans and 100,000 Filipino patriots resisting the Japanese invasion, the air units received nothing—not even mail—from the few submarines that delivered token help to the Army troops (and one Marine regiment). Three P-40Es were sent to Mindanao by coastal steamer from Australia in March 1942, and that was the only help the men of the 24th Pursuit Group and 19th Bomb Group *ever* received.

By 7 January 1942 MacArthur's forces were backed up on the Bataan Peninsula and on Corregidor and its satellite fortresses guarding Manila Bay. There the "Battling Bastards of Bataan" fought so fiercely that enemy attacks ceased from mid-February until April while the invader brought in additional forces—eventually, a total of six divisions.

The pilots of the 24th PG flew their remaining P-40s from narrow dirt strips, bombing and strafing the enemy, until Bataan fell on 9 April, and the last two P-40s were flown 500 miles south to Del Monte Airfield on Mindanao. Gen. Lewis Brereton, who had assumed command of the FEAF the day before the Phillipines were attacked, had previously taken the surviving B-17s to Australia. The P-40 pilots who were left also escaped to Australia (more P-40s were lost to landing accidents on the tiny, improvised landing strips, and to lack of spare parts, than to enemy action). Gen. MacArthur and his family, by order of the President, had escaped to Australia in late December.

Under the command of Gen. Johnathan Wainwright, the American forces on Bataan retreated to the caves of Corrigedor, where they held out until 6 May. Then, out of food and ammunition, they had no option but to surrender. They had been doomed from the beginning, and everyone knew that—except the Battling Bastards of Bataan.

For a time it seemed that nothing could stop the Japanese. In less than three weeks after Pearl Harbor, the isolated American outposts on Wake and Guam fell, the British garrison at Hong Kong was overwhelmed, and powerful Japanese forces were converging on Malaya and the Netherlands Indies. Singapore and its 80,000 British troops were captured 15 February 1942; Adm. Hart lost all but four destroyers of his small Asiatic Fleet, while two of Britain's proudest ships, *Prince of Wales* and *Repulse* were sunk by enemy air attacks. By May, the Japanese were in control of Burma, Malaya, Thailand, French Indochina (Vietnam), and the Malaya Archipelago, and farther to the east had won strong positions on the islands of New Guinea, New Britain, and in the Solomons, flanking the approaches to Australia and New Zealand from the U.S.

Coral Sea and Solomons

America's goal in the Pacific during 1942 was to build up a base in Australia and secure the chain of islands leading to it. In January, a task force of division size was hastily assembled and sent to New Caledonia to guard its eastern approaches. In March and April, two additional U.S. infantry divisions, the 41st and 32nd, went to the Southwest Pacific. After the western anchor of the stepping-stone island chain was secured at New Caledonia, Army and Marine garrisons were sent to various other islands along the line, culminating with the arrival of the 37th Division in the Fiji Islands in June. Establishment of this critical lifeline to Australia may not have been possible had the Japanese timetable of conquest not been so seriously delayed by the stubborn defense of the Philippines.

In any case, these moves came none too soon because during the spring the enemy occupied Rabul, pushed into the southern Solomons, and established himself on the northeast coast of New Guinea—just across the narrow Papuan peninsula from Port Moresby. The stage was thus set for a major test of strength in the Pacific, with American forces thinly spread along an immense arc from Hawaii to Australia, and the Japanese securely in possession of vast areas north and west of the arc, prepared to strike in force at any point.

The first test came in May, when the Japanese attempted to take Port Moresby, the obvious staging base for an invasion of Australia. This resulted in the Battle of the Coral Sea, which lasted from the 4th through the 8th of May, the first naval engagement in history fought entirely with aircraft, the opposing ships never approaching within sight of one another.

U.S. Navy Task Force 17, commanded by Rear Adm. F. J. Fletcher, with the carrier *Yorktown* bombed enemy transports landing troops in Tulagi Harbor, then, joined by other Allied naval units including Rear Adm. A.W. Fitch's Task Force 11 with the carrier *Lexington* south of the Louisiades, stationed an attack group in the probable track of the enemy transports and moved north in search of the Japanese covering force. Carrier aircraft located and sank the Japanese escort carrier *Shoho* covering the convoy, while enemy carrier aircraft sank one U.S. destroyer and one fleet tanker. On the final day of the action, the enemy covering force was located and taken under an attack that damaged but did not sink the carrier *Shokaku*. Almost simultaneously, enemy carrier airplanes attacked Task Force 17, damaging the *Yorktown*, and then scored hits on the *Lexington* that set off uncontrollable fires which spelled her doom. The air battles attendant to the action took an unusually heavy toll of enemy aircraft and pilots, a fact more significant than realized at the time. The Japanese naval aviators of this period were well-trained and highly experienced. They would not soon be replaced.

Although the *Lexington* was a poor trade for the little *Shoho*, Port Moresby was denied the enemy, along with his only opportunity to invade Australia.

The Doolittle Raid

The Navy tells it this way: "From a position at sea 668 miles from Tokyo, the Carrier *Hornet* launched 16 B-25s of the 17th Bomb Group led by Lt. Col. J.H. Doolittle, USA, for the first attack on the Japanese homeland. *Hornet* sortied from Alameda 2 April, made rendezvous with *Enterprise* and other ships of Task Force 16 [Vice Adm. W.F. Halsey] north of the Hawaiian Islands, and proceeded across the Pacific to the launching point without making port."

That is, of course, accurate as far as it goes. Actually, Doolittle's raiders were forced to take off prematurely when the task force was spotted by a Japanese fishing boat. Halsey was unwilling to risk so much so deep in hostile waters. That meant that the radio beacons which were to guide the B-25s to safe landings in unoccupied China were not in operation. A storm over the planned landing area in China added to the problems of the B-25 crews, and

347

all except one crew either crash-landed in the water near the coast or bailed out.

Eight of the eighty fliers were captured by the Japanese; three were executed and one died in prison. One crew landed in Russia and was interned. Most of the rest were rescued by friendly Chinese and returned to the U.S.

The Doolittle raiders inflicted only token damage on Tokyo, but their attack, 18 April 1942, came at a time when all other war news for America was bad, and it provided a tremendous morale boost for civilians and the military alike. It also induced the Japanese to divert ships and planes from combat areas to guard the home islands.

The Battle of Midway

Several months prior to the attack on Pearl Harbor, U.S. Navy cryptographers had broken the Japanese diplomatic code, and since it was one of the family of codes used by the Japanese, Adm. Chester Nimitz, Commander of the Pacific Ocean Area, was soon reading portions of the Imperial Japanese Navy's messages. That, in turn, allowed Nimitz to plan an ambush of the Japanese Navy of truly classic proportions, as bold as it was brilliant, because the American force would be greatly outgunned.

The Japanese armada, comprising a total of 185 ships and 450 aircraft, was commanded by Adm. Isoroku Yamamoto and split into four forces: Two aircraft carriers with their support ships would steam north to attack the U.S. Navy base at Dutch Harbor, deliver occupation troops to Kiska and Attu in the Aleutians, and also serve as a diversionary force to lure the U.S. Navy into battle, while far to the south, Yamamoto's Midway Invasion Force—consisting of one carrier, troop transports, and escorting warships—approached Midway. In between would be Midway Strike Force One under Adm. Chuichi Nagumo, containing four aircraft carriers—*Akagi, Kaga, Hiryu,* and *Soryu*—two battleships, three battle cruisers, and twelve destroyers. Finally, Yamamoto himself would loiter about 300 miles behind Nagumo with one carrier, three battleships, and a host of escort vessels.

Waiting in the bushes some 350 miles northeast of Midway were U.S. Navy Task Forces 16 and 17, commanded by Rear Adm. Raymond Spruance and Frank Fletcher, respectively (with Fletcher in overall command), containing the carriers *Enterprise, Hornet* and *Yorktown,* along with a total of eight cruisers and 14 destroyers.

Planes from Midway located and attacked the Midway Invasion Force on 3 June 1942, when it was about 600 miles from Midway. When Nagumo's Strike Force was located the next day, planes from Midway joined the carrier-base aircraft laying in wait for Nagumo to attack his four-carrier force at 0930 hours on 4 June. These units suffered heavy losses. Only six of 41 TBD Devastator torpedo planes from the three U.S. carriers survived, but their determined attacks drew the enemy's combat air patrols down to the surface, expended their fuel and ammunition, and allowed the 51 Dauntlesses from *Enterprise* and *Yorktown* to dive-bomb unmolested by enemy fighters. That cost the Japanese the carriers *Akagi, Kaga,* and *Soryu.*

Later that afternoon, U.S. carrier aircraft struck Nagumo's force again, sinking his fourth carrier, *Hiryu.* Meanwhile, the enemy found and damaged the *Yorktown.* It was later sunk, along with a destroyer, by a Japanese submarine.

Japanese losses at Midway totaled two heavy and two light carriers, one heavy cruiser (*Mikuma*) sunk and another (*Mogami*) severely damaged, plus 258 airplanes and a large percentage of their experienced carrier pilots.

U.S. losses were 40 shore-based and 92 carrier aircraft, in addition to the *Yorktown* and the destroyer *Hammann.* This decisive defeat administered to the Japanese put an end to their dominance of the Pacific Ocean. The enemy had lost five of her nine carriers within a month.

Guadalcanal

The battle for Guadalcanal in the Solomons lasted from 7 August 1942 to 9 February 1943; air support for the U.S. Marines' first amphibious landing of WWII was provided by three carriers and Navy, Marine, and Army Air Forces airplanes operating from bases on New Caledonia and in the

New Hebrides. The carriers withdrew on 9 August but remained in the area to give overall support to the campaign, during which they participated in several of the naval engagements fought over the island.

Saratoga's air groups sank the Japanese light carrier *Ryujo* in the Battle of the Eastern Solomons fought 23-25 August. On 24 August, *Enterprise* was hit by carrier-based bombers and forced to retire for repairs. On 31 August, *Saratoga* was damaged by submarine torpedo and retired for repairs. Then, on 15 September, *Wasp* was sunk by a submarine while escorting a troop convoy to Guadalcanal.

Hornet, in Task Group 17, commanded by Rear Adm. G.D. Murray, hit targets in the Buin-Tonolei-Faisi area on 5 October, attacked enemy transports and supply dumps on Guadalcanal, and fought the battle of Santa Cruz 26-27 October in which she was sunk by air attack. In the final carrier actions of the campaign, *Enterprise* participated in the Battle of Guadalcanal 12-15 November and the Battle of Rennell Island 29-30 January 1943.

Ashore, Army, Navy, and Marine air provided direct support. Marine Fighter Squadron 223 and Scout Bombing Squadron 232, (delivered by the first of the new escort carriers, *Long Island*), began operations from Henderson Field on Guadalcanal on 20 August while fellow Marines guarded the field perimeter against enemy soldiers in the nearby jungle. Air Force and Navy aircraft arrived on the field within a week. The Army's 339th Squadron of the 347th Fighter Group (FG), originally equipped with P-39s, received P-38s in November 1942, and shortly afterwards a second squadron, the 70th, was Lightning-equipped. The P-38 Lightning was popular with most fighter pilots in the Southwest Pacific, primarily because of its two engines "over all that water." America's top ace in WWII, Richard Bong, flew the P-38 there while amassing most of his 40 victories (as did the number two ace, Thomas McGuire, 38 victories).

The 44th Fighter Squadron (FS) of the 15th FG arrived on Guadalcanal in January 1943, flying P-40 Warhawks to share Fighter One Base at Lunga Beach with Marine Fighting Squadron VMF 214, the famous "Black Sheep," flying F4U Corsairs.

Guadalcanal was declared secure on 9 February 1943, and a little over two months later—on 18 April—14 Lightnings of the 347th FG intercepted a pair of Japanese Betty bombers near Ballale, Bougainville, and shot them down, killing Adm. Yamamoto. Four P-38s went after the Bettys while the others dealt with the escorting Zeros. Capt. Thomas Lanphier was officially credited with downing the Betty carrying Japan's top military strategist. Decoding of an enemy message made the interception possible.

The U.S. 13th Air Force was formed in New Caledonia in June 1943 and USAAF planes in the Solomons were incorporated into it under the overall command of Adm. "Bull" Halsey.

Gen. Kenney's 5th AF was in New Guinea, operating from airstrips around Port Moresby with American and Australian air units, and was part of Gen. MacArthur's command. After the big Japanese air and naval base at Rabaul was neutralized by U.S. airpower in November 1943, the 13th and 5th Air Forces were united as the reborn Far East Air Force in preparation for America's return to the Philippines.

Meanwhile, the U.S. Fleet was gaining strength daily as a virtual flood of escort carriers joined the fray. Originally called AVGs, then ACFs, the Jeep carriers—flight decks built atop a variety of hulls such as former merchantmen—were finally designated as CVEs in mid-1943. The big fleet carriers, with the designation CVS, also saw their ranks swelled as shipyards back home expanded along all of America's coasts, working 24-hour days and seven-day weeks. Between 7 December 1941 and the day the Japanese surrendered—14 August 1945—30 new fleet (attack) carriers were commissioned, along with 75 escort carriers, with 35 of the latter going to Great Britain under the terms of Lend-Lease.

North Africa

The invasion of northwest Africa on 8 November 1942 by U.S. forces under the command of Gen. Dwight Eisenhower was Churchill's idea. It was obvious that the Allies would not be strong enough to invade Europe before 1944. In the meantime, if Russia fell, as seemed entirely possible, Hitler's Europe could prove well-nigh impregnable by then. A second front would stiffen Russia's resolve, perhaps take some German troops out of Russia, and at the same time relieve the threat that Field Marshal Erwin Rommel's German-Italian forces posed to Britain's supply of Mideast oil.

The Desert War had see-sawed back and forth across North Africa for more than two years, and in July 1942, when Roosevelt and Churchill agreed upon the North African invasion—Operation Torch—Rommel had pushed the British 8th Army into Egypt, less than 70 miles from Alexandria.

Eisenhower insisted on landing on the Atlantic coast of French Morocco. That left him nearly a thousand miles from Tunis, where he hoped to be by Christmas, but he feared that a landing further east on the Mediterranean coast could cut him off from supply if the Germans sealed the Straits of Gibraltar. He also hoped that the Vichy French would refuse to fight the Americans on behalf of their German conquerors.

Meanwhile, just five days before the invasion, Gen. Bernard Montgomery's British 8th Army broke out of its defensive position at El Alamein in Egypt and, moving swiftly beneath skies controlled by Air Marshal A.W. Tedder's Desert Air Force (which contained the newly-formed U.S. 9th AF), was pursuing Rommel into Libya.

U.S. Navy carrier aircraft from the *Ranger* and escort carriers *Sangamon, Suwannee,* and *Santee* of Task Group 34.2, commanded by Rear Adm. E.D. McWhorter, of the Western Naval Task Force, covered the landings of Army troops near Casablanca, Oran, and Mehedia. There was some resistance from the confused French, and at least one large air battle was fought between Navy Wildcats and French pilots flying the export version of the Curtiss P-36 (an action which the official records almost totally ignore). The French had small inclination to fight Americans, however, and all threw down their arms within three days.

On that date—11 November—the first P-38 Lightning mission was flown from African soil by pilots of the 48th FS, 14th FG. Two P-38 groups, the 1st and 14th, had flown from Land's End in England with a stop at Gibralter. The 48th's aircraft were on the field at Tafaraoui on the 11th, the 49th on the 18th. The 1st FG followed on the 20th.

The 33rd FG's 78 P-40 Warhawks were flown from the deck of the escort carrier *Chenango* on the 10th to operate from a field at Port Lyautey; on the 13th, Navy Patrol Squadron 73 arrived at Port Lyautey from Iceland to begin anti-submarine operations over the western Mediterranean. In addition, B-25 Mitchells of the 12th and 17th BGs, A-20s of the 47th BG, and the Spitfire-equipped 31st FG were in action before Christmas as part of Gen. Jimmy Doolittle's hastily-organized 12 AF. The 325th FG, originally equipped with P-40s, would follow in April, 1943.

Meanwhile, the 79th, 324th, and veteran 57th Warhawk groups in the 9th AF were fighting with Tedder's Desert Air Force, leading British and Commonwealth troops across Libya to meet Eisenhower's forces in Tunisia—with Rommel squeezed between them.

Although Marshal Rommel was poorly supplied, there was a lot of fight left in the Desert Fox and his Afrika Korps. The key to his defeat was successful interdiction of his supply line across the Mediterranean. And the key to *that*, of course, was Allied airpower.

The American drive toward Tunis became stalled in the mud of seasonal rains due to the Germans' quick reaction to Torch, spotty resupply (almost everything Eisenhower needed had to come all the way from the U.S.) and, at first, inefficient use of the 12th AF. Apparently, Eisenhower did not trust Doolittle's judgment, and instead of allowing Doolittle to concentrate the American fighter strength so that it could gain control of the air over northwest Africa, the Lightnings and Warhawks were thinly spread to serve the immediate needs of local ground commanders. With his fighter strength being frittered away piecemeal, Eisenhower recognized that he was doing something wrong and sent to England for Gen. Carl "Tooey" Spaatz.

By February 1943, Spaatz had brought all U.S. air units in Africa into the 12th AF, including two additional fighter groups, the 81st and 82nd (P-47s and P-38s), along with the 97th, 99th, and 301st BGs, and merged the 12th AF with the British air units in the area into the Northwest African Air Forces, commanded by Gen. Spaatz but part of Air Chief Marshal Tedder's Mediterranean Air Force. With a unified chain of command, consisting of air officers, Allied airpower in the Mediterranean Theater of Operations (MTO) soon took charge of the air.

The Palm Sunday Massacre

On 7 April 1943 the British 8th Army linked up with the Americans in Central Tunisia, and closed the jaws of their giant pincer to surround Rommel's forces and back him against the sea on the Cape Bon peninsula in northeastern Tunisia. Since the British Navy had long since denied Rommel meaningful supply by sea, the Germans had assembled a large fleet of transport aircraft for that purpose but it was bared to the Allied air forces when the Luftwaffe was defeated over North Africa, and would be almost completely destroyed in a single blow.

It happened on Palm Sunday, 18 April 1943, and that action has been known ever since as the Palm Sunday Massacre. All that day, NAAF fighters had been patrolling the sea watching for the German transports. Then, at 1650 hours, three squadrons of the 57th FG (64th, 65th, 66th)—34 Warhawks in all, along with 12 Warhawks from the 324th FG's 314th Sqdn—took over the patrol.

The Warhawks, flying from their base at El Djem, rendezvoused over Hergla with 12 RAF Spitfires that would provide top cover while the P-40s stayed low, flying four abreast, and stair-stepped between 4000 and 12,000 feet. Near dusk, the P-40s found "100-plus" trimotor transports—some Savois, but mostly Ju 52s—just off the water about six miles west of Cape Bon, escorted by "30-plus" Me 109 Messerschmitts and Macchi Mc 202s. The Warhawks pounced on them and, in a ten-minute battle, shot down 58 Ju 52s, 14 Mc202s, four Me 109s, two Me 110s, and damaged 29 other aircraft. Two Warhawk pilots each got five of the Junkers, and eight other Warhawk pilots each destroyed three enemy transports. Six Warhawks were lost.

The enemy made several more attempts to slip 20-plane formations into Tunis, but lost twelve of them on the 19th and all 20 of them, along with ten fighters, on the 22nd. When the Germans tried to slip in some transports at night, Desert Air Force night Beaufighters were waiting, and Rommel's last hope for supply by air was gone.

By early May, the stubborn enemy was forced from the Tunisian hills, and on 13 May 1943, 270,000 German and Italian troops pinned on the Cape Bon Peninsula surrendered. Field Marshal Rommel escaped to Italy.

The Italian Campaign

At the end of the North African Campaign, the U.S. 12th AF contained (along with the previously mentioned air groups) the 47th, 48th, 310th, 319th, and 320th BGs, flying B-17s, B-24s, and B-25s. Gen. Lewis Brereton's 9th AF had four Warhawk groups, the 57th, 79th, 324th, and 325th; two medium BGs, the 12th and 322nd; and two heavy BGs, the 98th and 376th. Altogether, the 9th and 12th Air Forces could count slightly over 2600 aircraft in the MTO, including transports and photo-reconnaissance machines. The RAF had slightly over 1000 aircraft in the Desert Air Force, and the Free French 94, mostly Warhawks supplied by the U.S.

With North Africa secured, the Allies turned their attention northward to what Prime Minister Churchill had referred to as "Europe's soft underbelly," Italy, which would be invaded by way of Sicily. As a prelude, the Italian-held island of Pantelleria, 50 miles off the North African coast, was neutralized by airpower, including the all-black 99th FS (which had no parent group at the time) flying P-40s. Pantelleria was the first bit of enemy territory in history to surrender to the U.S. Air Forces with no surface forces moving against it.

The Sicilian Campaign lasted five weeks, 10 July to 17 August 1943. Almost 3000 Allied ships put ashore the British 8th Army and the U.S. 7th Army at widely separated points, while the NAAF and the Desert Air Force, ranging as far as Naples, maintained complete control of the air.

To speed reinforcement, the Allies on two successive nights flew in American and British paratroopers. On both nights the Navy gunners on U.S. ships standing offshore shot down a number of the C-47s and the troop gliders the Gooney Birds were towing. The half-dozen aircraft that went down under "friendly" gunfire on the first night could have been charged to whatever foul-up was claimed, and then forgiven. There was *no* defense for the downing of eight more on the following night. (The identity of the Navy officer responsible was never revealed. Later, in 1944, U.S. Army antiaircraft batteries in Luxembourg, under the command of Gen. Omar Bradley, shot down so many British and American airplanes that the antiaircraft units were taken from Bradley's control and placed under the command of Air Forces Gen. Hoyt Vandenberg.)

Even as the Allies had been preparing to invade Sicily, the Italian people and their government had become increasingly disenchanted with the war. Under the impact of the loss of North Africa, the invasion of Sicily, and the first bombing of Rome, the Italian king forced Dictator Benito Mussolini to resign. The Italians wanted out of the war, but were virtual prisoners of the German forces in Italy. The Italian government nevertheless agreed to surrender, a fact Gen. Eisenhower announced on the eve of the Allied landings in Italy. Interestingly, Italian fighter pilots who remained sympathetic to the Axis cause were incorporated into a Luftwaffe fighter squadron (II/JG 77) with their Macchi C205s to form Italian Fighter Group. Since another group of Italian pilots had flown their airplanes to Sicily to join the Americans when Italy's surrender was announced, the possibility of Italian vs. Italian, flying identical fighter aircraft, thus existed. The Free Italians flew their first mission on 16 October 1943 when they served as top cover for 82nd FG Lightnings dive-bombing German shipping in Levkas Channel.

The Allied invasion of Italy began on 3 September 1943 and German forces in Italy under Rommel and Kesselring occupied Rome, digging in north of a strong defensive line that roughly divided the country in half. The ground war in Italy was stalemated throughout the winter, although Allied airpower gained a tenuous control of the air by 1 October.

At the end of 1943, Gen. Spaatz was named overall commander of the U.S. Strategic Air Forces in Europe, and Gen. Doolittle took over the 8th AF from Gen. Ira Eaker, who in turn was promoted to boss of the (renamed) Mediterranean Allied Air Forces (MAAF). MAAF contained, in addition to RAF units, Gen. Cannon's 12th AF and Gen. Twining's 15th AF. The 12th relinquished bomber groups to Gen. Nathan Twining's new 15th AF, formed in Italy 1 November 1943, and the 12th AF became strictly a tactical air force. The 9th AF, by then under command Gen. Hoyt Vandenberg, and which had fought beside the 12th from Africa to Italy, was moved to England in October 1943 in anticipation of the invasion of Europe, then planned for 1 May 1944. Gen. Eisenhower left Italy in January 1944, to go to England, and was succeeded by Britain's Field Marshal Sir Henry Wilson.

During February 1944, the 15th AF in Italy and the 8th AF in England began to coordinate mass air attacks deeper and deeper into Germany. The ground war in Italy was bogged down with the Germans in strong defensive positions in mountainous terrain while the 12th AF fighter-bombers steadily eroded the enemy's supply lines. The P-38 groups were transferred to the 15th AF to provide long-range escort for the heavies going to Germany.

The German positions were not breached until May 1944, but then the enemy withdrew to the Apennines in northern Italy and remained there until his lines of supply were again severed by the 12th AF, and one million hungry Germans, lacking the means to further resist, surrendered just a few days before the war ended in Europe.

It is not clear which side could claim a strategic victory in Italy—the Allies, who pinned down such a sizable enemy force that otherwise might have been freed to bolster German resistance to the Allied invasion of Europe, or the Germans, who kept almost as many Allied troops engaged in Italy and similarly diverted from the showdown battles in

France and Germany. In any case, Europe's underbelly did not prove to be very soft.

Cross-Channel Attack

The combined strategic air offensive against Germany began early in 1944, and by late March Allied airpower was dominant over Europe, air superiority having largely been won by the 22 P-47 Thunderbolt groups in the 8th and 9th AFs. (It may be true that the P-51 Mustang was the best air superiority fighter of the war, but the first Mustangs did not reach England until late in 1943, and they were not in action in large numbers until after the showdown air battles of February and March 1944. And while Mustangs were the first U.S. fighters to escort bombers over Berlin [4 March 1944], P-38s of the 20th and 55th Fighter Groups were actually over Berlin the day before to rendezvous with heavies that were recalled due to weather.)

The essential prelude to the cross-channel invasion of Europe from England was the strategic air offensive agreed upon by President Roosevelt and Prime Minister Churchill at a conference in Casablanca early in 1943. The Americans would bomb Germany by day, the British by night, striking at the enemy's industrial base. The buildup of crews and machines had been slowed by the requirements of the North African and Italian Campaigns, as well as the demands of the Pacific War, but the air offensive that began early in 1944 would not abate until the enemy capitulated on 7 May 1945.

The invasion began at 0145 hours 6 June 1944, when 9000 paratroopers of the U.S. 82nd and 101st Airborne Divisions began plunging from some 800 9th AF C-47s over France's Cherbourg Peninsula inland from the Normandy beaches. Concurrently, 200 RAF transports dropped 5000 British paratroopers, while 200 gliders followed with heavy weapons, guided to open fields by flare paths set up by the paratroopers. The paratroopers' job was to seize key bridges and otherwise secure routes of egress for the seaborne forces that began coming ashore at 0630, just after sunrise. With daylight, too, came tactical air, P-38s stooging high above the landing sites to keep the air above the beaches free of enemy aircraft, while other fighters ranged inland at low altitude to paralyze all enemy troop movements and prevent reinforcement and supply to resisting German forces in the area. General Eisenhower had learned his lessons well in North Africa and Italy, and the role of U.S. tactical aircraft in the invasion was performed to textbook perfection.

Later, advancing across France, Gen. George Patton would add a chapter to that book when, on 25 July, he sent his 3rd Army through a breach blasted in enemy defenses along the St. Lo-Periers Road by tactical aircraft (with some aid from 8th AF B-17s), and raced ahead, depending *entirely* on U.S. tactical fighters and medium bombers to protect his exposed southern flank—a mission the fliers performed so well that 20,000 German troops surrendered *directly* to the Air Force without ever engaging the 3rd Army.

Victory in Europe

A week after D-Day (the date of the Normandy landings; the operation itself was code-named Overlord), Hitler began sending his *Vergeltungswaffe*—vengeance weapons—against London. The first of these was the V-1 buzz bomb, a small, unmanned, gyro-stabilized aircraft powered with a pulse-jet engine and carrying a ton of explosives. The buzz bombs had a range of 150 miles at a speed of 400 mph. Eight thousand of them were fired, first at London, then at Antwerp, from bases in France and Holland. About 4000 of the V-1s were intercepted—1847 by fighter planes directed by radar, 1866 by anticraft artillery, and 244 flew into barrage balloon cables around London before Allied bombers snuffed out V-1 production and destroyed the flying bombs' launching sites. By September 1944, the V-1 was no longer a serious threat. However, Hitler had another, more terrifying weapon, the V-2.

The V-2 was first fired against London and liberated Paris on 8 September 1944. It was a short-range ballistic missile, fueled with liquid oxygen and alcohol, that reached a speed of 3600 mph and a peak altitude of 60 miles. It carried slightly less than a ton of explosive in its warhead, and had a maximum range of 220 miles. Like the Me 262 jet fighter, the V-2 did not receive Hitler's backing during its development stages, and it appeared too late to change the course of the war. Nevertheless, the 2500 buzz bombs and perhaps 100 V-2s that struck England killed almost 10,000 civilians and destroyed 20,000 buildings.

Meanwhile, American strategic bombers systematically eliminated the enemy's oil industry. By the spring of 1945, gasoline production in Hitler's Europe had dropped to seven percent of normal capacity and the Nazi war machine became almost immobile. The mechanized German Army was afoot; panzer units were crippled, and new fighter airplanes produced with supreme effort in underground plants were grounded with empty tanks.

Allied ground forces, moving swiftly beneath skies completely controlled by their own aircraft, swept over the Reich to meet Russian forces at the Elbe River on 25 April 1945. Five days later, Adolf Hitler committed suicide amid the rubble of Berlin. A week after that a prostrate Germany, her cities in ruin, surrendered.

The American advance had been halted by Eisenhower short of Berlin because U.S. forces had outrun their lines of supply. The Russian advance across Poland and into Germany was owed in no small way to $11 billion in American Lend-Lease materials, including 400,000 Jeeps and trucks, 12,000 armored vehicles (including 7000 tanks, enough to equip 20 U.S. armored divisions)—1.75 million tons of food, and 14,000 aircraft, all of it delivered at great cost in ships and men through U-boat infested waters.

At the end of the fighting in Europe the Allies had 4.5 million men in nine armies (five of them American) six tactical air commands (four American), and two strategic air forces (one American). The Allies had 28,000 combat aircraft, of which 14,845 were American.

From the time of the Normandy landings on 6 June 1944, the U.S. lost 135,576 dead in Western Europe; the other Allies, approximately 60,000. The Germans lost 263,000 killed in the west during that same period, and more than 3 million dead throughout five years of war.

The Allies had dropped 2,755,540 tons of bombs on Hitler's Fortress Europe during five years of war. Of that total, U.S. bombers were responsible for 1,500,000 tons, mostly during the final ten months of fighting.

China-Burma-India

On 19 April 1942, Army Air Forces Colonel Caleb V. Haynes arrived in India in a B-24 Liberator, leading a flight of 13 B-17s and a C-47 that he had brought across the South Atlantic. Haynes was carrying secret orders to bomb Tokyo from a base in China, but the Doolittle raiders had struck Tokyo the day before, and the plan to bomb the enemy's principal city a second time was dropped.

General Lewis Brereton, in India to form the U.S. 10th AF from whatever he could lay his hands on, grabbed Haynes' bombers and then sent Haynes to Dinjan (in the Assam Valley near India's northeastern border with Burma) to organize the Assam-Burma-China Ferrying Command, an airlift operation that would supply fuel and ammunition to Col. Claire Chennault's American Volunteer Group (AVG), popularly known as the Flying Tigers. Haynes was given two C-47s for the job, and promised more. His pilots were a mixed group of USAAF pilots with "midnight" order changes and some Pan American Airways personnel (Pan Am was part owner of China National Airways).

These were the first "Hump" pilots, the ones who pioneered the high northern routes over the Himalayas for those who followed in ever-increasing numbers to supply not only the Flying Tigers, but the 14th AF that grew in their place, as well as Chinese forces fighting under Generalissimo Chiang Kai-shek. The purpose was to keep China in the war on the Allied side and thus tie down many thousands of Japanese, both soldiers and airmen.

The Japanese had cut the tortuous Burma Road that had been Chiang Kai-shek's main supply artery from India (both Burma and India were part of the British Empire until after WWII).

In April and May, the ABC Ferrying Command

was able to maintain a trickle of supplies to Chunking, and by Decmber, when it had received substantial numbers of Curtiss C-46s to greatly increase its effectiveness, it became a part of the newly-created Air Transport Command (ATC), or the Assam Trucking Company, as Hump crews called it. By mid-1945, the ATC was delivering 71,000 tons of supplies per month. A total of 1314 airmen died flying the Hump while taking more than three quarters of a million tons of cargo to China during 38 months of operations.

The Flying Tigers

The American Volunteer Group was put together during the summer of 1941 by retired Air Corps Capt. Claire Chennault with the consent of President Roosevelt. Chennault had been in China since 1937 trying to organize an air force for Chiang Kai-shek to counter Japanese air attacks. In desperation, Chennault returned to the U.S. in December 1940 and, with the support of Navy Secretary Frank Knox, Secretary of Treasury Henry Morgenthau, and Presidential Advisor Thomas Corcoran, received Roosevelt's permission to recruit U.S. military reservists to form a fighter group that would fly under the flag of China. Their airplanes would be 100 Curtiss H81-A2s, the export version of the P-40B, which the British agreed to give up in exchange for a like number of later model P-40s.

The AVG contained pilots and support personnel from the Army, Navy, and Marines, as well as a few civilians. All resigned their military commissions in order to go to China, but a year later all were gratefully received back into the U.S. military forces. Chennault personally trained the AVG pilots in southern Burma during the summer and fall of 1941, schooling them in air combat tactics that took advantage of the P-40's best qualities while exploiting known weaknesses in enemy fighter aircraft. AVG pilots flew their first combat mission 10 December 1941, just two days after the attack on Pearl Harbor.

The Flying Tigers, apparently so named by a *Life* magazine correspondent because of the tigershark mouth painted on their airplanes' radiator scoops (copied from RAF 112 Sqdn then in North Africa, which had copied it from Luftwaffe II/ZG 76 Group), shot down 217 enemy aircraft and claimed another 43 as probables during their first ten weeks in action. They lost 16 airplanes and four pilots during that period.

In almost every action the Tigers were greatly outnumbered, but Chennault's knowledge of the enemy and the tactics he imparted to his pilots were factors that largely obviated the enemy's numerical superiority. Due to lack of supply, the AVG never possessed more than 55 flyable aircraft at any given time, while the Japanese had about 600 combat airplanes in the area.

The AVG flew from primitive airstrips in East-Central Burma and from fields near Kunming in Southwestern China. There were 305 of them, including mechanics and other support personnel, although 64 of them either quit and went home or were discharged by Chennault. Among the latter was Gregory "Pappy" Boyington, who later became the top Marine ace with 28 victories. The Tigers were in action less than eight months, but were officially credited with 297 confirmed air victories, 240 unconfirmed, plus 40 enemy aircraft destroyed on the ground. Four AVG pilots died in air combat; six were killed while strafing, and three were taken prisoner by the Japanese. Ten were killed by enemy bombs or in operational accidents.

Robert H. Neale was the top AVG ace with 16 victories. Many of the 26 AVG aces added to their scores after leaving the Tigers. Most rejoined the service with which they had previously flown.

The AVG was disbanded on 4 July 1942 and the USAAF 23rd FG born in its place. The 23rd FG, with a nucleus of former Tigers and at first commanded by ex-Tiger Bob Neale, was originally part of Chennault's rag-tag China Air Task Force, a stepchild of the U.S. 10th Air Force based in India. Chennault had accepted re-induction into the USAAF as a brigadier general; his CATF began operations with the poorly-equipped 23rd FG and a group of B-25s commanded by Caleb Haynes. Early in 1943, the CATF became the U.S. 14th AF, and it remained in China, slowly gaining new aircraft, until war's end.

Meanwhile, the U.S. 10th AF in India supported the Allied troops serving under Lord Mountbatten, which were fighting the Japanese in Burma. The 10th AF had seven combat groups: the 7th BG flying B-24s, the 12 and 341st BGs equipped with B-25s, and the 33rd, 51st, 80th, and 311th FGs, the first three of which were flying P-40s, while the 311th had P-51s.

Two orphan squadrons of P-38s, the 449th and 459th, fought in the China-Burma-India Theater. They were formed in mid-1943 in North Africa from the P-38 groups there after Rommel's defeat in Tunisia. The pilots were volunteers who agreed to fly their Lightnings to India and join the 10th AF. The 459th FS was given a home by the Warhawk-equipped 80th FG in Upper Assam, India. The 449th FS was gratefully accepted as the 23rd FG's fourth squadron in Kunming, China, which by then was part of the 14th AF.

The Aleutians

The 11th AF fought the "forgotten war" in the Aleutian Islands, where the principal enemy was *not* the Japanese. These crews saw a surprising amount of action, and suffered a high casualty rate, but more were lost to the world's worst weather than to enemy guns.

At the time of the Battle of Midway, the Japanese landed about 3000 troops on the Aleutian Islands of Attu and Kiska as part of their diversionary thrust into the North Pacific that was designed to draw the weakened U.S. Navy into a showdown fight it could not win. The U.S. Navy refused to take the bait, and dry-gulched the enemy at Midway instead. However, the Japanese troops could not be allowed to remain on American soil, posing a threat to Alaska and the North American Continent. Therefore, a squadron of P-38s, designated the 54th FS, was hastily put together to join the 11th and 18th Fighter Squadrons of the 28th Composite Group that had been in Alaska since the previous December.

The 11th and 18th, flying P-40s and P-39s, respectively, had enjoyed the company of RCAF Squadrons No. 14 and 111. The Canadians, flying Kittyhawks, returned to Canada when the P-38s arrived. A few B-24s and B-17s made up the rest of the 11th AF for the time being. In September 1942, the three fighter squadrons would be designated the 343rd FG, as more B-24s arrived, along with some B-25s. Later in the war, the group's P-40s and P-39s would be replaced with P-38s.

The P-38 was the only fighter for that part of the world for the same reasons that it was favored in the South Pacific: It was a long-range machine and had two engines. These were indeed attractive features over the inhospitable Aleutians, which stretched across the icy waters of the North Pacific for 1500 miles.

The P-40s had seen action, flying from Otter Point on Umnak Island in the Eastern Aleutians, when the Japanese attacked the naval base at Dutch Harbor concurrent with the Midway engagement. But the closest enemy occupation forces were on Kiska, some 850 miles west of Umnak, and Attu was 225 miles beyond Kiska. Both targets were beyond range of the P-40s and P-39s.

The invading Japanese were themselves a bit short of airpower; they had a dozen or so Rufe fighters (the A6M2-N Zero on floats), a few Kawasaki Ki-97 flying boats, and were evidently supplied by Betty bombers (Mitsubishi G4M1) flying from bases in the Kuriles, Japan's northernmost home islands.

The 11th AF B-24s and P-38s raided the enemy, weather permitting, while airstrips were built closer to his positions—first at Adak, 375 miles west of Umnak, and then at Amchitka, just 75 miles from Kiska.

On 11 May 1943 a U.S. Navy task force formed around the escort carrier *Nassau* and the battleship *Pennsylvania* supported the landing of the U.S. Army's 7th Division on Attu, and remained in the area until 30 May while the enemy was slowly ferreted out of strong defensive positions in the mountainous terrain. Air Force Col. W.O. Eareckson, an experienced Aleutian pilot, directed the 11th AF aircraft from a command post aboard the *Pennsylvania*.

Two months later, when American forces went ashore on Kiska, they discovered that the Japanese there had slipped away in the fog.

With the Aleutians secured, the Liberators and Mitchells of the 11th AF turned their attention to the enemy's bases in the Kuriles; the fighters moved westward to Shemya to stand guard—and sweat the weather.

Victory in the Pacific

MacArthur's return to the Phillippines was the payoff for a year-long combined offensive by Adm. Nimitz' Central Pacific forces that had captured bases from Tarawa through the Gilberts and Marshalls to the Marianas, and by Gen. MacArthur's South Pacific forces that had fought from Port Moresby and Hollandia to the island bases of Wakde and Morotai, just 250 miles from Mindanao. These two forces came together at Morotai for the invasion of the Phillippines at Leyte on 20 October 1944.

Throughout the twin campaigns, Nimitz' forces had been spearheaded by his carrier-based aircraft, along with the 7th AF and shore-based Marine air units joined by the 13th AF. MacArthur's forces had been able to move beneath Kenney's 5th AF, but as plans went forward to retake the Phillippines, the 7th and 13th Air Forces were merged with Kenney's 5th AF to achieve the rebirth of the U.S. Far East Air Forces.

The FEAF would clear the skies ahead of the U.S. 6th and 8th Armies as they fought their way northward to Manila, but as the invading armies were establishing themselves in the Leyte Gulf area against strong enemy resistance, the U.S. Navy was fighting the Battles of Leyte Gulf, a series of actions that lasted from 10 October to 30 November 1944.

The opening blow of the naval showdown was struck by Task Force 38 under Vice Adm. M.A. Mitscher when it raided enemy airfields on Okinawa and the Ryukyus. This force, built around 17 carriers, then hit airfields on northern Luzon and Formosa, destroying 438 aircraft in the air and 366 on the ground in five days. Between 18 and 23 October, the Army's amphibious landings were directly supported by Task Group 77.4, which had a core of 18 escort carriers and was commanded by Rear Adm. T.L. Sprague.

The Japanese decided to risk much of their seapower in an attempt to defeat the U.S. naval forces supporting the Allied landings in the Southern Phillippines. The enemy fleet, in three elements, was met by the U.S. 7th Fleet under the command of Vice Adm. T.C. Kinkaid, as well as Vice Adm. Halsey's Fast Carrier Task Force of the 3rd Fleet.

In separate but related actions, Halsey, on 24 October, traded the escort carrier *Princeton* for the 63,000-ton battleship *Musashi,* while on the following day Kinkaid accounted for two enemy battleships and three destroyers in what was called the Battle of Surigao Strait. That same day the third element of the enemy force engaged Sprague's T.G. 77.4 in a battle off Samar; the Japanese lost three heavy cruisers, but sank the escort USS *Gambier Bay* and three U.S. destroyers.

Meanwhile, the Fast Carrier Force was in action off Cape Engano in a major engagement that cost the enemy the heavy carrier *Zuikaku* and the light carriers *Chiyoda, Zuiho,* and *Chitose.* However, off Leyte, *kamikaze* pilots appeared for the first planned suicide attacks of the war to sink the escort carrier *St. Lo* and damage the *Sangamon, Suwannee, Santee, White Plains, Kalinin Bay,* and *Kitkun Bay.* Then, as remnants of the Japanese Fleet limped home through the Central Philippines, its air units gone, U.S. carrier-based planes sank a light cruiser and four destroyers on 26-27 October to bring enemy battle losses to 26 major combatant ships totalling over 300,000 tons.

The Fast Carrier Force continued to pound Japanese shipping and airfields in the Northern Philippines, and at month's end U.S. carrier-based aircraft could claim a total of 1046 enemy planes destroyed for that month. Task Force 38 would get 770 more Japanese aircraft in the actions of late October and through November, along with three heavy cruisers, four destroyers, and 20 merchant ships, while suffering *kamikaze* attacks that damaged the carriers *Intrepid, Franklin, Belleau Wood, Essex, Cabot,* and the new *Lexington.*

Luzon was invaded during the first three weeks in January 1945, and again, TG 77.4 and TF 38 were attacked by *kamikazes*; though suffering damage, they lost no ships, but accounted for 600 enemy aircraft and 325,000 tons of Japanese shipping.

While Allied troops were mopping up in the Philippines, Adm. Nimitz was ordered to take Iwo Jima, a volcanic island in the Bonins about 750 miles from Tokyo. Rear Adm. Durgin's TG 52, put together around one heavy and eleven escort carriers, and Mitscher's TF 58 with eleven heavy carriers and five escort carriers supported the operation. On 19 February 1945, two Marine divisions, assisted by minor Army units and later a third Marine division, went ashore to face approximately 23,000 Japanese troops manning perhaps the most impenetrable defensive positions in the Pacific. The island was secured on 16 March at the cost of 4590 Marine dead and the escort carrier *Bismark Sea* which was sunk by *kamikaze* attack. The new *Saratoga* was severely damaged and the *Lunga Point* lightly damaged by *kamikazes*.

The high cost of Iwo Jima was more or less justified during ensuing months, since possession of this base cut in half the distance to Tokyo from U.S. air bases in the Marianas, allowing P-51 fighter escorts for the B-29 Superfortresses pummeling Japan, and also provided a much-needed emergency airfield for damaged or fuel-starved B-29s returning to the Marianas from missions over the enemy's home islands. However, the Pentagon hyped the truth, claiming that no less than 2400 Superforts made emergency landings on Iwo Jima, and therefore 25,000 U.S. airmen were saved from ditching in the ocean as a result of the Marines' sacrifice. Actually, fewer than 1200 B-29s served in the Pacific.

The Okinawan Campaign, 18 March to 21 June 1945, was the last, and for naval forces the most violent of the major amphibious operations of WWII. The invasion of the Ryukyus began with seizure of the Keramas, just 15 miles west of Okinawa and about 400 miles southeast of Japan, when troops of the U.S. 10th Army went ashore supported by Durgin's TG 52. Then Okinawa was invaded on 1 April by four Army and two Marine divisions. But despite American air superiority, many *kamikazes* slipped through to hit U.S. warships. Both Durgin's 18-carrier TG, operating south of Okinawa, and Mitscher's TF 58, standing between Okinawa and Japan, were subjected to repeated mass *kamikaze* assaults, at times as many as 400 suicide planes at once. During the 83 days of the Okinawan Campaign, *kamikazes* sank 35 U.S. ships and damaged 288.

The island was at last secured at a cost of 12,500 dead and missing American; 110,071 of the enemy. The U.S. lost 763 aircraft, while Japan lost 1,733. Opposition from Japanese naval forces was brief. An enemy task force of one light cruiser, eight destroyers, and the world's largest battleship, the *Yamoto,* made what was to be the final effort by the Japanese Navy. Only four enemy destroyers survived attacks by American carrier-based aircraft. Of the several records for continuous operations in an active combat area, the most outstanding was logged by the *Essex* with 79 consecutive days.

At the end of the Okinawan Campaign, the Japanese Navy no longer existed. The air assault against the enemy's home islands would continue for another four months, and TS 38, then under Vice Adm. J.S. McCain, would send its carrier-based aircraft over Japan to attack airfields and other military targets. But the triumphant U.S. Navy may as well have gone home. It had done its job, and LeMay's B-29s would do theirs.

Superfortresses against Japan

The B-29 offensive against Japan was carried out by the 20th and 21st Bomber Commands operating as the 20th Air Force directly under the Joint Chiefs of Staff as a global air force. A false start was made in mid-June, 1944, when the 58th Bomb Wing of the 20th Bomber Command began operations from Chengtu in west-central China, about 125 miles northwest of Chungking. But everything needed to support the effort had to be flown over the Himalayas from India, and that foreclosed the possibility of a large-scale operation. By October, bases taken from the enemy in the Marianas were ready to handle B-29s and Superfort strength was concentrated there on five big airfields, two each on Guam and Tinian and one on Saipan.

B-29 raids against Japan resumed on 24 November 1944 from the Marianas, but that raid, by 111 Superforts, and those that followed during the winter proved relatively ineffective against the mass of lightly-constructed and widely dispersed small shops that fed component parts to the enemy's war machine, while too many B-29s were going into the Pacific out of fuel short of their base after bucking the prevailing westerlies 1500 miles to Tokyo. (Iwo Jima, halfway between the Marianas and Japan, became available as an emergency field in March 1945.)

In January 1945, Maj. Gen. Curtis E. LeMay was sent to the Pacific to take command of the Superforts. LeMay had been a flier since 1929; he had gone to England in 1942 as a colonel and boss of the 305th BG. When ordered to the Pacific, he commanded a bombardment division in Europe, had personally led many air strikes against Germany, and had gained the sobriquet of "Iron Butt" (the actual term was usually less delicate. It did not bother LeMay. "I'm not over here to win friends," he said, "I'm here to win a war").

By early March LeMay determined that conventional bombs, released at 30,000 feet, were not punishing the Japanese as they had done on more substantial targets in Germany. He ordered the B-29s loaded with incendiaries, stripped out their guns to save weight, and told his crews to bomb at night from 7000 feet. It seemed a bold gamble.

However, the Japanese were unprepared for such a tactic: Their large-bore antiaircraft guns were set to fire at targets five miles high, and they had few night-fighter aircraft.

LeMay's plan worked. Between the 9th and 19th of March, 1945, 1489 Superfort loads of incendiaries created fire storms that completely destroyed more than 30 square miles in the cities of Tokyo, Nagoya, Osaka, and Kobe. Only 21 Superforts were lost in those ten raids, primarily because of the extra fuel each carried in place of its defensive guns and ammunition.

The B-29s were temporarily diverted to missions over Okinawa, then returned to the firebombing of Japan. By the end of July they had destroyed more than 100 square miles of the enemy's largest cities, as well as portions of 60 lesser cities. All told, 2,333,000 structures were levelled, 241,000 people killed, and 313,000 injured.

At the beginning of August, enemy resistance was so feeble that LeMay announced in advance to the Japanese where he would strike next in order to cut civilian casualties, and urged his superiors in Washington to delay the planned invasion of Japan because the enemy could not possibly hold out much longer. He did not know that America possessed the A-bomb.

The Japanese did indeed offer to discuss surrender, but they chose to send the message through the Soviets (who remained at peace with Japan until after the first A-bomb was dropped, and then declared war in order to grab Manchuria, the Kuriles, and North Korea). The Soviets did not forward the peace proposal to President Truman as requested.

Whether or not the use of atomic weapons was justified in the Pacific War, President Truman did order their use because to shorten the conflict by a few weeks, or even a single day, would save American lives. Therefore, at 0245 hours 6 August 1945, the *Enola Gay*, a B-29 of the 509th Composite Group, 20th AF, took off from Tinian in the Marianas, accompanied by two sister ships—one to take photographs and one to record special instrument readings—and set course for Hiroshima with an atomic bomb in its bays. The *Enola Gay*, named for the mother of her commander, Col. Paul Tibbets, loosed her terrible weapon above the Japanese city at 0816 hours. Four square miles of Hiroshima were obliterated and, eventually, casualties were put at 78,150 dead, 13,083 missing, and 37,425 injured.

Three days later, the B-29 *Bock's Car* loosed the second A-bomb over Nagasaki, killing 73,884 people. That silenced Japan's small but powerful cadre of military die-hards who had appeared to favor national suicide rather than accept unconditional surrender, and the Japanese formally sued for peace 14 August 1945.

The surrender documents were signed aboard the battleship *Missouri* in Tokyo Bay 2 Sep 1945, a ceremony that would have more appropriately taken place on the flight deck of an aircraft carrier.

U.S. Army Air Forces Personnel and Direct Appropriations 1941-1945

1941	152,125	$ 2,173,608,961
1942	764,415	23,049,935,463
1943	2,197,114	11,317,416,790
1944	2,372,292	23,655,998,000
1945	2,282,259	1,610,717,000

The explosive dismantling of American airpower at the end of WWII is reflected in the personnel and money figures for 1946: 455,415; $517,100

Military Aircraft Production* Calendar Years 1941-1945

1941	19,445
1942	47,675
1943	85,433
1944	95,272
1945	46,865

---

*Total Army and Navy; includes Lend-Lease airplanes

U.S. Navy Combat Aircraft 1941-1945

Manufacturer	Designation	Contract	Procured
	Attack		
Douglas	AD series	1944	3,180
Martin	AM-1	1944	152
Douglas	BTD-1	1941	30
	Fighter		
Grumman	F6F	1941	12,275
Eastern (F4F)	FM-1, 2	1941	5,927
Grumman	F7F	1941	364
Grumman	F8F	1943	1,263
McDonnell	FH	1943	61
Ryan	FR	1943	69
	Patrol		
Lockheed	P-2	1943	1,036
North American	PBJ	1942	706
Lockheed	PBO	1941	20
Consolidated	PB4Y-1	1942	977
Consolidated	PB4Y-2	1943	739
Martin	P4M	1944	21
Lockheed	PV series	1942	2,162
	Observation		
Consolidated	OY-1	1943	306
Curtiss	SC	1943	577

Note: Many Navy/Marine combat aircraft of WWII were contracted prior to 1941. See lists at end of Chapter 12.

U.S. Army Air Forces Aircraft 1941-1945*

Model	Manufacturer	Cal Yr 1st Del	Number Procured	Engine/hp
Fighter				
P-35A	Republic	1941	60**	P&W 1200
P-36G	Curtiss	1942	30***	Wright 1200
XP-38A	Lockheed	1941	(1)	2 Allison 1150
P-38B, C	designations not used			
P-38D	Lockheed	1941	36	2 Allison 1150
P-38E	Lockheed	1941	210	2 Allison 1150
P-38F	Lockheed	1942	566	2 Allison 1325
P-38G	Lockheed	1942	1,242	2 Allison 1325
P-38H	Lockheed	1943	601	2 Allison 1425
P-38J	Lockheed	1943	3,170	2 Allison 1425
P-38K	Lockheed	1944	(1)	2 Allison 1425
P-38L-LO	Lockheed	1944	3,811	2 Allison 1475
P-38L-VN	Vultee	1945	113	2 Allison 1475
P-38M	Lockheed	1945	(75)	2 Allison 1475
P-39C (P-45)	Bell	1941	20	Allison 1150
P-39D	Bell	1941	863	Allison 1150
XP-39E	Bell	1942	3	Allison 1325
P-39F	Bell	1942	229	Allison 1150
P-39G and H	designations not used			
P-39J	Bell	1942	25	Allison 1100
P-39K	Bell	1943	210	Allison 1325
P-39L	Bell	1943	250	Allison 1325
P-39M	Bell	1943	240	Allison 1200
P-39N	Bell	1943	2,095	Allison 1200
P-39Q	Bell	1944	4,905	Allison 1200
P-40B	Curtiss	1941	131	Allison 1040
P-40C	Curtiss	1941	193	Allison 1040
P-40D	Curtiss	1941	22	Allison 1150
P-40E	Curtiss	1941	2,320	Allison 1150
XP-40F	Curtiss	1941	(1)	Merlin 1300
P-40F	Curtiss	1942	1,311	Merlin 1300
P-40G (P-40)	Curtiss	1941	(45)	Allison 1040
P-40H	designation not used; J cancelled			
P-40K	Curtiss	1942	1,300	Allison 1325
P-40L	Curtiss	1943	700	Merlin 1300
P-40M	Curtiss	1942	600	Allison 1200
P-40N	Curtiss	1944	5,215	Allison 1200
XP-40Q	Curtiss	1945	1	Allison 1425
P-40R	Curtiss	1944	(300)	Allison 1200

361

Model	Manufacturer	Cal Yr 1st Del	Number Procured	Engine/hp	
P-43A-1	Republic	1941	125	P&W	1200
P-43B	Republic	1942	(150)	P&W	1200
P-45 (P-39C)	Bell	1941	(80)	Allison	1150
P-47B	Republic	1941	171	P&W	2000
P-47C	Republic	1941	602	P&W	2000
P-47D	Republic	'42-'44	12,602	P&W	2000
XP-47E, F	Republic	1942	(2)	P&W	2000
P-47G-CU	Curtiss	1942	354	P&W	2000
XP-47H	Republic	1943	(1)	Chrysler	2300
XP-47K, J, L, M	Republic	1944	1	P&W	2100
P-47M	Republic	1944	130	P&W	2100
XP-47N	Republic	1944	1	P&W	2100
P-47N	Republic	1944	1,816	P&W	2100
XP-48	Douglas	1941	not built		
XP-49	Lockheed	1942	1	2 Continental	1350
XP-50 (XF5F-1)	Grumman	crashed prior to acceptance			
XP-51	N. American	1941	2	Allison	1150
P-51	N. American	1941	150	Allison	1150
P-51A	N. American	1942	310	Allison	1200
XP-51B	N. American	1942	(2)	Merlin	1300
P-51B	N. American	1942	1,988	Merlin	1380
P-51C	N. American	1942	1,750	Merlin	1380
P-51D	N. American	1944	7,956	Merlin	1490
TP-51D	N. American	1944	10	Merlin	1490
XP-51F	N. American	1944	3	Merlin	1490
XP-51G	N. American	1944	2	Merlin	1500
P-51H	N. American	1944	555	Merlin	1380
P-51J	N. American	1945	2	Allison	1500
P-51K	N. American	1944	1,337	Merlin	1490
P-51M	N. American	1945	1	Merlin	1400
XP-52	Bell	1941	canceled		
XP-53	Curtiss	1941	canceled		
XP-54	Vultee	1942	2	Lycoming	2300
XP-55	Curtiss	1942	4	Allison	1275
XP-56	Northrop	1942	2	P&W	2000
XP-57	Tucker	1941	canceled		
XP-58	Lockheed	1943	1	2 Allison	2600
XP-59	Bell	1941	canceled		
XP-59A	Bell	1943	3	2 G.E. jet	2000[1]
YP-59A	Bell	1943	13	2 G.E. jet	2000
P-59A	Bell	1944	20	2 G.E. jet	2000
P-59B	Bell	1944	30	2 G.E. jet	2000
XP-60	Curtiss	1942	1	Merlin	1300
XP-60A	Curtiss	1942	3	Allison	1425

Model	Manufacturer	Cal Yr 1st Del	Number Procured	Engine/hp	
P-60A	Curtiss	1943	1	P&W	2000
XP-60B, C, D, E,	YP-60E	1943	(4)	various	
XP-61	Northrop	1942	2	2 P&W	2000
YP-61	Northrop	1942	13	2 P&W	2000
P-61-1, 5	Northrop	1943	80	2 P&W	2000
P-61A-10	Northrop	1943	120	2 P&W	2000
P-61B	Northrop	1943	450	2 P&W	2000
P-61C	Northrop	1944	41	2 P&W	2100
XP-61D, E, F	Northrop	1945	(3)	2 P&W	2100
XP-62	Curtiss	1943	1	Wright	2300
XP-63	Bell	1942	2	Allison	1325
XP-63A	Bell	1943	1	Allison	1325
P-63A-1, 10	Bell	1943	1,725	Allison	1325
RP-63A	Bell	1943	100	Allison	1325
P-63C-1, 5	Bell	1943	1,227	Allison	1325
RP-63C-2	Bell	1944	200	Allison	1325
P-63D	Bell	1945	1	Allison	1425
P-63E-1	Bell	1945	13	Allison	1425
P-63F	Bell	1945	2	Allison	1425
RP-63G	Bell	1945	32	Allison	1425
P-64	N. American	1941	6	Wright	875
XP-65	Grumman	1941	canceled		
P-66	Vultee	1942	144	P&W	1200
XP-67	McDonnell	1943	1	2 Cont	1350
XP-68	Vultee	1942	not built		
XP-69	Republic	1942	canceled		
XP-70 (A-20)	Douglas	1942	(1)	2 Wright	1600
P-70	Douglas	1942	(59)	2 Wright	1600
P-70A, B	Douglas	1943	(40)	2 Wright	1600
XP-71	Curtiss	1942	canceled		
XP-72	Republic	1945	1	P&W	3450
P-73	not used				
P-74	not used				
XP-75	Fisher	1943	2	Allison	2600
XP-75	redesign Fisher	1944	6	Allison	2600
P-75A-1	Fisher	1945	6	Allison	2600
P-76	Bell	1942	canceled		
XP-77	Bell	1944	2	Ranger	520
XP-78 (XP-51B)	N. American	1942	(2)	Merlin	1300
XP-79	Northrop	1943	not procured		
XP-79B	Northrop	1945	1	2 West	1150[1]
XP-80	Lockheed	1944	1	Goblin	3000[1]
XP-80A	Lockheed	1944	2	G.E.	4000[1]
YP-80A	Lockheed	1944	13	G.E.	4000

Model	Manufacturer	Cal Yr 1st Del	Number Procured	Engine/hp	
P-80A	Lockheed	1945	917	G.E.	4000
XP-81	Vultee	1945	2	2 G.E. jet	4000
XP-82	N. American	1945	2	2 Merlin	1380
XP-82A	N. American	1945	1	2 Allison	1500
P-82B	N. American	1945	20	2 Merlin	1380
XP-83	Bell	1945	2	2 G.E. jet	4000
XP-84	Republic	1945	3	G.E. jet	5200

Attack

Model	Manufacturer	Cal Yr 1st Del	Number Procured	Engine/hp	
A-24 (SBD-3)	Douglas	1941	168	2 Wright	1000
A-24A	Douglas	1942	170	2 Wright	1000
A-24B	Douglas	1942	615	2 Wright	1200
A-25A	Curtiss	1943	900	2 Wright	1700
XA-26	Douglas	1943	1	2 P&W	2000
XA-26A	Douglas	1943	1	2 P&W	2000
XA-26B	Douglas	1943	1	2 P&W	2000
A-26B	Douglas	1943	1,356	2 P&W	2000
A-26C	Douglas	1945	1,091	2 P&W	2100
A-26D, E	cancelled				
A-26F	Douglas	1945	1	2 P&W	2100
A-27 (AT-6)	N. American	1941	10	Wright	785
A-28 (Hudson)	Lockheed	1941	52	2 P&W	1050
A-28A	Lockheed	1942	450	2 P&W	1200
A-29 (Hudson)	Lockheed	1942	416	2 Wright	1200
A-29A (C-63)	Lockheed	1942	384	2 Wright	1200
A-29B (Photo)	Lockheed	1942	(24)	2 Wright	1200
A-30 (Baltimore)	Martin	1941	281	2 Wright	1600
A-30A	Martin	1942	894	Wright	1700
A-31 (Vengeance)	Vultee	1941	100	Wright	1600
A-31	Northrop	1942	200	Wright	1600
XA-31A, B, C	Vultee	1942	1	various	
YA-31C	Vultee	1943	(5)	B-29 engine test	
XA-32	Brewster	1944	1	P&W	2100
XA-32A	Brewster	1944	1	P&W	2100
A-33	Douglas	1942	31	Wright	1200
A-34	Brewster	1942	canceled		
A-35 (A-31)	Vultee	1942	99	Wright	1600
A-35B	Vultee	1942	831	Wright	1700
A-36A (P-51)	N. American	1943	500	Allison	1325
XA-37	Hughes	1943	canceled		
XA-38	Beech	1945	2	2 Wright	2300
XA-39	Fleetwings	1943	canceled		
XA-40 (XSB3C)	Curtiss	1943	canceled		

Model	Manufacturer	Cal Yr 1st Del	Number Procured	Engine/hp	
XA-41	Vultee	1944	1	P&W	3000
XA-42	Douglas	1945	2	2 Allison	1325

Observation

Model	Manufacturer	Cal Yr 1st Del	Number Procured	Engine/hp	
O-52	Curtiss	1941	203	P&W	600
O-53	Douglas	1941	canceled		
YO-54 (105)	Stinson	1941	6	Cont.	80
YO-55	Ercoupe	1941	1	Cont.	65
O-56	Lockheed	1941	canceled		
YO-57	Taylorcraft	1942	4	Cont.	65
O-57	Taylorcraft	1942	70	Cont.	65
O-57A	Taylorcraft	1942	336	Cont.	65
YO-58	Aeronca	1942	4	Cont.	65
O-58	Aeronca	1942	50	Cont.	65
O-58A	Aeronca	1942	20	Cont.	65
O-58B	Aeronca	1942	335	Cont.	65
YO-59 (J-3)	Piper	1942	4	Cont.	65
O-59	Piper	1942	140	Cont.	65
O-59A	Piper	1942	649	Cont.	65
XO-60 autogiro	Kellett	1942	7	Jacobs	300
YO-60	Kellett	1943	(6)	Jacobs	300
XO/YO-61	Aga Aviation	1943	canceled		
O-62	Vultee	1942	275	Lycoming	185
XO-63	Interstate	1942	1	Franklin	100

Reconnaissance

Model	Manufacturer	Cal Yr 1st Del	Number Procured	Engine/hp	
F2A (UC-45)	Beech	1942	(13)	2 P&W	450
F2B	Beech	1944	42	2 P&W	450
F-3A (A-20)	Douglas	1944	(46)	2 Wright	1600
F-4 (P-38)	Lockheed	1942	99	2 Allison	1150
F-4A	Lockheed	1942	20	2 Allison	1325
F-5A(P-38)	Lockheed	1942	181	2 Allison	1325
F-5B	Lockheed	1943	200	2 Allison	1325
F-5C	Lockheed	1943	(128)	2 Allison	1325
XF-5D	Lockheed	1943	(1)	2 Allison	1325
F-5E	Lockheed	1943	(713)	2 Allison	1425
F-5G	Lockheed	1945	(63)	2 Allison	1475
F-6A (P-51)	N. American	1942	(57)	Allison	1150
F-6B	N. American	1943	(35)	Merlin	1300
F-6C	N. American	1944	(91)	Merlin	1400
F-6D	N. American	1944	(136)	Merlin	1490
F-6K	N. American	1945	(163)	Merlin	1490
XF-7 (B-24)	Convair	1943	(1)	4 P&W	1200
F-7-FO (B-24)	Ford	1943	data incomplete		

Model	Manufacturer	Cal Yr 1st Del	Number Procured	Engine/hp	
F-7A	Convair/Ford	1944	(86)	4 P&W	1200
F-7B	Convair/Ford	1944	data incomplete		
F-8 (D.H. Mosquito)	deHavilland	1943	40	2 Merlin	1300
F-9 (B-17)	several	1943	(75+)	4 Wright	1200
F-10 (B-25)	N. American	1943	(10)	2 Wright	1700
XF-11	Hughes	1945	2	2 P&W	3000
XF-12	Republic	1945	1	4 P&W	3000
F-13A (B-29)	Boeing	1945	data incomplete		
XF-14 (YP-80)	Lockheed	1945	(1)	G.E. jet	4000
XF-15 (P-61)	Northrop	1945	(1)	2 P&W	2000
XFA-26C	Douglas	1945	(1)	2 P&W	2000

Liaison

Model	Manufacturer	Cal Yr 1st Del	Number Procured	Engine/hp	
L-1 (O-49)	Stinson	1942	(142)	Lycoming	295
L-1A	Stinson	1942	(182)	Lycoming	295
L-1B	Stinson	1942	(3)	Lycoming	295
L-1C,D,E,F	Stinson	1942	(25)	Lycoming	295
L-2	Taylorcraft	1942	(74)	Cont.	65
L-2A	Taylorcraft	1942	(476)	Cont.	65
L-2B	Taylorcraft	1943	490	Cont.	65
L-2C thru L2L	Taylorcraft	1943	8	Cont/Lyn/Fkln	65
L-2M	Taylorcraft	1943	900	Cont.	65
L-3 (O-58)	Aeronca	1942	(54)	Cont.	65
L-3A	Aeronca	1942	(20)	Cont.	65
L-3B	Aeronca	1942	875	Cont..	65
L-3C	Aeronca	1943	490	Cont.	65
L-3D	Aeronca	1942	10	Franklin	65
L-3E	Aeronca	1942	10	Cont.-8	65
L-3F	Aeronca	1942	1	Cont.-8	65
L-3G	Aeronca	1942	2	Lycoming	65
L-3H	Aeronca	1942	1	Lycoming	65
L-3J	Aeronca	1942	2	Cont.	65
L-4 (O-59)	Piper	1942	(144)	Cont.	65
L-4A (J-3)	Piper	1942	(948)	Cont.	65
L-4B	Piper	1942	(981)	Cont.	65
L-4C	Piper	1942	10	Lycoming	65
L-4D	Piper	1942	5	Franklin	65
L-4E	Piper	1942	16	Cont.	75
L-4F	Piper	1942	45	Cont.	75
L-4G	Piper	1942	41	Lycoming	100
L-4H	Piper	1943	1,801	Lycoming	65
L-4J	Piper	1944	1,680	Lycoming	65
L-5	Vultee	1942	1,731	Lycoming	185
L-5A	Vultee	1942	(688)	Lycoming	185

Model	Manufacturer	Cal Yr 1st Del	Number Procured	Engine/hp
L-5B	Vultee	1942	679	Lycoming 185
L-5C	Vultee	1944	200	Lycoming 185
L-5E	Vultee	1944	558	Lycoming 185
L-5G	Vultee	1945	115	Lycoming 190
XL-6 (XO-63)	Interstate	1942	(1)	Franklin 100
L-6	Interstate	1943	250	Franklin 100
L-7A (Monocoupe)	Universal	1943	19	Franklin 90
L-8A	Interstate	1943	8	Cont. 65
L-9A (Voyager)	Stinson	1943	8	Franklin 90
L-9B	Stinson	1942	12	Franklin 90
L-10 (SCW)	Ryan	1942	1	Warner 145
L-11 (31-50)	Bellanca	1942	1	P&W 600
L-12 (SR-5)	Stinson	1944	2	Lycoming 300
L-12A (SM-7B)	Stinson	1944	2	Lycoming 300
XL-13	Vultee	1945	2	Franklin 245
YL-14	Piper	1945	5	Lycoming 130

Bombardment

Model	Manufacturer	Cal Yr 1st Del	Number Procured	Engine/hp
B-17D	Boeing	1941	42	4 Wright 1200
B-17E	Boeing	1942	512	4 Wright 1200
B-17F-BO	Boeing	1942	2,300	4 Wright 1200
B-17F-DL	Douglas	1942	600	4 Wright 1200
B-17F-VE	Vega	1942	500	4 Wright 1200
B-17G-BO	Boeing	1943	4,035	4 Wright 1200
B-17G-DL	Douglas	1943	2,395	4 Wright 1200
B-17G-VE	Vega	1943	2,250	4 Wright 1200
B-17H (Rescue) various, converted Gs		1944	(approx. 130)	4 Wright 1200
XB-19	Douglas	1941	1	4 Wright 2200
XB-19A	Douglas	1943	(1)	4 Allison 2600
B-24A	Consolidated	1941	9	4 P&W 1200
XB-24B	Consolidated	1941	(1)	4 P&W 1200
B-24C	Consolidated	1941	9	4 P&W 1200
B-24D-CO	Convair	1941	2,415	4 P&W 1200
B-24D-CF	Convair	1942	303	4 P&W 1200
B-24D-DT	Douglas	1941	10	4 P&W 1200
B-24E-CF	Convair	1942	144	4 P&W 1200
B-24E-DT	Douglas	1942	167	4 P&W 1200
B-24E-FO	Ford	1942	430	4 P&W 1200
XB-24F-CO	Convair	1942	(1)	4 P&W 1200
B-24G-NT	N. American	1943	430	4 P&W 1200
B-24H-CF	Convair	1943	738	4 P&W 1200
B-24H-FO	Ford	1943	1,780	4 P&W 1200
B-24H-DT	Douglas	1943	582	4 P&W 1200
B-24J-CO	Convair	1943	2,792	4 P&W 1200

Model	Manufacturer	Cal Yr 1st Del	Number Procured	Engine/hp
B-24J-CF	Convair	1943	1,558	4 P&W 1200
B-24J-FO	Ford	1943	1,587	4 P&W 1200
B-24J-NT	N. American	1943	536	4 P&W 1200
B-24J-DT	Douglas	1943	205	4 P&W 1200
XB-24K-FO	Ford	1943	(1)	4 P&W 1350
B-24L-CO	Convair	1944	417	4 P&W 1200
B-24L-FO	Ford	1944	1,250	4 P&W 1200
B-24M-CO	Convair	1944	916	4 P&W 1200
B-24M-FO	Ford	1944	1,677	4 P&W 1200
XB-24N-FO	Ford	1944	1	4 P&W 1350
YB-24N-FO	Ford	1945	7	4 P&W 1350
XB-24P-CO	Convair	1945	(1)	4 P&W 1200
XB-24Q-FO	Ford	1945	(1)	4 P&W 1200
B-25A	N. American	1941	40	2 Wright 1700
B-25B	N. American	1941	119	2 Wright 1700
B-25C	N. American	1941	1,619	2 Wright 1700
B-25D	N. American	1942	2,290	2 Wright 1700
XB-25E, F, G	N. American	1942	(1)	2 Wright 1700
B-25G[2]	N. American	1942	405	2 Wright 1700
B-25H[2]	N. American	1943	1,000	2 Wright 1700
B-25J	N. American	1943	4,318	2 Wright 1700
B-26	Martin	1941	201	2 P&W 1850
B-26A	Martin	1941	139	2 P&W 1850
B-26B	Martin	1942	1,883	2 P&W 2000
B-26C	Martin	1942	1,235	2 P&W 2000
XB-26D, E	Martin	1942	(1)	2 P&W 1850
B-26F	Martin	1943	300	2 P&W 2000
B-26G	Martin	1944	893	2 P&W 2000
TB-26G	Martin	1945	57	2 P&W 2000
XB-26H	Martin	1945	(1)	2 P&W 2000
XB-27	Martin	canceled		
XB-28 (B-25 mod)	N. American	1941	1	2 P&W 2000
XB-28A	N. American	1942	1	2 P&W 2000
XB-29-BO	Boeing	1943	3	4 Wright 2200
YB-29-BW	Boeing	1943	14	4 Wright 2200
B-29-BW	Boeing	1943	1,620	4 Wright 2200
B-29-BA	Bell	1943	357	4 Wright 2200
B-29-MO	Martin	1944	536	4 Wright 2200
B-29-BN	Boeing	1944	1,119	4 Wright 2200
B-29B-BA	Bell	1944	311	4 Wright 2200
XB-30	Lockheed	not built		
XB-31	Douglas	not built		
XB-32	Convair	1943	3	4 Wright 2200
B-32	Convair	1944	115	4 Wright 2200

Model	Manufacturer	Cal Yr 1st Del	Number Procured	Engine/hp
XB-33	Martin	1942	canceled	
B-34	Vega	1942	200	2 P&W 2000
XB-35	Northrop	1944	2	4 P&W 3000
YB-35	Northrop	1945	2	4 P&W 3000
B-37 (O-56)	Lockheed	1945	18	2 Wright 1700
XB-39-GM (B-29)	Boeing	1944	(1)	4 Allison 2600
XB-40 (B-17)	Vega	1943	(1)	4 Wright 1200
XB-41 (B-24)	Convair	1943	(1)	4 P&W 1250

Transport

Model	Manufacturer	Cal Yr 1st Del	Number Procured	Engine/hp
C-32A	Douglas	1942	24	2 Wright 740
UC-36A (Model 10)	Lockheed	1942	15	2 P&W 450
UC-36B	Lockheed	1942	4	2 P&W 600
UC-36C	Lockheed	1942	7	2 P&W 450
UC-40D	Lockheed	1942	10	2 P&W 450
UC-43 (D-17S)	Beech	1942	207	P&W 450
UC-43A (D-17R)	Beech	1942	13	Wright 440
UC-43B (D-17S)	Beech	1942	13	P&W 450
UC-43C (F-17)	Beech	1942	38	Jacobs 300
UC-43D (E-17)	Beech	1942	31	Jacobs 285
UC-43E (C-17)	Beech	1942	5	Wright 440
UC-43F (D-17A)	Beech	1942	1	Wright 350
UC-43G (C-17B)	Beech	1942	10	Jacobs 285
UC-43H (B-17R)	Beech	1942	3	Wright 440
UC-43J (C-17L)	Beech	1942	3	Jacobs 225
UC-43K (D-17W)	Beech	1942	1	P&W 600
C-45A	Beech	1941	20	2 P&W 450
C-45B	Beech	1942	223	2 P&W 450
UC-45C	Beech	1942	2	2 P&W 450
UC-45D	Beech	1942	2	2 P&W 450
UC-45E	Beech	1942	6	2 P&W 450
UC-45F	Beech	1943	1,137	2 P&W 450
C-46	Curtiss	1941	25	2 P&W 2000
C-46A	Curtiss	1942	1,041	2 P&W 2000
C-46A-CK	Curtiss	1943	438	2 P&W 2000
C-46A-CS	Curtiss	1943	12	2 P&W 2000
C-46A-HI	Higgins	1943	2	2 P&W 2000
XC-46B, C	Curtiss	1943	(2)	2 P&W 2000
C-46F	Curtiss	1944	234	2 P&W 2000
C-46G	Curtiss	1945	1	2 P&W 2100
C-46H, K	Curtiss	1945	canceled	
XC-46L	Curtiss	1945	3	2 Wright 2500
C-47	Douglas	1941	965	2 P&W 1200
C-47A	Douglas	1942	2,954	2 P&W 1200

Model	Manufacturer	Cal Yr 1st Del	Number Procured	Engine/hp
C-47A (Oke City)	Douglas	1942	2,299	2 P&W 1200
C-47B	Douglas	1943	300	2 P&W 1200
C-47B	Douglas	1943	3,065	2 P&W 1200
XC-47C (amphib)	Douglas	1943	(1)	2 P&W 1200
C-48 (DC-3)	Douglas	1941	1	2 P&W 1200
C-48A	Douglas	1941	3	2 P&W 1200
C-48B	Douglas	1942	16	2 P&W 1200
C-48C	Douglas	1942	16	2 P&W 1200
C-49 (DC-3)	Douglas	1941	6	2 Wright 1200
C-49A	Douglas	1941	1	2 Wright 1200
C-49B	Douglas	1941	3	2 Wright 1200
C-49C	Douglas	1941	2	2 Wright 1200
C-49D	Douglas	1941	11	2 Wright 1200
C-49E	Douglas	1942	22	2 Wright 1100
C-49F	Douglas	1942	9	2 Wright 1200
C-49G	Douglas	1942	8	2 Wright 1200
C-49H	Douglas	1942	19	2 Wright 1200
C-49J	Douglas	1943	34	2 Wright 1200
C-49K	Douglas	1943	23	2 Wright 1200
C-50 (DC-3)	Douglas	1941	4	2 Wright 1100
C-50A	Douglas	1941	2	2 Wright 1100
C-50B	Douglas	1941	3	2 Wright 1100
C-50C	Douglas	1941	1	2 Wright 1100
C-50D	Douglas	1941	4	2 Wright 1100
C-51 (DC-3)	Douglas	1941	1	2 Wright 1100
C-52 (DC-3)	Douglas	1941	1	2 P&W 1200
C-52A	Douglas	1941	1	2 P&W 1200
C-52B	Douglas	1941	2	2 P&W 1200
C-52C	Douglas	1941	1	2 P&W 1200
C-53 (DC-3)	Douglas	1941	193	2 P&W 1200
XC-53A	Douglas	1942	(1)	2 P&W 1200
C-53B	Douglas	1942	8	2 P&W 1200
C-53C	Douglas	1943	17	2 P&W 1200
C-53D	Douglas	1943	159	2 P&W 1200
C-54 (DC-4)	Douglas	1941	24	4 P&W 1350
C-54A	Douglas	1942	207	4 P&W 1350
C-54B	Douglas	1943	220	4 P&W 1350
C-54C (sleeper)	Douglas	1943	(1)	4 P&W 1350
C-54D	Douglas	1943	350	4 P&W 1350
C-54E	Douglas	1944	75	4 P&W 1350
C-54G	Douglas	1945	76	4 P&W 1450
C-54H, J,	canceled			
C-54K	Douglas	1945	1	4 Wright 1425
C-55 (C-46)	Curtiss	1941	1	2 Wright 1700

Model	Manufacturer	Cal Yr 1st Del	Number Procured	Engine/hp
C-56 (Lodestar)	Lockheed	1941	1	2 Wright 1100
C-56A	Lockheed	1942	1	2 P&W 875
C-56B	Lockheed	1942	13	2 Wright 1200
C-56C	Lockheed	1942	12	2 P&W 875
C-56D	Lockheed	1942	7	2 P&W 875
C-56E	Lockheed	1943	2	2 Wright 1200
C-57	Lockheed	1941	13	2 P&W 1200
C-57B (no A)	Lockheed	1943	7	2 P&W 1200
C-57C, D,	Lockheed	1942	(4)	2 P&W 1200
C-58 (B-18)	Douglas	1942	(2)	2 Wright 1000
C-59	Lockheed	1941	10	2 P&W 875
C-60	Lockheed	1941	36	2 Wright 1200
C-60A	Lockheed	1942	325	2 Wright 1200
XC-60B, C	Lockheed	1943	(1)	2 Wright 1200
UC-61 (24W)	Fairchild	1941	163	Warner 165
UC-61A	Fairchild	1942	512	Warner 165
UC-61B	Fairchild	1942	1	Warner 165
UC-61C	Fairchild	1942	1	Ranger 175
UC-61D	Fairchild	1942	3	P&W 300
UC-61E	Fairchild	1942	3	Ranger 175
UC-61F	Fairchild	1942	2	Ranger 145
UC-61G	Fairchild	1942	2	Warner 145
UC-61H	Fairchild	1942	1	Warner 145
UC-61J	Fairchild	1942	1	Ranger 150
UC-61K	Fairchild	1943	306	Ranger 200
YC-62	Waco	canceled		
C-63	Lockheed	canceled		
YC-64	Noorduyn	1942	7	P&W 600
C-64A	Noorduyn	1943	746	P&W 600
C-64B (floats)	Noorduyn	1943	6	P&W 600
XC-64 (Skycar)	Stout	1942	1	Franklin 90
C-66	Lockheed	1942	1	2 P&W 1200
UC-67 (B-23)	Douglas	1942	(12)	2 Wright 1600
C-68 (DC-3)	Douglas	1942	2	2 Wright 1200
C-69 (Model 49)	Lockheed	1942	19	4 Wright 2200
C-69C (no A, B,)	Lockheed	1945	1	4 Wright 2200
UC-70 (DGA-15)	Howard	1942	11	P&W 450
UC-70A	Howard	1942	2	Jacobs 300
UC-70B	Howard	1942	4	Jacobs 300
UC-70C	Howard	1942	1	Wright 350
UC-70D	Howard	1942	2	Jacobs 285
UC-71 (7W)	Spartan	1942	16	P&W 400
UC-72 (SRE)	Waco	1942	12	P&W 400
UC-72A (ARE)	Waco	1942	1	Jacobs 300

Model	Manufacturer	Cal Yr 1st Del	Number Procured	Engine/hp	
UC-72B (EGC)	Waco	1942	4	Wright	350
UC-72C (HRE)	Waco	1942	2	Lycoming	300
UC-72D (VKS)	Waco	1942	2	Cont.	240
UC-72E (ZGC)	Waco	1942	4	Jacobs	285
UC-72F (CUC)	Waco	1942	1	Wright	250
UC-72G (AQC)	Waco	1942	1	Jacobs	300
UC-72H (ZQC)	Waco	1942	5	Jacobs	285
UC-72J (AVN)	Waco	1942	3	Jacobs	300
UC-72K (YKS)	Waco	1942	2	Jacobs	225
UC-72L (ZVN)	Waco	1942	1	Jacobs	300
UC-72M (ZKS)	Waco	1942	2	Jacobs	285
UC-72N (YOC)	Waco	1942	1	Jacobs	285
UC-72P (AGC)	Waco	1942	2	Jacobs	300
C-73 (247)	Boeing	1942	27	2 P&W	600
C-74 (DC-7)	Douglas	1943	14	4 P&W	3000
C-75 (307)	Boeing	1942	5	4 Wright	1100
YC-76 (all wood)	Curtiss	1942	11	2 P&W	1200
YC-76A/C-76	Curtiss	1943	14	2 P&W	1200
UC-77 ('29 model)	Cessna	1942	4	Wright	300
UC-77A, B, C	Cessna	1942	7	various	
UC-78	Cessna	1942	1,287	2 Jacobs	225
UC-78A	Cessna	1942	17	2 Jacobs	225
UC-78B	Cessna	1942	1,806	2 Jacobs	225
UC-78C	Cessna	1943	327	2 Jacobs	225
C-79 (Ju.52)	Junkers[3]	1942	1	3 B.M.W.	800
UC-80	Harlow	1942	4	Warner	145
UC-81 (SR-8)	Stinson	1942	5	Lycoming	245
UC-81A (SR-10)	Stinson	1942	2	Lycoming	290
UC-81B (SR-8E)	Stinson	1942	1	Wright	350
UC-81C (SR-9C)	Stinson	1942	3	Lycoming	260
XC-81D	Stinson	1943	1	P&W	450
UC-81E (SC-9F)	Stinson	1942	4	P&W	450
UC-81F (SR-10)	Stinson	1942	8	P&W	450
UC-81G (SR-9D)	Stinson	1942	3	Wright	285
UC-81H (SR-10)	Stinson	1942	1	Wright	350
UC-81J (SR-9E)	Stinson	1942	10	Wright	350
UC-81K (SR-10)	Stinson	1942	5	Lycoming	260
UC-81L (SR-8C)	Stinson	1942	2	Lycoming	240
UC-81M (SR-9C)	Stinson	1942	1	Lycoming	290
UC-81N (SR-9B)	Stinson	1942	1	Lycoming	260
XC-82	Fairchild	1945	1	2 P&W	2100
C-82A	Fairchild	1944	200	2 P&W	2100
C-82N	N. American	1945	3	2 P&W	2100
UC-83 (J5)	Piper	1942	1	Cont.	75

Model	Manufacturer	Cal Yr 1st Del	Number Procured	Engine/hp	
C-84 (DC-3)	Douglas	1942	4	2 Wright	1200
UC-85 (Orion)	Lockheed	1942	1	P&W	550
UC-86 (24-R)	Fairchild	1942	9	Ranger	175
XUC-86A, B	Fairchild	1943	(2)	Ranger/Franklin	
C-87 (B-24D)	Convair	1942	276	4 P&W	1200
C-87A	Convair	1943	6	4 P&W	1200
UC-88	Fairchild	1942	2	Wright	350
UC-89 ('28 model)	Hamilton	1942	1	P&W	525
UC-90 (8A)	Luscombe	1942	1	Cont.	65
UC-90A	Luscombe	1942	1	Lycoming	65
C-91 (SM-6000)	Stinson	1942	1	3 Lycoming	260
UC-92	Funk	1942	1	Lycoming	75
C-93A	Budd	1942	canceled		
UC-94 (C-165)	Cessna	1942	3	Warner	165
UC-95 (BL-65)	Taylorcraft	1942	1	Lycoming	65
UC-96 (F-71)	Fairchild	1942	3	P&W	450
XC-97	Boeing	1945	3	4 Wright	2200
C-98 (Clipper)	Boeing	1942	4	4 Wright	1500
XC-99	Convair	1945	1	6 P&W	3000
UC-100 (Gamma)	Northrop	1942	1	Wright	700
UC-101 (Vega)	Lockheed	1942	1	P&W	600
UC-102	Rearwin	1942	2	Ken-Royce	90
UC-102A	Rearwin	1942	1	Ken-Royce	120
UC-103 (G-32)	Grumman	1942	2	Wright	890
C-104A	Lockheed	1943	canceled		
XC-105 (XB-15)	Boeing	1943	(1)	4 P&W	1000
C-106	Cessna	1943	2	2 P&W	600
UC-107 (Skycar)	Stout	1943	test only		
XC/YC-108 (B-17)	Boeing	1944	(4)	4 Wright	1200
XC-109-FO (B-24)	Ford	1943	(1)	4 P&W	1200
C-109 (tanker)	various	1944	(40+)	4 P&W	1200
C-110 (DC-5)	Douglas	1944	3	2 Wright	1100
C-111 (14)	Lockheed	1944	3	2 Wright	760
XC/YC-112 (C-54)	Douglas	1945	1	4 P&W	2100
XC-113 (XC-46)	Curtiss	1945	(1)	2 P&W	plus jet
XC-114 (C-54)	Douglas	1945	1	4 Allison	1620
XC-115 (C-54)	Douglas	1945	canceled		
YC-116 (XC-114)	Douglas	1945	(1)	4 Allison	1620
C-117A (C-47)	Douglas	1945	17	2 P&W	1200

Primary Trainer

Model	Manufacturer	Cal Yr 1st Del	Number Procured	Engine/hp	
PT-13D	Stearman	1942	318	Lycoming	220
PT-14A (UPF-7)	Waco	1942	1	Cont.	220

Model	Manufacturer	Cal Yr 1st Del	Number Procured	Engine/hp	
PT-19A	Fairchild	1941	3,181	Ranger	200
PT-19A	Aeronca	1942	477	Ranger	200
PT-19A	St. Louis	1942	44	Ranger	200
PT-19B	Fairchild	1942	774	Ranger	200
PT-19B	Aeronca	1942	143	Ranger	200
PT-20	Ryan	1941	30	Menasco	125
PT-20A	Ryan	1941	(27)	Kinner	132
PT20B	Ryan	1941	(3)	Menasco	125
PT-21 (STM)	Ryan	1941	100	Kinner	132
PT-22 (ST3)	Ryan	1941	1,023	Kinner	160
PT-22A	Ryan	1942	25	Kinner	160
XPT-23 (PT-19)	Fairchild	1942	(1)	Cont.	220
PT-23	Fairchild	1942	2	Cont.	220
PT-23-AE	Aeronca	1942	375	Cont.	220
PT-23-HO	Howard	1942	199	Cont.	220
PT-23-SL	St. Louis	1942	200	Cont.	220
PT-23-FE	Fleet	1942	93	Cont.	220
PT-23A-HO	Howard	1942	150	Cont.	220
PT-23A-SL	St. Louis	1942	106	Cont.	220
PT-24 (Tiger Moth)	deHavilland	1942	200	Gipsy	140
YPT-25	Ryan	1942	5	Lycoming	185
PT-26	Fairchild	1942	670	Ranger	200
PT-26A	Fleet	1942	807	Ranger	200
PT-26B	Fleet	1943	250	Ranger	200
PT-27 (PT-17)	Boeing	1942	300	Cont.	220

Basic Trainer

Model	Manufacturer	Cal Yr 1st Del	Number Procured	Engine/hp	
BT-12	Fleetwings	1942	24	P&W	450
BT-13A	Vultee	1941	6,607	P&W	450
BT-13B	Vultee	1942	1,775	P&W	450
BT-14A (BT-9)	N. American	1941	(27)	P&W	400
BT-15	Vultee	1941	1,693	Wright	450
XBT-16 (BT-13A)	Vidal	1942	(1)	P&W	450
XBT-17	Boeing	1942	1	P&W	450

Advanced Trainer

Model	Manufacturer	Cal Yr 1st Del	Number Procured	Engine/hp	
AT-6A	N. American	1941	1,549	P&W	600
AT-6B	N. American	1941	400	P&W	600
AT-6C	N. American	1942	2,970	P&W	600
AT-6D	N. American	1942	4,388	P&W	600
XAT-6E	N. American	1944	(1)	Ranger	575
AT-6F	N. American	1944	956	P&W	600
AT-7	Beech	1941	577	2 P&W	450

Model	Manufacturer	Cal Yr 1st Del	Number Procured	Engine/hp	
AT-7A	Beech	1942	7	2 P&W	450
AT-7B	Beech	1942	9	2 P&W	450
AT-7C	Beech	1943	549	2 P&W	450
AT-8	Cessna	1941	33	2 Lycoming	295
AT-9	Curtiss	1941	491	2 Lycoming	295
AT-9A	Curtiss	1942	300	2 Lycoming	295
AT-10	Beech	1942	1,771	2 Lycoming	295
AT-10-GF	Globe	1942	600	2 Lycoming	295
AT-11	Beech	1941	1,582	2 P&W	450
AT-11A	Beech	1942	(36)	2 P&W	450
AT-12	Republic	1941	50	P&W	1050
XAT-13	Fairchild	1942	1	2 P&W	600
XAT-14	Fairchild	1942	1	2 Ranger	520
XAT-15	Boeing	1942	2	2 P&W	600
AT-16	Noorduyn	1942	1,500	P&W	600
AT-17	Cessna	1942	450	2 Jacobs	245
AT-17A	Cessna	1942	223	2 Jacobs	245
AT-17B	Cessna	1942	466	2 Jacobs	245
AT-17C	Cessna	1942	60	2 Jacobs	245
AT-18	Lockheed	1942	217	2 Wright	1200
AT-18A	Lockheed	1942	83	2 Wright	1200
AT-19 (Reliant)	Vultee	1943	500	Lycoming	280
AT-20 (Avro)	Federal	1943	50	2 Jacobs	330
AT-21	Fairchild	1942	106	2 Ranger	520
AT-21-BL	Bellanca	1942	39	2 Ranger	520
AT-21-MC	McDonnell	1942	30	2 Ranger	520
AT-22 (C-87)	Convair	1943	5	4 P&W	1200
AT-23A (B-26)	Martin	1943	208	2 P&W	2000
AT-23B	Martin	1943	350	2 P&W	2000
AT-24A, B, C, D	N. American	1944	B-25 mods for training		
P-322 (P-38)	Lockheed	1943	138	2 Allison	1150[4]

Helicopters

Model	Manufacturer	Cal Yr 1st Del	Number Procured	Engine/hp	
XR-1	Platt-LePage	1944	1	P&W	440
XR-1A	Platt-LePage	1945	1	P&W	450
XR-2 (YG-1C)	Kellett	1941	(1)	Jacobs	300
XR-3 (YG-1B)	Kellett	1941	(1)	Jacobs	225
XR-4	Sikorsky	1942	1	Warner	165
YR-4A	Sikorsky	1942	3	Warner	180
YR-4B	Sikorsky	1943	27	Warner	180
R-4B	Sikorsky	1944	100	Warner	200
XR-5	Sikorsky	1944	5	P&W	450
YR-5A	Sikorsky	1944	26	P&W	450

Model	Manufacturer	Cal Yr 1st Del	Number Procured	Engine/hp
R-5A	Sikorsky	1945	34	P&W 450
XR-6	Sikorsky	1944	1	Lycoming 225
XR-6A	Sikorsky	1944	5	Franklin 240
YR-6A-NK	Nash	1944	26	Franklin 240
R-6A-NK	Nash	1945	193	Franklin 240
XR-8	Kellett	1945	1	Franklin 240
XR-8A	Kellett	1945	1	Franklin 240
XR-9	G&A	not delivered		
XR-10	Kellett	1945	2	P&W 450

*Figures in parentheses indicate modification or redesignation of previously listed aircraft
**Built for export to Sweden. Accepted by USAAF after German occupation of Sweden
***Built for export to Norway Seized by USAAF

1. Jet engines rated in pounds of thrust.
2. 75mm nose cannon.
3. War prize.
4. Lightning Mk Is for Britain. Total of 143 ordered. Five apparently sent to England. Remainder used for training in U.S. No superchargers; propellers rotated in same direction, and no armor.

Note: Lend-Lease airplanes are included in the above lists, all of which were delivered with USAAF serial numbers. The Lend-Lease Act became law in March 1941, and under its provisions Russia received 2091 Curtiss P-40s of various models, most of the 2947 Bell P-63A and P-63C models built, and about 5000 of the P-39N and P-39Q models.

Among the aircraft supplied to Britain under Lend-Lease were 2552 P-40's (Tomahawks and Kittyhawks), and the British transferred 100 of those to the Soviets.

Interestingly, a Soviet propaganda film about WWII, shown on TV in March 1984 by Ted Turner's cable network, did not mention the extent to which the U.S. equipped the Soviet Air Force as Russia desperately struggled to halt the German invasion. It did claim 66,000 German aircraft destroyed in those battles, which is at least five times the true number.

Total German aircraft production, 1 September 1939 through 1945, was 113,514. Total aircrew losses, both combat and training, for the same period were 80,558 killed and missing; 31,258 wounded or injured. Of these, 70,030 were killed or missing in combat.

The Germans produced 20,001 FW 190 and 30,480 Me 109 fighters (some writers prefer "Bf" 109 since the Messerschmitts were mostly built by *Bayerisch Fleugzeugwerke*). Hitler had no strategic air force, and no suspicion that he needed one. The 15,000 Junkers Ju 88s were the principal Luftwaffe bomber. The Luftwaffe's principal night fighter and fighter-bomber was the Me 110, of which 5762 were built. Certain versions of the FW 190 were configured as fighter-bombers. Only 4881 of the famed Ju 87 Stuka dive bombers were produced.

The U.S. aircraft industry turned out 300,000 airplanes 1941-1945 inclusive. About 175,000 of that total were combat aircraft of 18 principal types. Total dollar value was placed at $45 billion. Average price of a P-51 or P-47 was near $75,000. The P-40s were down to $44,000 when production ended in 1944.

The U.S. lost 18,000 aircraft in the war against Germany and Italy, while the British lost 22,000, and the Germans 57,000.

A total of 79,265 American airmen were lost in combat in WWII.

Japanese aircraft production 1941-1945 inclusive totalled 69,888 units, of which 52,242 were combat airplanes. The Japanese had about 3500 first-line combat airplanes at the time of the Pearl Harbor attack, but so great were their losses—50,955, combat and training—during the ensuing 44 months of the Pacific War that at no time did Japan have more than 8000 airplanes committed to combat duty. When the end came, Japan did have approximately 7000 airplanes, including trainers, held in reserve for *kamikaze* missions against the anticipated invasion.

The USAAF claimed a total of 10,343 Japanese aircraft destroyed in all theaters of the Pacific War. Navy and Marine fliers claimed slightly over 15,000, according to U.S. records. Japanese official figures list 23,835 Japanese Army planes lost, of which 16,255 went down in combat. Japanese Navy planes expended totaled 27,120, of which 10,370 were lost in combat. In other words, the Japanese lost a total of 26,625 airplanes in combat in WWII, according to their records.

U.S. Navy and Marine pilots sank 161 Japanese warships and 447 merchant ships. American submarines, however, accounted for 63 percent of all Japanese merchant shipping sent to the bottom.

Curtiss P-40C of the 33rd PS, 8th PG, which went to Iceland in August 1941 to preempt possible German seizure of that strategically located island. The P-40s flew from the deck of the carrier *Wasp*. (USAF)

Donovan Berlin (L), designer of the Curtiss P-36/P-40, with Maj. Gen. H.H. Arnold, Chief USAAC, and Burdette Wright (R), VP and General Manager of Curtiss-Wright's Airplane Division at Buffalo, NY, September 1940. (Donovan Berlin)

The Curtiss AT-9 Jeep was a two-place, multi-engine pilot trainer during WWII. Engines were 300-hp Lycomings. (USAF)

Junkers Ju 87 dive bomber was very effective in the conquest of Europe, where it did not have to contend with enemy fighters. Lightly armed, the Ju 87 Stuka had a maximum speed of 217 mph at 16,000 feet; engine was the Junkers Jumo of 900 hp. (USAF)

This Japanese photo, taken at the beginning of the attack on Pearl Harbor, shows a Kate climbing away as its torpedo explodes against the Oklahoma. Ships around Ford Island are: 1) the cruiser *Raleigh*; 2) target ship *Utah*; 3) cruiser *Tangier;* 4) battleship *Nevada;* 5) battleship *Arizona* with repair ship *Vestal* on far side; 6) battleships *Tennessee* and *West Virginia;* 7) battleships *Maryland* and *Oklahoma*; 8) the oiler *Neosho*; 9) battleship *California*; 10) seaplane tender *Curtiss*; H) hospital ship *Solace,* and K) attacking *Kates*. (USN)

Explosions within the battleship *Arizona* particlally hide the doomed *Oklahoma* and *West Virginia* while the *California* (L) burns and sinks. (Movietone News)

At 1215 hours on 7 December an explosion rips the bow from the *USS Shaw* at Pearl Harbor while other fires continue to rage out of control. No U.S. carriers were anchored at Pearl when the Japanese struck. (USN)

One of the 62 P-40s destroyed on the ground at Wheeler Field during the Japanese surprise attack on Pearl Harbor. The attacking force consisted of 352 aircraft from six aircraft carriers. (USN)

The PBY series of Navy long-range patrol aircraft was in production from 1935 into 1945. A total of 2387 was built, with 636 of those going to America's allies during WWII. Most were built by Consolidated; the PBY-5 was produced by Boeing of Canada as the PB2B, and also by the Naval Aircraft Factory as the PBN. (USN)

Consolidated B-24 Liberator approaches for landing at Midway above nesting gooney birds. The goonies, named by U.S. servicemen, were actually members of the albatross family. More Liberators were built than any other heavy bomber in WWII. In addition to Consolidated (Convair), B-24s were also manufactured by Ford, North American, and Douglas. (USAF)

Lt. Ben S. Brown was one of the 24th PG pilots who flew patched-up P-40s resisting the Japanese in the Philippines until Bataan fell. He is pictured here at a training base in Florida later in the war. (USAF)

The first P-38s to see combat were F-4 photo-reconnaissance versions sent to Australia in April 1942 for operations over New Guinea. Late in June, Lightnings of the 1st and 14th FGs flew the Atlantic to England, while a squadron was hastily put together for operations in the Aleutians. (USAF)

The U.S. Navy took delivery of 2290 Grumman TBF Avengers between September 1941 and September 1945, 458 of which went to America's allies under provisions of the Lend-Lease Act. (USN)

Lt. Edward "Butch" O'Hare, flying a Grumman F4F Wildcat fighter from the carrier *Lexington*, shot down five enemy airplanes off Bougainville on 10 March 1942. (USN)

The Grumman F4F-4 Wildcat was the main Navy fighter throughout the first year of the war in the Pacific. Its P&W Twin Wasp of 1200 hp gave the Wildcat a top speed of 318 mph. Eastern Aircraft (GM) built F4F-4s as the FM-1. (USN)

Battle of the Coral Sea. Crew is unaware that the *Lexington* is doomed. Fires below will soon spread out of control. (USN)

The North American B-25 Mitchell was the most widely used medium bomber of WWII, with almost 10,000 produced. More than 2000 went to Russia, Britain, China, and Brazil. The U.S. Marines received 706, which were designated PBJ-1s. Pictured is an early model similar to those flown by the Doolittle Raiders. Armament varied widely on the several versions; the "hard nosed" G and H models were fitted with a 75 mm cannon in addition to as many as eight forward-firing .50-caliber machine guns. This original B-25 had a maximum speed of 322 mph; cannon and additional machine guns reduced that to 275 mph. Engines were Wright Double Cyclones, R-2600s, of 1700 hp. (Stubbs Collection)

Crew prepares to abandon the *Yorktown* at conclusion of the Battle of Midway as *Yorktown's* air group, returning from strikes against the enemy fleet, diverts to *Enterprise* and *Hornet*. Surviving Japanese pilots had no such alternative; the enemy lost all four of his aircraft carriers committed to this critical battle. (USN)

Grumman F4F Wildcat returns to the escort carrier *Suwanee* during the Battle of Rennell Island, 29 January 1943; this was the final carrier action of the larger battle for Guadalcanal, which lasted from 7 August 1942 to 9 February 1943. (USN)

The war in North Africa's western desert had been in progress for two years when American forces landed there in November 1942. Here an RAF salvage crew retrieves a Tomahawk IIB (P-40) forced down in the wasteland. (Curtiss-Wright)

Flight A of RAF 112 Squadron, Desert Air Force, was the first Allied unit to decorate its aircraft with the tiger shark mouth, September 1941. It was originally seen on the Luftwaffe's II/ZG 76 "Shark" Gruppe, which flew Me 110s during the Battle of Britain in the summer of 1940. (Imperial War Museum)

A P-40 Warhawk, 64th FS, 57th FG, 9th AF. The sketchily-equipped 9th AF was part of the RAF Desert Air Force, fighting through Libya and Tunisia before joining in the liberation of Italy under U.S. commanders. (USAF)

Lt. Charles B. Hall of the Warhawk-equipped all-black 99th FS, the first black airman to shoot down a German plane in WWII. Hall ended the war with three confirmed victories, two FW190s and an Me 109. The 99th belonged to no group when it went to North Africa early in 1943. Later, in Italy, enough black pilots had been trained to form the 332nd FG, commanded by Col. (later General) Benjamin O. Davis, Jr. A total of 992 black pilots were trained at Tuskegee, Alabama, during the war; 450 went overseas, and the 332nd FG claimed 111 air victories. (Charles Hall)

The certificate of a strictly unofficial organization for those Allied airmen who went down in North Africa and managed to return to their units. One P-38 pilot walked 250 miles from inside enemy-held territory. (Col. A.J. Knight)

Dense smoke engulfs the runway of an American Beaufighter base in Italy after one of the night fighters crash-landed. The radar-equipped two-place Bristol Beaufighter was powered with a pair of bristol Hercules radials of 1590 hp, and had a maxiumum speed of 325 mph. Armament was four 20mm cannon and six .303-inch wing guns. The USAAF had but one squadron of these British-built aircraft as the RAF replaced them with deHavilland Mosquitos. (USAF)

Hitler and his fighter aircraft boss, Lt. Gen. Adolf Galland. In background are Chief of German Armed Forces, Field Marshal Wilhelm Keitel (L) and Field Marshal Erhard Milch, State Secretary and boss of arms production. (National Archives)

A Douglas C-47 of the 9th Troop Carrier Command, painted with invasion stripes, air-snatches a WACO CG-4 troop glider. (USAF)

A Focke-Wulf 190 fighter, attacking a formation of 9th AF Martin B-26s as the Marauders released their bombs, narrowly missed this string of bombs as he passed beneath his intended victims. (USAF)

A Douglas A-20 of the 410th BG made a belly landing in friendly territory after suffering crippling hits during a bombing mission, 27 December 1944. (USAF)

The Focke-Wulf 190 was produced in many versions and with a wide variety of armament. It was unquestionably the best German fighter of the war, and a worthy adversary of the P-51. Top speed was 405 mph at 19,000 feet in early versions; 435 mph for the 1944 FW 190D. (Gaston Botquin)

An Me 109E, "Battle of Britain" version of the prolific German fighter. Maximum speed was 354 mph at 16,000 feet with the Daimler-Benz inverted V-12 of 1100 hp, 450 mph at 20,000 feet fitted with the 1800-hp engine in the K model. More than 30,000 Messerschmitt 109s and Czech and Spanish variants were built, with production continuing in the latter two countries after the war. (Arno Abendroth)

A Douglas C-47 of the 439th Troop Carrier Group, 9th AF, went down "somewhere in France" while on a supply mission to advancing U.S. troops. (USAF)

Eighth AF B-17s enroute to attack the Messerschmitt works at Brunswick are set upon by an enemy fighter (upper left) that has evaded the bombers' fighter escort. (USAF)

Thunderbolt pilot Maj. Everett Stewart had this Me 110 in his gunsight when Lt. John Coleman, intent on the same target, dropped in front of Stewart and shot down the enemy aircraft. (USAF)

A Junkers Ju 88 is destroyed by a U.S. fighter pilot not identified by the AF. Action took place near the Dummer Lake area, Germany. (USAF)

Capt. Robert S. Johnson of Lawton, Oklahoma, was the second ranking American ace in the ETO with 28 confirmed air victories. Johnson flew P-47 Thunderbolts. (USAF)

The Republic P-47 Thunderbolt carried the USAAF fighter burden over Europe alone until joined by two P-38 groups and a handful of P-51s late in 1943. The "Jug" was fast and extremely rugged, with eight .50-caliber wing guns. The Jug's engine was the 2000-hp P&W Double Wasp; maximum speed was 406 mph at 27,000 feet. The 1944 D model had a bubble canopy, 2500 hp, and a speed of 429 mph at 30,000 feet.

Lockheed P-38 Lightnings shootup an enemy locomotive as part of the U.S. air campaign to immobilize the German Army. At the time of the invasion of Europe by Allied forces, there were seven P-38 groups in the 8th and 9th Air Forces. (USAF)

The first of Hitler's Vengeance Weapons was the V-1 buzz bomb, or "Doodler," as the British called it. Powered with a pulsejet engine and gyro-stabilized, some 8000 were fired against London and Antwerp from bases in France and Holland, about half of which were shot down by Allied fighters and antiaircraft fire. The V-1 carried a ton of explosives and flew at 350 mph. This one was found intact in France by advancing U.S. troops. (USAF)

The Northrop P-61 Black Widow night fighter entered combat in the ETO shortly after D-Day; the 422nd NFS, 9th Tactical Air Command, was the first P-61 unit to become operational. The 425th and 414th followed. The Black Widow had a top speed of 369 mph at 20,000 ft. *Jukin Judy* is in center above; *Lovely Lady* on right. Both were 422nd machines. (USAF)

This Curtiss P-40N Warhawk of the 80th FG was flown by Col. Ivan McElroy while the group was stationed at Nagahuli, Upper Assam, India, 1944. (USAF)

Typical summer afternoon in the Aleutians. These Warhawks of the 343rd FG, 11th AF, shared the "Forgotten War" with a squadron of P-38s and one of P-39s, aided for a time by two squadrons of Royal Canadian Air Force Kittyhawks. This photo was made at Alexi Point Airbase, Attu, after the Japanese had been forced out in 1943. (USAF)

A Lockheed P-38F-5 Lightning of the 54th FS, 343rd FG, in the Aleutians. Of the original 31 Lightning pilots who went to the Aleutians in the summer of 1942, only ten survived, most lost to the "world's worst flying weather."

Pilots of the 23rd FG, China Air Task Force (later, 14th AF), which replaced the Flying Tigers on 4 July 1942. L to R, Maj. John R. Alison, eight victories; Maj. David "Tex" Hill, a former Flying Tiger with 14 victories (later a Colonel and group commander with 18¼ victories), Maj. Albert "Ajax" Baumler, six victories, and Lt. Mack Mitchell. (USAF)

During the biplane era these were the "Three Men on a Flying Trapeze," the Army Air Corps' aerobatic exhibition team, distant ancestor of the present-day Thunderbirds. This trio performed with their wingtips tied together with short lengths of rope! Pictured here in China late in 1943, they are, L to R: Lt. Col. John H. Williamson, Maj. Gen. Claire Chennault, and Capt. W. D. McDonald, of the China National Aviation Corp., CNAC, the Pan Am-sponsored Chinese airline (in China, "The Middle Kingdom Space Machine Family"), which helped form the ABC Ferry Command to fly the Hump. (USAF)

On 31 August 1944, a month of almost continuous action was begun by the four carrier groups of Task Force 38 in support of a series of amphibious landings by U.S. forces on islands within striking distance of the Philippines. Here, Grumman TBFs return to their carrier following attacks on enemy targets in the Bonin Islands. (USN)

The Vought F4U Corsair entered combat with VMF-124 on Guadalcanal, 11 February 1943. The Marines got this first-rate fighter because the Navy was slow to accept it for carrier operations because of its 90-mph landing speed and poor pilot-visibility forward in nose-high landing attitude. VMF-124 at last reported aboard the *Essex* late in 1945. Shown here is the AU-1 variant; specialized for the ground-attack role, it was a post-WWII development that saw action in Korea. (USN)

Douglas SBD Dauntless from Task Force 58 over Duhlan and Eaton Islands enroute to the Japanese stronghold at Truk, 29 April 1944. The unremarkable Dauntless had a remarkable record in the Pacific War, beginning with the sinking of four enemy carriers at Midway. (USN)

Maverick Marine pilot Maj. Gregory "Pappy" Boyington briefs pilots of VMF-214, the collection of misfits known as the Black Sheep Squadron, on Guadalcanal. Boyington had gone to China with Chennault and scored six victories as a Flying Tiger, but Chennault gave him a dishonorable discharge. As a Marine pilot, Boyington ran his score to 28 before being shot down and becoming a prisoner of the Japanese. (USMC)

The Japanese Mitsubishi A6M5, Model 52, Type 0—the famed Zero-sen. This version, of which 1701 were built, had a top speed of 330 mph at 15,000 feet. Allied code names were "Zeke" for earlier models, "Hamp" for this one. (USAF)

Grumman F6F Hellcat entered combat 31 August 1943 with VF-9. Fitted with the P&W Double Wasp of 2000 hp, the first production Hellcats, F6F-3s, had a top speed of 375 mph at 17,300 feet. A total of 12,276 were built, most of them F6F-5s. The example shown here is in postwar USN reserve markings.

Japanese Army bomber, Mitsubishi Ki-21, Type 97 "Sally," had a maximum speed of 250 mph at 10,000 feet, 2200 pound maximum bomb load, and 14,00 pound gross weight. A total of 1800 were produced. The Sally saw extensive service in the China-Burma-India Theater of War. (USN)

The Curtiss C-46 Commando was the largest twin-engine airplane built up to its time. Originally the CW-21, intended as a pressurized airliner, it was rushed into production—largely untested—due to the pressing need, and was the main transport used over the high northern Himalayas carrying supplies from India to China. (USAF)

Although formal surrender documents were not signed until 2 September 1945, the Japanese had sued for peace on the 14th of August, and this Japanese delegation arrived at Kunming on 27 August to arrange the peaceful departure of Japanese troops from China. The aircraft is a Nakajima L2D3, Type 0, "Tabby," a copy of the Douglas DC-3/C-47. A total of 450 was built. (USAF)

Chapter 14

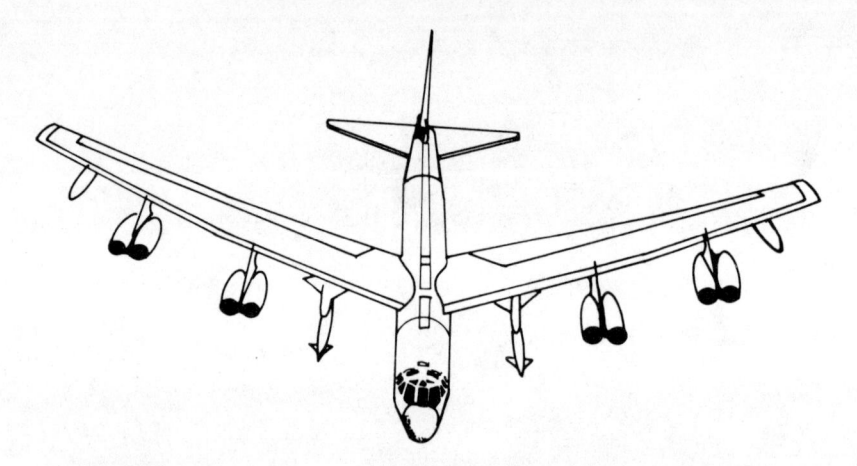

The Era of Confrontation

From the hour of Japan's capitulation ending the Second World War, there was acute pressure from the public, the Congress, and the troops themselves for demobilization of America's mighty military forces. Detailed plans for an orderly return of more than six million servicemen to civilian life were upset by the clamor, and Army personnel alone was reduced by half during the first six months of 1946. By 30 June 1947 the Army was a volunteer body of 684,000 ground troops and 306,000 airmen. The Navy was down to 484,000, the Marines, 92,000. There had been eight million Americans under arms just two years earlier.

National Security Act

The exigencies of war had demonstrated the need for unified control of our armed forces, and in July 1947, President Truman signed the National Security Act, which became effective 18 September that year, creating the Military Establishment, the Central Intelligence Agency, the National Security Council, and making the Air Force a separate branch of the military equal to the Army and Navy. The Army and Navy each kept its organic air service. In practice, this meant no restrictions on Navy/Marine aircraft types, although the Army's only fixed-wing aircraft would eventually be limited to liaison craft (and Beech King Airs for the brass to ride around in), the Army's air fleet consisting mostly of helicopters. In 1949, the Military Establishment became the Department of Defense, containing the Army, Navy, and Air Force; the civilian secretaries of the three services answered to the Secretary of Defense who had cabinet rank.

There was no doubt by that time that the hard-won peace was to be an uneasy one. Winston Churchill, in a speech at Fulton, Missouri early in 1946, had warned that the Soviets were lowering an "Iron Curtain" across the European continent. It drew East Germany, Poland, Hungary, Rumania, Bulgaria, Yugoslavia, and Albania behind that curtain. In Asia, besides insisting on full control in Northern Korea, the Soviets had turned Manchuria over to the Chinese Communists under Mao Tse-tung and were encouraging Mao in his renewed effort to wrest power from Chiang Kai-shek. The Soviets were clearly embarked on a policy of expansion. They refused to cooperate with Allied attempts to help the defeated Axis countries rebuild their shattered cities and economies, and labeled the Marshall Plan—$16 billion in aid to the countries of Western Europe—"an instrument of American Imperialism."

SAC Formed

During the two-year period between the war's end and creation of the U.S. Air Force, Gen. Carl Spaatz, one of America's ablest air commanders, held things together and presided over the birth of the Strategic Air Command, a nuclear air strike force with global capability, which was seen as the Free World's most effective deterrent to Soviet armed aggression.

SAC was officially born 21 March 1946 with Gen. George Kenney as its first commander. Kenney was given 36,800 men, 18 active air bases, and 600 aircraft, 250 of which were bombers—B-17s, B-29s, and B-25s. Converted B-29s became the first of the modern aerial tankers, serving the Boeing B-50, essentially a B-29 with heavier skin and more powerful engines.

Air Force leaders were divided over the question of a proper strategic bomber. Most favored the Convair B-36, a mammoth machine with six pusher engines of 3000 hp each, which was actually a prewar design. But Kenney argued for the B-50 because he could have more of them. The issue was further clouded for a time by Jack Northrop's four-engine XB-35 Flying Wing bomber, which promised a nonrefuelled range of 8000 miles with a 16,000 pound bomb load. The XB-35 first flew in June 1946, followed by an eight-jet version, the YB-49, which made its maiden flight late in 1947. Ultimately, the flying wings seemed to pose aerodynamic problems that could take years to solve and the project was dropped. Two of the three YB-49s built crashed.

Boeing B-52 Genesis

In the meantime, Boeing offered a series of design proposals to the Air Force for the B-52, beginning early in 1946 with Boeing Design 462. From the outset, the B-52 was planned as the global bomber that would replace the B-36 in the '50s, assuming that the B-36 was produced. For two and a half years the B-52 design studies concentrated on a four-engine prop-jet—partly because a prop-jet would be more fuel-efficient, and partly because no pure jet engine of sufficient power was then under development. None of the prop-jet designs promised the performance the Air Force felt that it would need ten years down the road. But late in 1948 a design study of a super bomber fitted with eight P&W J-57 turbojets (an engine in an early stage of development) resulted in a lot of broad smiles in the halls of the Pentagon. A few months earlier, the question as to whether or not the Air Force should have such a super bomber had been decided for us by the Soviets when they attempted to force the Western Allies out of Berlin.

Berlin Airlift

In a meeting between the Allied heads of state at the Russian port city of Yalta in February 1945, President Roosevelt, tired and ill just two months before his death, agreed to a joint sovereign authority over a defeated Germany, with U.S., British, French, and Soviet forces occupying separate zones, while an Allied Control Council, made up of the commanders of the four occupation armies, would decide all matters of national import relating to Germany's immediate future.

The Yalta Conference gave the Russians control of territory they could not have dreamed of otherwise possessing, and it soon became clear that they regarded every inch of it as their own freedom, the joint Control Council notwithstanding. The Western Allies were intent on "de-Nazifying" Germany, presiding over the establishment of a German-run free government, and restoration of the German economy—none of which pleased the Soviets. They were particularly incensed at the U.S. for its offer to help Germany rebuild with Marshall Plan funds.

With all of Eastern Europe firmly in the Soviet orbit, the Russians decided to remove the one remaining outpost of freedom behind the Iron

Curtain—that part of Berlin that was administered by the Western Allies. Berlin was 110 miles inside the Russian zone, and the Western Allies were guaranteed access to their part of the city by rail, waterway, the autobahn, and three 20-mile-wide air corridors.

On 24 June 1948 the Russians tore up rail lines and blew bridges to halt all surface traffic across their zone of occupation in an attempt to isolate Berlin. But there was nothing they could do about the air corridors short of shooting down any Western Allied aircraft that chose to make use of those aerial freeways. That, of course, would be an unmistakable act of war.

Within 48 hours, Gen. Curtis LeMay, Commander of the USAF in Europe, organized an airlift to fly essential supplies into Berlin, while Gen. Hoyt Vandenberg, AF Chief of Staff since Spaatz' retirement two months earlier, ordered all available Air Force transport aircraft to Frankfurt to support "Operation Vittles." During the 13 months the Berlin airlift was in operation, the Military Air Transport Service delivered 2,231,000 tons of life-giving supplies to the beleaguered city, averaging 700 flights daily.

The Soviets could find no way to salvage their ill-conceived ploy. They merely lifted the blockade.

Power in the Pod

Development of the turbojet aircraft engine took a surprisingly long time when one considers that the principle of jet propulsion has been understood for centuries, and that the first patent for such an engine was granted in 1914. That patent went to Dr. Henri Coanda, a Rumanian, who exhibited a jet-propelled airplane at the 1910 Paris Airshow. Coanda's machine had a ducted fan faired into its nose that was turned by a gasoline engine. He wrecked it attempting to combine its test flight with his first self-administered flying lesson.

In 1928, Frank Whittle in Great Britain wrote an exam thesis on the subject jet propulsion for aircraft; two years later at age 23, Whittle obtained his first patents on a gas turbine engine. Working independently in Germany was Hans von Ohain, who applied for his first patents in 1935.

Although Whittle was able to form Power Jets, Ltd., in 1936 to develop his design, not until 1939 did the British Government give him an order for an engine, concurrently commissioning Gloster Aircraft Company to build a special airframe to flight-test the new powerplant. In Germany, Dr. Ernst Heinkel had already added a gas turbine engine division to his aircraft plant and, working with Von Ohain, produced the first turbojet airplane, the He 178, which made its maiden flight, in secrecy, 27 August 1939 near Rostock. Five days later, Hitler marched into Poland, and Germany's lead in this field would not be known to the world until our airmen encountered the Me 262s over Europe late in the war. This airplane had first flown in July 1942, but Hitler and his army generals were not convinced of its potential by German air commanders until it was too late to produce it in meaningful numbers.

The Me 163 rocket fighter, designed by Dr. Alexander Lippisch, first flew in 1941 and, produced in small numbers, it began attacking American bombers in 1944. The He 162 jet fighter and the Arado 234 jet bomber were also in production in Germany at the end of WWII.

The Italians flew the Caproni-Campini N-1 jet in 1940, but it was merely an updated version of the 1910 Coanda—a gasoline engine powering a compressor—and with a top speed of 140 mph the N-1 clearly had no future.

In the U.S. prior to WWII, there was no apparent interest in either rocket or gas turbine engines for airplanes. When someone suggested to Gen. Arnold in 1940 that the expertise of America's rocket pioneer, Dr. Robert Goddard, be put to work by the Air Corps, Goddard was asked to work on small rockets used for extra takeoff boost on airplanes, a system known as JATO, for jet assisted takeoff.

American manufacturers were pushed into the jet engine business after Gen. Arnold saw the Whittle engine in England in mid-1941 and managed to borrow one, which was copied (and improved upon) by General Electric.

Bell Aircraft was given a contract to build a

suitable airframe, and on 1 October 1942 Bell test pilot Robert M. Stanley flew America's first jet-powered airplane on its initial flight. It was the Bell XP-59 Airacomet. A total of 66 were built. America's first combat-ready jet fighter was the Lockheed P-80 Shooting Star, which appeared in 1945.

Speed of Sound Exceeded

The development of turbojet and rocket engines promised speeds several times that which had been achieved in prop-driven airplanes if aerodynamicists could find a way to penetrate the "sound barrier." As early as 1941, test pilots had encountered the spectre of compressibility in dives from high altitudes in the Lockheed P-38. The airplane's nose tucked under to near vertical and the controls refused to respond until the airplane entered denser air at a lower altitude.

The use of high-velocity wind tunnels allowed engineers to see the shock waves created by airplane models as the airflow neared the speed of sound (the speed of sound varies with altitude and temperature—it is 760 mph at sea level and 59 degrees F; near 660 mph at 36,000 feet), and the problem was identified if not solved.

It was solved with a series of supersonic research aircraft beginning with the Bell X-1. The X-1, fitted with four rocket engines built by Reaction Motors, Inc., was the first airplane in history capable of exceeding the speed of sound in level flight. Flown by Air Force Capt. Charles "Chuck" Yeager (after Bell's chief test pilot refused the job), and dropped from a B-29 at 20,000 feet, the X-1 first accelerated through the sound barrier 14 October 1947 to reveal that there was no barrier to a properly designed airplane. Nothing dramatic happened to tell Yeager that he had crossed the line into supersonic flight. Capt. Yeager (later brigadier general) exceeded the speed of sound a dozen or more times in 30 flights during the 13 months he was associated with the project. The X-1's four rocket engines provided 6000 pounds of thrust, burning 4000 pounds of alcohol and liquid oxygen (LOX) per minute for 2½ minutes.

The X-1 flights were made at Muroc Dry Lake in California desert, "Muroc" being the name of the Corum brothers (who homesteaded the site) spelled backwards. In 1951 it becames Edwards Air Force Base, named for Capt. Glen W. Edwards, who was killed there in 1948 test-flying the Northrop XB-49 flying wing bomber.

In December 1953, Yeager took the Bell X-1A to Mach 2.5 (1650 mph) at 70,000 feet. That same year the North American F-100 Super Sabre claimed the official world's speed record at 754.98 mph, the Super Sabre being the first turbojet to exceed the speed of sound in level flight.

Another War

In the fall of 1949, shortly after the Soviets successfully detonated their first A-bomb, the communist insurgents in China defeated Chiang Kai-shek and, early in 1950, the new Chinese Communist Government and Soviet Russia signed a "mutual assistance" treaty.

Meanwhile, the communist government in North Korea, installed by the Soviets two years earlier, fostered an internal rebellion of sorts in South Korea, where a democratic form of government had been left behind by the U.S. occupation forces. When that insurgency failed to gain any popular support, the North Koreans decided to take South Korea by force.

The Korean communists had every reason to believe that an invasion of the south would succeed. They had a well-equipped army of 135,000, plus some 200 Russian aircraft, while South Korea's army of 95,000 awaited equipment promised, but not yet delivered, from America. Many analysts believe that the communists were misled into thinking that they had practically been invited to march across the 38th Parallel into South Korea by President Truman's Secretary of State Dean Acheson, who publicly defined the U.S. "defense line" in Asia as running south from the Aleutian Islands to Japan, to the Ryukyu Islands, and then to the Philippines. That delineation left Taiwan (Formosa), where Chiang Kai-shek and his followers had fled, and Korea, outside the line.

Before daylight on Sunday, 25 June 1950, the

North Korea People's Army moved swiftly south into the Republic of Korea in a surprise attack that carried the invader into the South Korean Capital of Seoul on the 28th.

In Washington, where a 14-hour time difference made it 24 June when the North Koreans crossed the 38th parallel, the first report of the invasion arrived that night. Early on the 25th the U.S. requested a meeting of the United Nations Security Council, and President Truman ordered Gen. MacArthur, at MacArthur's Far East Command headquarters in Tokyo, to supply Republic of Korea (ROK) forces with ammunition and equipment, evacuate American dependents from Korea, and report on how to best assist the Republic further. The President also ordered the U.S. 7th Fleet from Philippine waters to Japan.

Then, on the 27th, the U.N. Security Council passed a resolution to ". . . furnish such assistance to the Republic of Korea as may be necessary to repel the armed attack and to restore international peace." At that, Truman sent the USAF and the 7th Fleet to Korea, and by the 5th of July American ground forces were in South Korea to fight under the blue-and-white flag of the United Nations. Gen. MacArthur commanded all of U.N. forces, which immediately included troops from Great Britain, Australia, Turkey, and later those of 17 other nations besides the U.S. and South Korea.

The U.S. Navy was in action in Korean waters by 3 July, when the carrier *Valley Forge* with Air Group 5, along wth Britain's *Triumph*, operating in the Yellow Sea, launched air strikes on airfields, supply lines, and transportation facilities in and around Pyongyang, northwest of Seoul. These strikes put the F9F Panther and Douglas AD Skyraider in combat for the first time, and also were the occasion for the first Navy kills in air combat over Korea, when VF-51 F9F pilots Lt. j.g. L.H. Plog and Ensign E.W. Brown shot down two Yak-9s near Pyongyang.

The first Air Force victories were scored on 27 June, when five Yaks attacked five P-82 Twin Mustangs over Kimpo Airport at Seoul. Lts. William G. Hudson and Charles B. Moran, along with Maj. James W. Little, each downed a Yak. Hudson, flying with Staff Sgt. Nyle S. Mickley, was officially credited with the first American air combat victory over Korea.

The Navy began arriving in strength during the fourth week in July. On the 22nd, the escort carrier *Badoeng Strait* docked at Yokosuka, Japan, with the First Marine Air Wing aboard, followed four days later by the *Sicily*, and on 1 August the *Philippine Sea*. This was the beginning of carrier deployment to the combat area that, by war's end, totalled eleven attack carriers, one light carrier, and five escort carriers sent into action—several of them for two or three tours.

The Korean conflict was an undeclared war; President Truman characterized it as a "police action," and the U.N. forces, about 90 percent of which were American, sought only to push the North Koreans back across the 38th Parallel and to demonstrate that such aggressions would not be tolerated.

U.S. airpower played a major role in the Korean War, and it was a war largely fought by retread airmen, the reserves, and National Guard units. Of the 334 ships put in service during the war only ten were new—the others came from mothballs, as did 3300 Navy aircraft.

Actually, there were *two* wars—the first one with North Korea, the second with the Chinese Communists (ChiComs)—and both were fought on a peninsula, somewhat resembling Florida in shape, about 575 miles long and 95 miles wide at its narrowest. The two Koreas are separated by the 38th Parallel, and the South Korean capitol of Seoul is about 35 miles south of that dividing line roughly midway up the peninsula.

The initial North Korean thrust carried all the way to the southern end of the country by 15 September 1950, with the United Nations troops backed into a pocket around Pusan. Then, supplied by the Navy, a counterattack by the U.N. forces, combined with a daring amphibious landing at Inchon just south of Seoul, trapped the invaders and cut their supply line. About 30,000 North Koreans escaped through the eastern mountains, but the

North Korean army was soundly defeated. U.N. forces then crossed the 38th Parallel and continued to the Yalu River, which marks North Korea's border with Manchuria, arriving there on 2 November 1950.

The next day China sent 300,000 troops storming across the Yalu, and MacArthur's U.N. forces fell back below Seoul before halting the ChiCom advance in late January 1951.

A U.N. counterattack swept into North Korea once again, slowly moving northward until, in November 1951, the U.N. forces dug in along a line that was roughly 20 miles north of the 38th Parallel and which provided favorable defensive positions.

For the next 20 months there would be no significant change in this line while the ChiComs and North Koreans negotiated a settlement, communist-style, seeking to salvage an advantage from their unsuccessful military adventure. However, Dwight Eisenhower became President of the United States in January 1953. Following a firsthand look at the situation, Eisenhower sent a message to Moscow, Peiping, and Pyongyang, saying that if satisfactory progress toward an armistice was not immediately forthcoming, ". . . we intend to move decisively and without inhibition in our use of weapons, and will no longer be responsible for confining hostilities to the Korean peninsula."

That achieved the desired result, and a cease-fire agreement was signed 27 July 1953. The Republic of Korea remained free.

At the beginning of the war, the U.S. FEAF could count 535 combat aircraft, including 365 F-80s and 32 F-82s. Within a month, 764 National Guard F-51s were called to active duty, 145 of them rushed to Korea aboard the carrier *Boxer*. These fighter airplanes, along with the Navy F9Fs, and F4Us of VMF-214 and VMF-323, operating from the *Sicily* and *Badoeng Strait*, completely annihilated the North Korean Air Force during the first few months of the war, while two groups of B-29s from SAC's 15th AF destroyed the relatively few strategic targets in North Korea and then flew tactical missions against rail and communications facilities. The U.N. ground forces had marched the length of Korea to the Yalu beneath skies owned by their own aircraft, but when the Chinese entered the war they committed more than 1300 aircraft, at least 500 of which were MiG-15 fighters. America's response to that was the 4th Figher-Interceptor Wing, equipped with 105 North American F-86A Sabres, which arrived in Korea in Decemver 1950.

The ChiComs eventually had more than 1000 MiGs on Manchurian bases (off limits to U.N. attack), and often sent massive formations numbering 100 or more MiGs over Korea. More often than not they chose not to fight, or made a single diving pass on small formations of F-86s and then ran for sanctuary across the Yalu. On those occasions when the MiG pilots dared to fight they were soundly beaten.

The Far East Air Force controlled 1536 aircraft at the end of the Korean War, including seven squadrons of Marines and three foreign air units. FEAF flew a total of 720,980 sorties (a sortie is one flight, one aircraft), and dropped 476,000 tons of explosives. Marine air units operating under FEAF command flew 107,303 sorties and loosed 82,000 tons of munitions. Land-based "friendly foreigns" flew 44,873 sorties; the U.S. Navy flew 167,552 sorties, expending 120,000 tons of rockets, napalm, and bombs.

The U.N. air units destroyed 976 enemy aircraft, 1,327 tanks, 82,920 vehicles, 963 locomotives, 10,407 railway cars, 1,153 bridges, 118,231 buildings, 65 tunnels, numerous barges, boats, gun positions, and 184,808 enemy troops.

The FEAF lost 1466 aircraft, 139 in aerial combat. The Marines lost a total of 436 aircraft (none in aerial combat), and the friendly foreigns lost six airplanes in aerial combat. The FEAF downed 900 enemy aircraft, plus 168 probables. The Marines accounted for 35 enemy aircraft, including 15 MiGs, and Navy pilots shot down 16 airplanes, four of which were MiGs. The friendly foreigns had three air victories.

The Sabres were the MiG killers, downing a total of 810 enemy airplanes of which 792 were MiGs, while losing 78 Sabres.

The 38 jet aces in Korea destroyed a total of

305 planes in air combat; a check of U.S. pilots with MiG victories revealed that they had an average of 18 missions in WWII.

Army aviation flew 86,000 forward missions over Korea, 7000 of which were front-line evacuation flights.

At the end of the Korean War, Col. Francis S. Gabreski was America's top living ace with a total of 37½ victories, 6½ of those scored over Korea.

When the final casualty report for the 37 months of fighting was prepared, total United Nations casualties reached over 550,000, including almost 95,000 dead. U.S. losses numbered 142,091 of whom 33,629 were killed, 103,248 wounded, and 5,178 missing or captured. U.S. Army casualties alone totaled 27,704 dead, 77,596 wounded, and 4,658 missing or captured. The estimate of enemy casualties, including prisoners, exceeded 1,500,000 of which 900,000 were Chinese.

Vietnam

The American presence in Vietnam was owed to the desire of several U.S. administrations—beginning with that of Harry Truman—to contain Soviet expansionism. America was in Vietnam for the same reason it was in Korea, and the North Vietnamese had the same sponsors that supported the North Koreans.

President Truman backed the French return to Indochina at the end of WWII because he believed that the French would block the spread of communism in Southeast Asia. Indochina—Laos, Cambodia, and Vietnam—had been ruled by the French since the mid-19th century. The Japanese had occupied the area during WWII, and when the French returned in 1945 they found that the communist guerila force known as the Viet Minh, which had weakly opposed them there since 1930, had grown powerful during the Japanese occupation. With massive aid from the Soviets and Chinese Communists, the Viet Minh forced the French out of Indochina in 1954.

The cease-fire agreement between the French and the Viet Minh was mediated by a multi-nation conference in Geneva, which provided for the indepedance of Laos, Cambodia, and Vietnam, with a political dividing line drawn across Vietnam at the 17th parallel in recognition of the centuries-old differences between the 18 million people of the north—traditionally ruled from Hanoi—and the 16 million people of the south whose emperors had ruled from Hue before the French arrived.

A provision of the Geneva Accords called for free elections throughout Vietnam by July 1956, at which time the north and south were to be united under a single government. That embarrassed the U.S. and its friends. The U.S. Government, then in the hands of President Eisenhower, refused to sign the Geneva Accords because Ike well knew that the Viet Minh would have terrorists in every South Vietnamese hamlet to ensure the election of their leader, Ho Chi Minh.

Meanwhile, the U.S. found itself providing aid to a series of corrupt governments in South Vietnam that had little to recommend them except that they opposed communism. President Eisenhower sent 400 military advisors, with some military equipment, when the French pulled out the last of their forces in 1956, to join some 280 Americans who had been there since 1954.

A revolt within the Viet Minh left them unable to react when the South Vietnamese Government repudiated the Geneva Accords and refused to participate in the scheduled 1956 elections. By 1961, however, Ho Chi Minh had destroyed his enemies in the north and restructured his organization around veteran Red guerillas, which he called the National Liberation Front (Viet Cong), and then began a campaign of murder and abduction in South Vietnam.

President John Kennedy inherited the problem at that time; his response was to increase the number of U.S. military advisers in South Vietnam to 15,000. The Viet Cong continued to gain strength, however. By mid-1964, 115,000 were in South Vietnam, where they had murdered 5587 and kidnapped a *known* 26,504 people.

Then, when North Vietnamese torpedo boats attacked and slightly damaged the U.S. destroyer *Maddox* in the Gulf of Tonkin on 2 August 1964,

Lyndon Johnson, who had been in the White House since the previous November, told the Congress that there had been two such attacks (there was but one) and asked Congressional support for "all necessary action to protect our armed forces and to assist nations covered by the SEATO treaty," the SEATO Treaty being an agreement among certain interested nations* to give assistance to Laos, Cambodia, and South Vietnam if they were attacked and requested aid. The Congress obliged with a resolution which held that the President was authorized to "take all necessary steps, including the use of armed force, to assist any member or protocol state of the Southeast Asia Collective Defense Treaty requesting assistance in defense of its freedom." It was adopted by a vote of 416 to 0 in the House, 88 to 2 in the Senate.

Johnson had ordered air strikes on North Vietnam from U.S. 7th Fleet Attack Carriers *Constellation* and *Ticonderoga* the night before, and from that date increasingly committed U.S. military forces to a war in Vietnam.

It was a strange war, a war fought on the enemy's terms and at his convenience. Johnson, his Secretary of Defense Robert McNamara, and Deputy Secretary of Defense Cyrus Vance directed a war of attrition in the name of a policy known as "flexible response." It was a policy not of action, but of reaction. American airpower was tightly fettered, with important strategic targets in North Vietnam off limits. Tactical targets were elusive. The Viet Cong moved supplies into the south on bicycles down one of the "Ho Chi Minh Trails" inside Laos.

There was no front line; the enemy could be anywhere, and was often indistinguishable from civilians. U.S. ground forces were concentrated in fortified base camps and were dispatched on search-and-destroy missions by helicopter.

The Army's helicopters were perhaps the single most important weapon of the war. They provided great mobility to the troops, along with the advantage of surprise, while robbing the enemy of most opportunities to effect ambush attacks. The helicopter gunships, which were essentially rotary-winged fighters, gave the Army a kind of close air support that is not possible with fixed-wing aircraft. The Army's helicopters in Vietnam clearly established the value—the absolute *necessity*—of Army organic aviation under the direction of ground commanders.

As the Viet Cong and units of the regular North Vietnamese Army steadily increased their strength in South Vietnam, Johnson and his advisors matched the enemy's escalation of the war by sending more U.S. troops, eventually maintaining more than 500,000 Americans as well as a South Vietnamese Army of some 340,000, ad 48,000 South Koreans in South Vietnam.

Meanwhile the U.S. Army commanders on the scene claimed one victory after another based on the impressive number of enemy dead counted after each battle (most battles being at the company or platoon level). By the time Richard Nixon became President in January 1969, the enemy had lost half a million men, a figure readily confirmed by North Vietnamese Defense Minister Vo Nguyen Giap, but what Lyndon Johnson and his advisors failed to perceive—as did President Nixon in the beginning—was that Giap was prepared to lose that many *more*. At that time, American combat deaths were approaching the 33,000 mark in South Vietnam, and many Americans—perhaps most—were disenchanted with a war that had confused beginnings, no clearly defined purpose, and the promise of an inconclusive end at best.

Once President Nixon realized that enemy body counts reported by his generals in Vietnam bore small relevance to the eventual outcome of the conflict, and that communist-style negotiations for a cease-fire were being conducted primarily for the benefit of the world press, Nixon at last allowed concentrated air attacks on strategically important targets in North Vietnam (and, secretly, on enemy sanctuaries in "neutral" Cambodia).

That resulted in the "Eleven-Day War," which began on 18 December 1972. One hundred and twenty-nine B-52s, along with 400 fighters (in-

---

*The signatories to the SEATO Treaty were: United States; United Kingdom, France, Austrialia, New Zealand, Thailand, Philippines, and Pakistan.

cluding Navy A-6 Intruders from carriers in the Gulf of Tonkin and Air Force F-111s from bases in Thailand) struck military and industrial targets in the Hanoi area that day. Except for a 36-hour break over Christmas, the maximum-effort attacks continued until, on 30 December, the enemy signaled that he wanted to talk. With 80 percent of his electrical production gone, along with 25 percent of his petroleum resources—and the Americans apparently intent on finishing the job—the North Vietnamese gained a different perspective of the war.

A substantive agreement to end the war was reached by 27 January 1973, although Viet Cong activities in Cambodia and Laos prompted President Nixon to send the B-52s over those tortured nations a number of times until the last mission was flown 15 August 1973.

During the eight years that American airpower was (mis)used in Vietnam, 126,615 B-52 sorties were flown, with 55 percent of the targets in South Vietnam, 27 percent in Laos, 12 percent in Cambodia, and *six percent in North Vietnam*. A total of 17 B-52s were lost over North Vietnam, 15 of which went down during the Eleven-Day War, which cost 66 B-52 crewmen killed or captured. And this was in the face of the most concentrated defensive fire ever experienced by U.S. aircrews.

The Eleven-Day War was authorized just seven years too late.

The 385 Convair B-36s built equipped 33 SAC squadrons from mid-1948 until the B-52 replaced them in the late '50s, the last B-36 being retired in 1959. Power was six P&W Wasp Major engines of 3500 hp each, plus four GE J-47 turbojets of 5200 pounds thrust each. Maximum speed of 347 mph at 40,000 feet; range was 9500 miles with five tons of bombs. (Convair)

The Northrop XB-35 Flying Wing bomber appeared in 1946 fitted with four P&W Wasp Major engines driving pusher contrarotating propellers. Following a second XB-35 and a YB-35 came two YB-49s powered with eight Allison J-35 turbojets of 4000 pounds thrust each. Maximum speed was 530 mph. (Northrop)

A B-52 carrying a pair of Hound Dog missiles; 744 B-52s were built, with deliveries beginning in April 1953 and ending in March 1961. In 1984 approximately 300 remained in service, having been rebuilt and considerably updated. Cruising speed at 20,000 feet is .84 Mach. Unrefueled range of the H models with turbofan engines is 10,000 miles. Early models, mostly Ds, were reconfigured to carry "iron bombs" over Vietnam. (Boeing Company)

The Bell P-59 Airacomet was America's first jet airplane. Fitted with a copy of Britain's Whittle centrifugal-type turbojet engine, the XP-59 first flew 1 October 1942. In all, 66 were built, including three prototypes, 13 YP-59s, 20 P-59As, and 30 P-59Bs. Top speed was 413 mph at 30,000 feet. All were used for tests and training. (USAF)

The Bell X-1 was a research aircraft designed to furnish data on flight at and beyond the speed of sound. On 14 October 1947 Capt. (later General) Chuck Yeager attained a speed of 967 mph in the X-1 at 70,140 feet to become the world's first pilot of a manned aircraft to exceed the speed of sound. (Bell Aerospace Systems)

A North American X-15 being released from B-52 at 38,000 feet. The first of three X-15s, participating in a ten-year research program, began flying in mid-1959, and soon exceeded Mach 3 with test pilot Scott Crossfield at the controls. Altogether, eleven Air Force and NASA pilots flew the X-15s to altitudes as high as 354,200 feet, and to a speed of 6.73 Mach. One pilot and one X-15 were lost during the program. (North American Rockwell)

Originally designated the A-11 and YF-12, the Lockheed Blackbird was eventually to become the SR-71 as an operational high-speed reconnaissance aircraft. It first flew 26 April 1962. Powered with P&W J-58 turbojets, the Blackbird has a maximum speed in excess of Mach 3 above 80,000 feet. A two-placer, its unrefueled range is 3000 nm. (Lockheed-California)

Lockheed's P-80 was the USAAF's first service jet fighter, appearing in 1945. Fitted with the GE J-33 turbojet of 4000 pounds thrust, it had a maximum speed of 558 mph. Its stretched two-place version was the T-33, the standard Air Force trainer for many years. P-80s—by then called F-80s—saw combat in Korea. (Lockheed-California)

F-51 Mustangs aboard the escort carrier *Boxer* enroute to their second war. (Bart Bartimus)

Douglas B-26 Invaders of the 452nd Bomb Wing carrying five-inch rockets over Korea, February 1951. The Invader, originally the A-26 (redesignated B-26 in 1949 with the removal of the Martin B-26 Marauder from the inventory and the disuse of the A-for-Attack prefix), was the successor to Douglas' A-20 Havoc, and is unique in being the only USAF combat aircraft to have seen action in WWII, Korea, and Vietnam—as well as fighting on both sides in the Bay of Pigs invasion of 1961. (USAF)

The Marines' VMF-214, Pappy Boyington's "Black Sheep" Squadron, was back in action over Korea—still flying Corsairs. (Checker-nosed Corsair taxiing out carries the markings of VHF-312.) (USN)

Damaged Grumman F9F Panther of the attack carrier *Princeton's* air group returns to service after receiving rear half of another battle-damaged F9F. *Princeton* was part of Task Group 77, in Korean waters throughout the Korean War. (USN)

Troops of the First U.S. Marine Division are airlifted by CH-19 helicopter to a firing site near Panjong-ni, Korea, with their 4.5-inch rocket launchers 21 August 1952. (USMC)

Lt. William A. Todd of the 4th Fighter Interceptor Wing examines 37mm cannon damage to his F-86 Sabre following an air battle with Red Chinese MiG-15s, 30 October 1951. Two MiGs were shot down; all F-86s returned to base. Lt. Todd suffered minor hand injuries when a 20mm round shattered his canopy. (USAF)

Captured MiG-15 is test-flown by a USAF pilot. Mock combat with F-86 pilots allowed realistic evaluation of the MiG's strengths and weaknesses. Of its technological sophistication, one observer noted that the MiG was "evidently designed to be serviced and maintained by tractor mechanics." (USAF)

Royal Australian Air Force Mustangs, armed with napalm and rockets, prepare for combined mission with Mustangs of the U.S. 35th Fighter Interceptor Wing, 29 January 1951. (USAF)

Lockheed F-80 returns from a strafing mission with cannon damage following encounter with enemy tanks. In upper photo aircraft trails mud and water as it rebounds into the air after initial touchdown in rice paddy. Pilot was not injured. (USAF)

F-80 strafes tank, jeep, and North Korean motor pool. (USAF)

Vought F4U-4 Corsairs of VMF-323 aboard the escort carrier *Badong Strait* prepare to strike North Korean targets with rockets and napalm, 19 September 1950. (USN)

The first use of transport helicopters in combat came on 13 September 1951 when 247 Marines were airlifted to this site north of the 17th Parallel in North Korea. (USMC)

425

Lt. Benjamin Briggs, flying an F-100 Super Sabre, dive-bombs a Viet Cong base camp ten miles south of Saigon, 15 April 1966. (USAF)

Grumman A-6A Intruder, a low-altitude attack aircraft with all-weather capability. Integrated cockpit displays enable a crew to "see" target area at night or in obscuring weather. A total of 429 was built, with first deliveries beginning in 1960. (USN)

A Boeing KC-135 tanker refuels Air Force McDonnell F-4C Phantom fighters. The KC-135s have a takeoff weight of 300,000 pounds, 175,000 pounds of which is fuel, which can be transferred at the rate of 1000 gallons per minute. A total of 732 KC-135s was built between mid-1956 and 1965. (Boeing)

Douglas A-1E Skyraiders flown by pilots of the South Vietnamese Air Force (VNAF) return from a strike over the Mekong Delta. The Skyraider series began with the AD-1 in 1945 and remained in production through the AD-7 version in 1956, and had meanwhile been redesignated the A-1 series; 3180 were built for the Navy, a number of which were transferred to the USAF and VNAF in the mid-'60s. The U.S. Air Force used the Skyraider in combat until 1972. (USAF)

An A-4 Skyhawk returns to the nuclear carrier *Enterprise* in the South China Sea following its first mission over Vietnam, December 1965. Skyhawk's maximum speed was 710 mph. (USN)

Early morning launch of F-4 Phantoms from the attack carrier *Franklin D. Roosevelt* in the Gulf of Tonkin. The Navy took delivery of 2337 Phantoms, beginning in 1960; 2337 also went to the USAF. (USN)

Bell UH-1D Huey near An Khe, South Vietnam, 16 March 1966. Crew Chief Spec. 4 James Ralph mans M60 machine gun of this 1st Cavalry Division chopper. The Hueys were ubiquitous workhorses for the Army in Vietnam. (U.S. Army)

Republic F-105D Thunderchiefs over Vietnam. The F-105, a 1350-mph nuclear carrier capable of 970 mph at 4000 feet, was armed with a six-barrel 20mm Vulcan "Gatling Gun" cannnon firing 6000 rounds per minute. Georgia Air National Guard was the last unit to operate the Thunderchief, finally trading its two-seat "Wild Weasel" F-105Gs for F-4Ds in March 1984. (USAF)

Cessna A-37 was modified from T-37 Tweety Bird trainer; saw action in Vietnam with both USAF and VNAF in the ground attack or counterinsurgency (COIN) role. T-37 remains in service in the '80s as an Air Force primary jet trainer. (USAF)

Boeing B-52D, its reflective underside paint as a nuclear carrier replaced by black for service over Vietnam. B-52s carried 30 tons of iron bombs each, flying from Guam and Thailand, and surprised the experts when the earlier D models proved to be able to sustain more battle damage than the G model due to high-pressure hot-air activation of D models's operating systems. The G model has hydraulic and electrical operating systems. (USAF)

The Boeing-Vertol CH-47 Chinook, a tactical transport helicopter powered by two Lycoming T-55 turbojets, entered Army inventory in 1962. Payload is more than six tons, and normal cruising speed 150 mph. This one is delivering 105mm ammo to a field artillery unit near Phu My, Vietnam, 28 February 1967. (U.S. Army)

Flight deck safety man signals all-clear to catapult officer as Douglas A-4 Skyhawk is poised for launch from the attack carrier *Franklin D. Roosevelt* in the Gulf of Tonkin, 6 September 1966. (USN)

Bell AH-1G Hueycobra, a heavily armed helicopter fighter designed for close air support of Army ground forces, has 7.62mm "Gatling Gun" and rocket pods. The Army ordered 320 Cobras in 1966 after the prototype demonstrated a cruising speed of 200 mph. It entered combat in Vietnam in October 1967. (U.S. Army)

The General Dynamics F-111 resulted from Defense Secretary Robert McNamara's attempt to "mate a sports car with a six-ton truck," to quote Gen. Eaker. Originally conceived as an Air Force tactical fighter, McNamara ordered it modified to serve as both a Navy and Air Force fighter as well as a strategic bomber (replacing the B-70, which he cancelled). The Navy refused the bungled compromise (waiting for the airplane they really wanted, the Grumman F-14 Tomcat); the Air Force accepted it. SAC's proud emblem on this one indicates that it is an FB-111, which has the slightly longer wings of the cancelled F-111B (Navy) version. Based in Thailand, TAC F-111A crews flew a number of dangerous low-level bad-weather missions over North Vietnam. (USAF)

A product of Kelly Johnson's "Skunk Works" at Lockheed, the TR-1 is a tactical reconnaissance aircraft capable of 430 mph at 70,000 feet with a 3000 mile range. The TR-1 is 40 percent larger than the famed U-2 of the '60s which spawned it, and which saw its mission taken over by spy satellites. (Lockheed-California)

The world's largest aircraft, the 728,000-pound (gross) Lockheed C-5A Galaxy, first flew 30 June 1968. Designed to operate from 7500-foot runways, spreading its "footprints" over a 28-wheel landing gear, a fleet of these 500-mph super transports provides the Military Airlift Command with the ability to airlift large numbers of U.S. troops anywhere in the world on short notice. The improved C-5B version joins MAC in the mid-'80s. (Lockheed-Georgia)

Lockheed C-130H Hercules of the Oklahoma Air National Guard's 137th Tactical Airlift Wing which entered service in mid-1979. The "Herky Bird" prototype flew in 1954, and this aircraft has been in production since in several versions for the Air Force and foreign nations. The H model, built since 1966, is powered with four Allison T-56 turboprops, has a range of 4500 miles, cruises at 320 knots, and can take off in 3500 feet at 155,000 pounds gross weight. (Lockheed Georgia)

Chapter 15

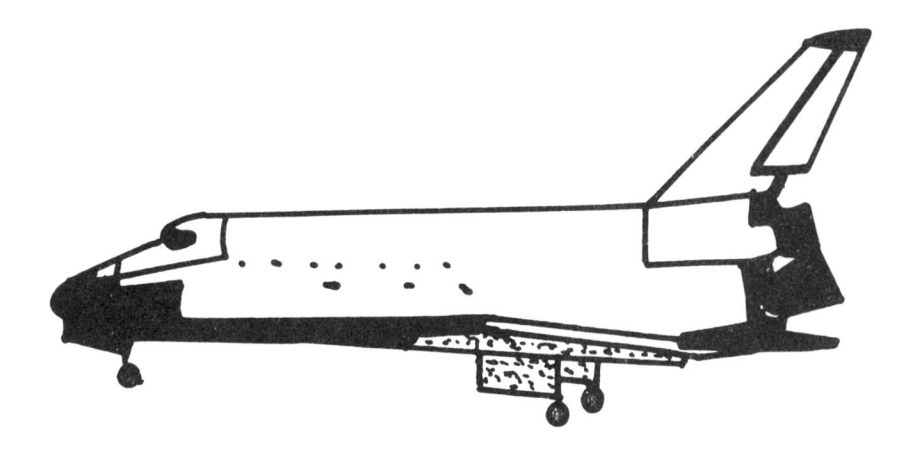

America into Space

Modern rocketry was fathered by an American, Robert H. Goddard. Goddard spent his life building a sound foundation for this new technology, then, like the Wright brothers, saw European experimenters take over initial development of his discoveries. Once again, the pupils surpassed the teacher, and when America at last began to recognize all that this new science portended, we appropriated the fruits of *their* labors to nourish an industry that should have grown up here in the first place.

At the end of WWII, U.S. intelligence people scrambled around Germany tracking down German rocket experts. We didn't get them all—the Soviets were looking for them, too—but we found the ones at the top of the list, particularly Gen. Walter Dornberger and Werner von Braun, along with several of their best technicians, and brought them to the United States. We also sought Maj. Gen. Wolfgang von Chamier-Glisczinski, but accepted the report that he had died in an American bombing raid on Peenemunde. When the Soviets orbited Sputnik I in 1957, we had reason to doubt that report.

In the U.S., von Braun was placed in charge of the test firing of some captured German V-2 rockets (called the A-4 by the Germans) at White Sands Missile Test Center near Alamagordo, New Mexico, And when our engineers asked von Braun about his early research and basic formulas, he seemed surprised. "Why, of course," he replied. "We started with the published papers of your Dr. Goddard!"

Dr. Robert Goddard was a rare person; he dedicated himself to a task that offered no tangible reward in his lifetime. Most people thought he was a nut.

Goddard was 17 years old when he decided that man would someday go to the moon. He always remembered the day—19 October 1899—and thereafter celebrated that date as the anniversary of his life's work.

After finishing high school in his hometown,

Worcester, Massachusetts, Goddard attended Worcester Polytechnic Institute where he authored a paper suggesting the use of radioactive material as a fuel for deep space flight. The idea was ridiculed.

Goddard worked as a teacher while earning his Master's degree, and then a PhD, after which he moved on to Princeton as a research fellow. But a year later he was told by his doctor that he was so seriously infected with tuberculosis that he had not long to live. He went to bed as ordered, but continued his computations.

A year later, in 1912, the disease was arrested and Goddard received patents on a multi-stage rocket system designed while bedridden.

During the next four years Goddard worked as an assistant professor at Clark University and spent much of his salary on experiments with small rockets; he authored a paper describing his experiments and what he had learned to date about rocket propulsion. He sent copies to several non-profit organizations, noting that he needed more money for additional research. That resulted in a $5,000 grant from the Smithsonian Institution early in 1917, along with the suggestion that Goddard design a small battlefield rocket for the U.S. Army.

By the time Goddard's infantry rocket had been proven in Army tests, WWI was over, and then the device was forgotten until, 25 years later, he reminded Army authorities that it should prove useful against Hitler's panzers. It was. Produced as the 2.36-inch antitank weapon, it became known to American GIs as the "Bazooka."

But it was during the period between World Wars One and Two that Goddard accomplished his most significant work. On 16 March 1926 he tested the world's first liquid-propelled rocket. It was 10 feet in length, and had only a 2½-second burn, which took it to an altitude of a mere 184 feet, but it represented Goddard's proof-of-concept vehicle, and as far as he was concerned *it* represented man's "first small step" into space.

Goddard tested two improved rockets during the next three years; then, on 17 July 1929, he tested a rocket that was far more successful than it at first appeared to be. It weighed 55 pounds, was 11½ feet in length, and was fueled with 14 pounds of gasoline and 11 pounds of LOX (liquid oxygen). It flew (from a converted windmill tower) a distance of 171 feet at about 90 feet of altitude. Most importantly, its gyro guidance system and on-board instrument package functioned perfectly. Unfortunately, the vehicle's thunderous and fiery flight, from the pasture of Aunt Effie Ward's farm near Auburn, Massachusetts, panicked a number of people and brought police, firemen, sheriff, and countless small boys with their dogs to investigate. As a result, newspapers jibed that the professor's "moon rocket" had missed its target by approximately 239,000 miles, and the state fire marshal forbid any more nonsense of that kind.

The ridicule paid off that time, because the news wire services picked up the story and among those who read it was a man who saw past the crude humor. He was a man who, just two years earlier, had demonstrated how one could parlay courage, vision, and good planning into unmatched success. His name was Charles A. Lindbergh.

Lindbergh went to see Goddard and was clearly impressed, because shortly afterwards he phoned to say that he had talked with Daniel Guggenheim about Goddard's work, and Guggenheim had agreed to furnish $50,000 for rocket research over a two-year period. If Goddard were able to report reasonable progress at the end of that time, more money would be waiting.

Goddard and his wife Esther (a former Clark University honor student whom he had married in 1924 after she volunteered to type his papers for him) moved to New Mexico. There, in a barren valley some 15 miles northwest of Roswell, Goddard continued his experiments, disturbing only those secretive four-footed inhabitants of the high desert.

By March 1935, a 75-pound Goddard rocket exceeded Mach 1 as it streaked beyond a mile in height and flew 9000 feet downrange. There were some failures, but Goddard learned something from each, and he maintained careful records. He fired rockets with a cluster of four engines, with gyro-controlled, gimbal-mounted tails, and with steering vanes positioned within the exhaust stream. In 1938 he made a series of demonstrations for the National Aeronautic Association, and although those representatives, as most others of the scientific world, were impressed, it is obvious that no one in America then knew quite what to do with Goddard's fire-breathing birds.

In Germany, however, there were men who *did* know—men who saw rockets as a form of extra-long-range artillery. They were members of the clandestine General Staff, and they especially liked the idea of rockets because such weapons could be made secretly, and outside conventional arms factories. The Treaty of Versailles forbid Germany to have an air force and severely limited the other military weapons she was allowed. The generals turned to Hermann Oberth, whose interest in rocketry was generally known and who, as it turned out, had been corresponding with Goddard on the subject. Gen. von Chamier-Glisczinski, a ballistics specialist, took charge of the Germans' first rocket research facility at Kummersdorf. Later, with Dornberger and the youthful von Braun aboard, a new facility was secretly opened at Peenemunde, on the Baltic Coast.

The Peenemunde complex became operational in 1937, and two years later, on the eve of WWII, a von Braun rocket attained an altitude of five miles. But Hitler then cancelled the priority for rocket research in order to concentrate Germany's resources on the buildup of his panzers and aircraft designed for close support of the highly mobile *Wehrmacht* in the conquest of Europe. It was therefore not until four years later—3 October 1943—that the 5½-ton A-4 rocket (V-2) was deemed a success when it left the Earth's atmosphere and flew a distance of 124 miles.

By that time, Hitler's drive into Russia had proven disastrous, Rommel had been defeated in North Africa, the Allies were moving north through Italy, the buildup of American airpower in Britain portended a massive assault on *Festung Europa,* and Hitler turned in desperation to the German rocket program. He was too late. There was no way that a sufficient number of A-4 rockets could be produced at that late date to alter the course of the war. Von Braun later said that the A-4 could have been ready at least a year earlier had not Hitler cancelled Peenemunde's first priority, and the A-10 trans-Atlantic rocket, capable of striking New York City, would have been operational by 1946.

But wars are not decided by the "what ifs." Victory favors the side that makes the fewest mistakes.

The First U.S. Rockets

During the years immediately following WWII, the U.S. rocket program was limited to what could be learned from the firing of the captured A-4/V-2s (while Convair, as early as 1947, produced some slightly improved versions of this missile), and a low-priority effort was made at developing short-range artillery rockets such as the Corporal, along with an antiaircraft missile, the Nike-Ajax, both of which became operational in 1953. The Air Force meanwhile concentrated on flying bombs, unmanned miniature jet airplanes such as the Matador and Snark—with one notable exception.

In 1951, the Redstone Intercontinental Ballistics Missile program was born and nourished by Maj. Gen. Donald L. Putt and Brig. Gen. John W. Sessums, Jr. This pair of Air Force generals, arguing that the Soviets were already working feverishly in this field, faced a lot of high-level opposition but managed to slowly gain adherents. In 1953, Air Force Secretary Harold Talbot and Air Force Chief of Research and Development Trevor Gardner joined this minority crusade, and then others in the executive branch and in the Congress began to listen. Finally, in 1954, President Eisenhower gave America's ICBM program top priority. That sparked a crash program for development of a 6000-mile ballistics missile booster.

Born of that activity was the 100-ton, 80-foot Atlas, built by Convair, which first flew in June 1957 and became operational in September 1959. By March 1962, 119 Atlases had been fired, with the Air Force terming 80 completely successful. Four months earlier, an Atlas booster sent a chimpanzee named Enos twice around the Earth in a prelude to the Project Mercury manned orbital flights. On 20 February 1962 an Atlas-Mercury launch vehicle boosted Astronaut John H. Glenn, Jr., into America's first manned orbit of the Earth.

As Putt and Sessums had warned, the Soviets had, from the end of WWII, made an all-out effort to develop ICBMs. When they succeeded in orbiting the 184-pound Sputnik I on 4 October 1957, it became clear that they had gained a substantial lead in missile booster capability. Sputnik II, weighing 1121 pounds, was lofted into orbit on 3 November carrying a dog named Laika. The U.S. managed to put its first satellite into earth orbit on 31 January 1958 when a Jupiter C booster sent 31-pound Explorer I into space to discover the Van Allen Radiation Belt.

In October 1958, the National Aeronautics and Space Administration (NASA) was created from that venerable wellspring of aviation technology, the National Advisory Committee for Aeronautics (NACA), and NASA became the official agency for all exploratory and scientific programs in space as well as in the air. Therefore, in 1961, when newly elected President Kennedy challenged the Soviets to a race for the moon, the U.S. had a proper organization to call its space shots, three years, experience with its Explorer and Discoverer series, and a bold and youthful engineering force daily designing space hardware that had to be named as it was created.

By 1968 man had ridden thunderous rockets into space 24 times—the Americans 16 times, the Soviets eight times—but none had yet left Earth orbit to venture into deep space. America's Mercury shots, with Redstone and Atlas boosters, carried a single astronaut during the early '60s. Then came the Gemini series, with two astronauts aboard, thrust into orbit by the mighty Titan. It was, however, the Apollo series that took man to the moon, and 15 years after the first Americans left their footprints on the moon's surface, no Soviet manned spacecraft had been farther into space than you can drive your car in half a day.

The Lunar Landing Program had been suggested by NASA back in July 1960, and feasibility studies by several aerospace companies were handed to President Kennedy in mid-May, 1961. The President presented a plan to Congress ten days later, along with his dramatic commitment to put Americans on the moon by the end of the decade. Congress promised the money and the first engineering contracts were let in 1962. More than 20,000 companies were involved in the program, and the first unmanned Apollo was tested on 26 February 1966 (Apollo was the name given the complete vehicle; its Saturn V booster had three stages). Several more unmanned test flights were made, plus a manned orbital mission with three astronauts aboard—Walter Schirra, Donn Eisele, and Walter Cunningham—before Apollos 8 and 10 flew to the moon and back prior to the actual moon landing by Apollo 11's lunar module 20 July 1969. Apollo 9 was the Earth orbit mission, that checked out the Lunar Excursion Module, the little space taxi that would actually land on the moon with two astronauts while the command module remained in moon orbit.

In the meantime, on 27 January 1967, three astronauts—Virgil Grissom, Edward White, and Roger Chaffee—died in an accident at Florida's Kennedy Space Center during a prelaunch test when the pure oxygen atmosphere in their Apollo command module was ignited by a short in the electrical system (three months later, Soviet Cosmonaut Komarov was killed on landing after an 18-orbit mission).

When man eventually travels into deep space to explore and perhaps colonize planets in another solar system (the closest sun besides our own is 4.5 light years away; it may or may not have a planetary system), his children will undoubtedly be required to memorize the date of 21 December 1968 as that point in time when man first shed the bonds of Earth's gravity to fly free into the cosmos. At 10:41 a.m. EST on that date, Air Force Colonel Frank Borman, Navy Captain James Lovell, and Air Force Major William Anders, in Earth orbit in the predawn darkness 118 miles above Hawaii, ignited Apollo 8's third stage booster for a 302-second burn to thrust their craft into translunar injection at a speed of 24,196 mph. Translunar injection (TLI) results when just the right amount of thrust is applied, at the proper moment during Earth orbit, to propel a spacecraft onto an intercepting course for the moon. It of course takes into account that the moon will be somewhere else by the time the spacecraft gets there (now that is "deflection shooting" on a grand scale!).

Apollo 8 then coasted for the next 53 hours, gradually slowed by the Earth's gravitational pull. Then they passed that point in space 214,000 miles from Earth where the pull of the moon's gravity became stronger than that of Earth's. By then slowed to 2217 mph, their velocity began to increase as they were drawn toward the moon.

Passing close to the moon the next morning, Christmas Eve, and turned so that the spacecraft's rocket engines faced forward, a 240-second burn slowed Apollo 8 from 5758 mph to 3643 mph to balance its speed with the moon's gravitational pull, which gave an elliptical lunar orbit 194.5 miles at apogee and 69.6 miles at perigee. (Actually, *apogee* and *perigee* are correctly applied only to the orbits of objects circling the Earth. In lunar orbit the high point is the *apocynthion,* and the low point the *pericynthion.*)

Apollo 8 circled the moon until shortly after noon on Christmas Day. Then a 198-second burn of its service module rockets accelerated the spaceship to an escape velocity of nearly 6000 mph and the three astronauts began their journey back to the "big blue marble" in the blackness of space that was their home.

Apollo streaked through the first thinly-spread molecules of earth's atmosphere 35 hours later. Re-entry, beginning at about 400,000 feet, was made at a seven-degree angle, 1600 miles from the landing zone, in order to dissipate at an acceptable rate as much of the spacecraft's 24,630-mph velocity as possible before encountering denser air at lower altitudes. Nevertheless, Apollo 8 arced Earthward like a giant meteor as air friction built up a temperature of nearly 5000 degrees F on its ablative heat shield. Inside the command module the temperature remained a comfortable 70 degrees F, although the astronauts were subjected to deceleration forces of near seven Gs.

Four and a half miles above the surface, Apollo's pair of 16-foot stabilizing parachutes opened, slowing the CM to 300 mph. A minute later the trio of orange-and-white main 'chutes were deployed, and at 10:51 a.m. EST on 27 December 1968, Apollo 8 splashed into the Pacific 1450 miles southwest of Hawaii and three miles from the carrier *Yorktown*, which was standing by with helicopters ready to retrieve the CM and its crew. The spacecraft had completed a 537,000-mile journey in 147 hours.

Apollo 9 flew in March 1969 to check out the lunar excursion module (LEM), or "moon taxi." Apollo 10 took the LEM for a dry run to the moon in May, and then on Wednesday, 16 July 1969, Apollo 11 left Planet Earth to carry out the first moon landing. In command was civilian Astronaut Neil Armstrong, a former Navy pilot. His crew were Air Force Lt. Col. Michael Collins, the Command Module pilot, and Air Force Col. Edwin Aldrin, Jr.

Four days later, on Sunday afternoon, 20 July, Armstrong and Aldrin separated from the command module in the LEM and, at 3:08 p.m. EDT Earth time, as Apollo 11 emerged from behind the moon's dark side, fired the LEM's braking rockets for half a minute to drop down into a moon orbit with a pericynthion of ten miles above the lunar surface.

Further use of the LEM's descent-stage rockets took the moon taxi out of orbit and to within 350 feet of the lunar surface by 4:15 p.m. Noting that the landing would be within a crater, Armstrong maneuvered the LEM to a more suitable spot within an area known to astronomers as *Mare Tranquillitatis* (Sea of Tranquility), and at 4:17 p.m. touched down gently.

The two astronauts rested and ate a leisurely supper. Then at 10:56 p.m. EDT 20 July 1969, Armstrong descended the LEM's ladder and set foot on the moon. "That's one small step for [a] man; one giant leap for mankind," he said to the millions of Earthlings who watched the event via live television.

Aldrin joined Armstrong to plant the American flag, set up some experiments, and gather rock and dirt samples. Then they returned to the LEM for a few hours of sleep before blasting off to rendezvous with Mike Collins, orbiting above in the command module.

Apollo 11 returned to Earth at 12:50 p.m. EDT on 24 July. The three astronauts were held in quarantine for three weeks while doctors looked for signs of any exotic malady that may have been brought back from the moon, then they were freed to face hordes of news reporters—that part of the mission for which they were *least* prepared.

Subsequently, American astronauts would spend much more time on the lunar surface for scientific studies and exploration, and if, as has been suggested, the moon one day becomes a sort of base camp for space safaris far beyond, those latter-day astronauts may find it amusing to examine the primitive hardware left there by Apollo crews—the descent stages of their LEMs, for example, and the electric car, Moon Rover.

The Mercury series space capsules of the early '60s, each of which carried a single astronaut, put America in space as the wood-and-cloth airplanes of 1910 put us into the air. Then Project Gemini, each capsule of that series carrying two astronauts into Earth orbit, represented great advances in all areas of the U.S. space effort and in a remarkably short time. In 1965, Geminis 6 and 7 rendezvoused in orbit, flew in formation a few feet apart, and proved the concept of space-docking that followed with the Skylabs in the early '70s.

When President Nixon assumed office, he drastically cut funding for the space program, and except for a single docking mission in 1975, American astronauts did not leave Earth again until April 1981, in the first of four tests of the new Space Shuttles.

The Space Shuttles

The shuttles, officially known as Space Transportation Systems (STS), were conceived as space trucks, primarily to serve the Air Force, but also to serve commercial customers. The program is directed by Air Force Lt. Gen. James Abrahamson, and the most interesting cargoes delivered into space by the shuttles will not be announced to the public. NASA would like to use the shuttles to build a space station, but if it does, that, too, will be a compromise. The Air Force will run it.

In the spring of 1984 two shuttles were operational, the *Columbia* and *Challenger*. A third, *Discovery*, made its first flight that August. Meanwhile, a second launch and recovery complex was being readied near Vandenberg Air Force Base in California. The West Coast home of the shuttles will allow cargoes to be placed in polar orbits. Those launched from the Kennedy Space Center in Florida will achieve equatorial orbits.

In order for the shuttle program to be "economically viable," NASA says that 24 flights per year with significant commercial cargoes are necessary. The business is there, for the time being in communications satellites, if the European Space Agency's unmanned Ariane booster does not prove too competitive. The Arianes are launched from French Guiana, and the sixth one was successfully fired in June 1983, just two days before STS-7— with female astronaut Sally Ride aboard—was launched.

But the manned shuttles are far more versatile than the Ariane, and STS-11 dramatically demonstrated some of that versatility 8 April 1984 when George Nelson and James van Hoften retrieved the satellite Solar Max from its orbit for maintenance and repair.

In any event, the shuttles are essential to the nation's security, and are the most cost-effective space vehicles so far built. Not only are the shuttles themselves—huge, pressurized cargo gliders— serviced after each trip into space to fly again, but the solid-propellant boosters, jettisoned 130 miles downrange, parachute into the ocean for retrieval and re-use.

Returning from space, each shuttle commander actually has many alternative landing runways available to him in a pinch, because he has sufficient maneuverability to deviate as much as 1500 miles from his planned landing approach. He will always prefer to land at the launch site, because otherwise the shuttle must be (expensively) transported piggyback on a Boeing 747 to its takeoff facility. Edwards AFB is presently the shuttle's first alternate landing field because of its 15,000-foot runway, and because a specially-built structure is located there to facilitate mounting the shuttles on the NASA 747 for cross-country ferry.

Unmanned Space Vehicles

While the manned spacecraft may be of more immediate use to us, the unmanned probes into deep space serve a longer-range and a more compelling function. These little instrument packages, roaming among the planets, help satisfy our compulsive need to know more about ourselves and our place in the universe. We launch them in the name of scientific investigation, but we seek greater truths than they are constructed to discover. True, we send them to sample the atmosphere of Jupiter, mark the edge of the heliosphere, and chart the carrier waves of gravitational force, along with a hundred other measurements, but ultimately we hope to learn from these clues how our solar system was formed, and then the universe. Whether it was all an accident, a "Big Bang" in which the universe somehow created itself out of nothing (the theory currently fashionable among many scientists), or whether there is a Supreme Intelligence—and therefore, purpose—behind it all is a question each of us must decide for ourselves, Charles Darwin and Carl Sagan notwithstanding.

Some of our space probes surprise us. Pioneer 10, built by TRW and launched by NASA 3 March 1972, hopefully to make the first reconnaissance of Jupiter if the 500-pound craft could survive passage through the asteroid belt beyond Mars. If, by any chance, Pioneer 10 got that far, it would also confirm the outer limit of the heliosphere, the bubble of particles blown into space by the solar wind, believed to be in the vicinity of Jupiter.

Pioneer 10 did indeed reach Jupiter—and then accelerated by that great planet's gravitational field, raced beyond, passing the orbit of Uranus and, in June 1983 Earth time, crossed Neptune's orbit headed out of our solar system into the realm of the stars. It had discovered the heliosphere to be far beyond its predicted boundary, and had sent back intriguing evidence of an unseen source of gravity—perhaps another planet beyond Pluto—as it left our solar system on its journey into infinity. It was traveling at 30,550 mph, three billion miles away, and scientists at the Ames Research Center in California who are monitoring Pioneer's signals expect to hear from it until sometime in 1986. By then, it may have confirmed—or confounded—our theories about the composition of interstellar space. It is not expected to approach another sun for 10,000 years.

Voyager Two flew past Jupiter to investigate Saturn and its rings in August 1981, then tracked toward Uranus. If its radio continues to function, it will send back data about Neptune when it reaches there in 1989.

Earlier, in 1962, Mariner Two gave us the first reports on Venus's hostile environment. Three subsequent U.S. unmanned spacecraft went to Venus. Mariner Four flew by Mars for a preliminary look in 1965, and Mariner Nine went to Mars in 1971 to fill in the details. The soil of the Red Planet really is red, but there are no canals—and no Martians.

Our next scientific space vehicle (if Congress funds it) will be the Galileo Probe, which is presently scheduled to be launched from the fourth space shuttle, *Atlantis,* from Earth orbit in May 1986. The Galileo Probe will go to Jupiter. It and its orbiter will separate before reaching its destination. The probe will parachute into Jupiter's stormy atmosphere in August 1988, and report its discoveries for an hour before being destroyed by the crushing pressure and intense heat in Jupiter's lower atmosphere. The Galileo orbiter, meanwhile, will spend 20 months circling the big planet, photographing and collecting additional data on Jupiter and its four major moons.

We have other means of searching the cosmos—a huge radiotelescope in New Mexico, for example, and the Infrared Astronomical Satellite. The latter, launched in January 1983, is an infrared "telescope," placed in Earth orbit in order to allow it to operate above the disturbing influences of Earth's atmosphere. Within months, it found "something" orbiting the star Vega (all stars are suns). H.H. Aumann of the Jet Propulsion Laboratory and Fred Gillett of Arizona's Kitt Peak Observatory, working at the infrared telescope's tracking station at Chilton, England, made the discovery. The infrared data cannot reveal the size or composition of Vega's satellites, but it does give us firm evidence that another sun out there—this one 150 trillion miles away—has a satellite system of some kind and, perhaps, planets like those of our own sun.

Extraterrestrials?

And, of course, the question that always comes up when laymen talk about space: Is there anyone else out there?

Well, consider: We live in the Milky Way Galaxy, which is an enormous pinwheel of suns, perhaps as many as ten billion of them. There are millions of other galaxies similar to our own. We would have to possess a very narrow concept of life and the universe to conclude that, of all those billions of suns, only our own has a planetary system.

There are many kinds of suns, ranging from the red giants to the white dwarfs. Ours is a yellow star, a very common type.

Draw your own conclusions.

U.S. Manned Space Flight Log

Mission	Crew	Date	
Mercury-Redstone 3	Alan Shepard	4/12/61	suborbital
Mercury-Redstone 4	Virgil Grissom	7/21/61	suborbital
Mercury-Atlas 6	John Glenn	2/20/62	3 orbits
Mercury-Atlas 7	Scott Carpenter	5/24/62	3 orbits
Mercury-Atlas 8	Walter Schirra	9/ 3/62	6 orbits
Mercury-Atlas 9	Gordon Cooper	5/15/63	22 orbits
Gemini-Titan 3	Virgil Grissom John Young	3/23/65	3 orbits
Gemini-Titan 4	James McDivitt Edward White	6/ 3/65	62 orbits
Gemini-Titan 5	Gordon Cooper Charles Conrad	8/21/65	128 orbits
Gemini-Titan 7	Frank Borman James Lovell	12/ 4/65	220 orbits
Gemini-Titan 6	Walter Schirra Thomas Stafford	12/15/65	17 orbits
Gemini-Titan 8	Neil Armstrong David Scott	3/16/66	6½ orbits
Gemini-Titan 9	Thomas Stafford Eugene Cernan	6/ 3/66	47 orbits
Gemini-Titan 10	John Young Michael Collins	7/18/66	46 orbits
Gemini-Titan 11	Charles Conrad Richard Gordon	10/12/66	47 orbits
Gemini-Titan 12	James Lovell Edwin Aldrin	11/11/66	63 orbits
Apollo-Saturn 7	Walter Schirra Donn Eisele Walter Cunningham	10/11/68	163 orbits moon orbit
Apollo-Saturn 8	Frank Borman James Lovell William Anders	12/21/68	7-day flight moon orbit
Apollo-Saturn 9	James McDivitt Russell Schweickart David Scott	3/ 3/69	10-day flight moon orbit
Apollo-Saturn 10	Thomas Stafford Eugene Cernan John Young	5/18/69	8-day flight
Apollo-Saturn 11	Neil Armstrong Michael Collins Edwin Aldrin	7/16/69	moon landing
Apollo-Saturn 12	Conrad, Gordon, Bean	11/14 to 11/24/69	

Mission	Crew	Date
Apollo-Saturn 13	Lovell, Swigert, Haise	4/11 to 4/17/70
Apollo-Saturn 14	Shepard, Roosa, Mitchell	1/31 to 2/9/71
Apollo-Saturn 15	Scott, Worden, Irwin	7/26 to 8/7/71
Apollo-Saturn 16	Young, Mattingly, Duke	4/16 to 4/27/72
Apollo-Saturn 17	Cernan, Evans, Schmitt	12/7 to 12/19/72
Skylab 2	Conrad, Kerwin, Weitz	5/25 to 6/22/73
Skylab 3	Bean, Garriott, Lousma	7/28 to 9/25/73
Skylab 4	Carr, Gibson, Pogue	11/16/73 to 2/8/74
ASTP	Stafford, Brand, Slayton	7/15 to 7/24/75
Shuttle 1	Young, Crippen	4/12 to 4/14/81
Shuttle 2	Engle, Truly	11/12 to 11/14/81
Shuttle 3	Lousma, Fullerton	3/22 to 3/30/82
Shuttle 4	Mattingly, Hartsfield	6/27 to 7/4/82
Shuttle 5	Brand, Ovenmyer, Allen, Lenoir	11/11 to 11/16/82
Shuttle 6	Weitz, Bobko, Peterson, Musgrave	4/4 to 4/9/83
Shuttle 7	Crippen, Hauck, Ride, Fabian, Thagard	6/18 to 6/24/83
Shuttle 8	Truly, Brandenstein, Gardner, Bluford, Thornton	8/30 to 9/5/83
Shuttle 9	Young, Shaw, Garriott, Parker, Lichtenberg, Merbold	11/28 to 12/8/83
Shuttle 10	McCandless, Brand, Stewart, Gibson, McNair	2/10 to 2/18/84
Shuttle 11	Crippen, Scobee, Hart, Nelson, van Hoften	4/6 to 4/13/84

Dr. Robert H. Goddard, Father of Modern Rocketry, began his work before WWI; he fired the world's first liquid-fueled rocket on 16 March 1926. Aided by Charles Lindbergh's intervention on his behalf, Goddard, with Guggenheim grants, made significant progress during the '30s. (USIA)

America's first intercontinental ballistic missile, the Atlas, first flew in June 1957, and became operational under the direction of the Strategic Air Command in September 1959. The Atlas had a range of 9000 miles and was built by General Dynamics. (General Dynamics)

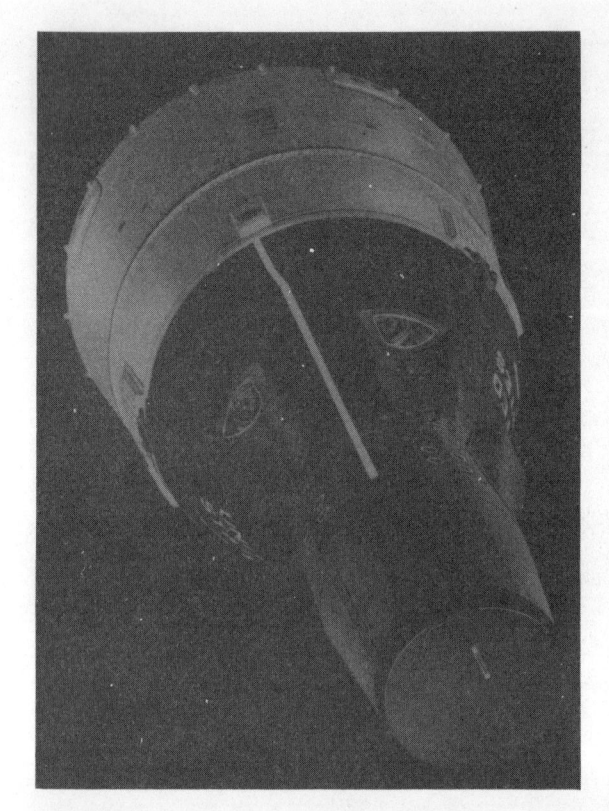

Gemini command module, built by McDonnell Aircraft Corp. America's Gemini Program of two-man Earth-orbit space missions began in March 1965; by the time Gemini 12 flew in November 1966, the Soviet's demonstrated space technology had been eclipsed in all areas of accomplishment by Americans. (McDonnell-Douglas)

This is how a Lockheed-built Agena Target Vehicle appeared to Gemini astronauts as they approached for docking in Earth orbit. Space docking, extravehicular activity, and orbit manipulation were among new space skills acquired during Gemini flights. (Lockheed Missiles & Space Company)

The Saturn V second stage S-II was built by North American and its five J-2 hydrogen-fueled engines were designed by Rocketdyne. The J-2s generated 200,000 pounds of thrust each. (North American Rockwell)

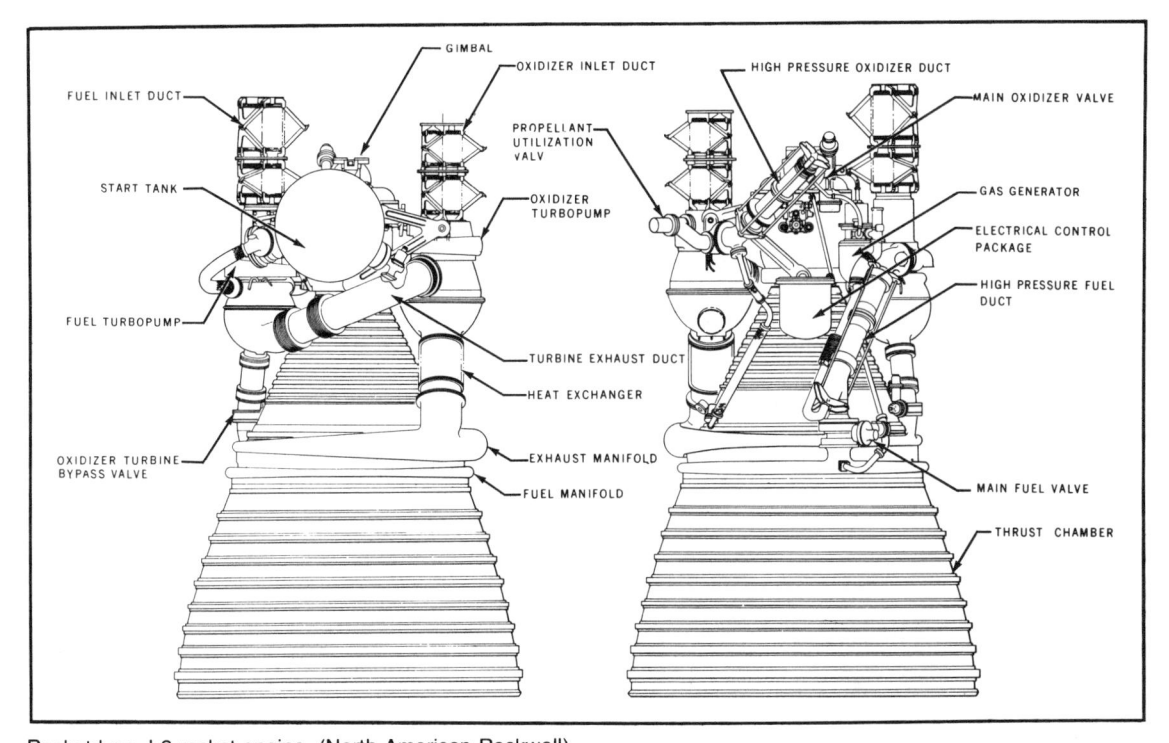

Rocketdyne J-2 rocket engine. (North American Rockwell)

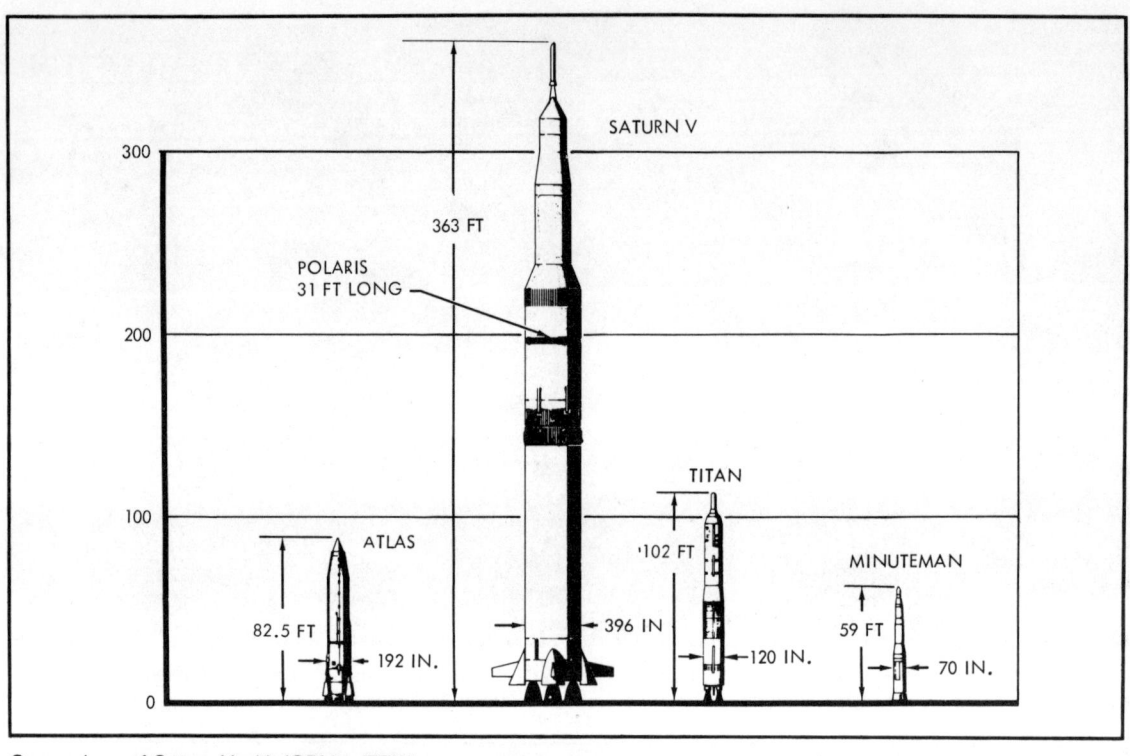

Comparison of Saturn V with ICBMs. (TRW)

This photo, taken from atop the assembly building, shows Apollo-Saturn at Kennedy Space Center as the 363-foot space vehicle and its launch umbilical tower move out of the assembly building for the launch pad on the crawler-transporter. (North American Rockwell)

LAUNCH ESCAPE SYSTEM

BOOST PROTECTIVE COVER

COMMAND MODULE

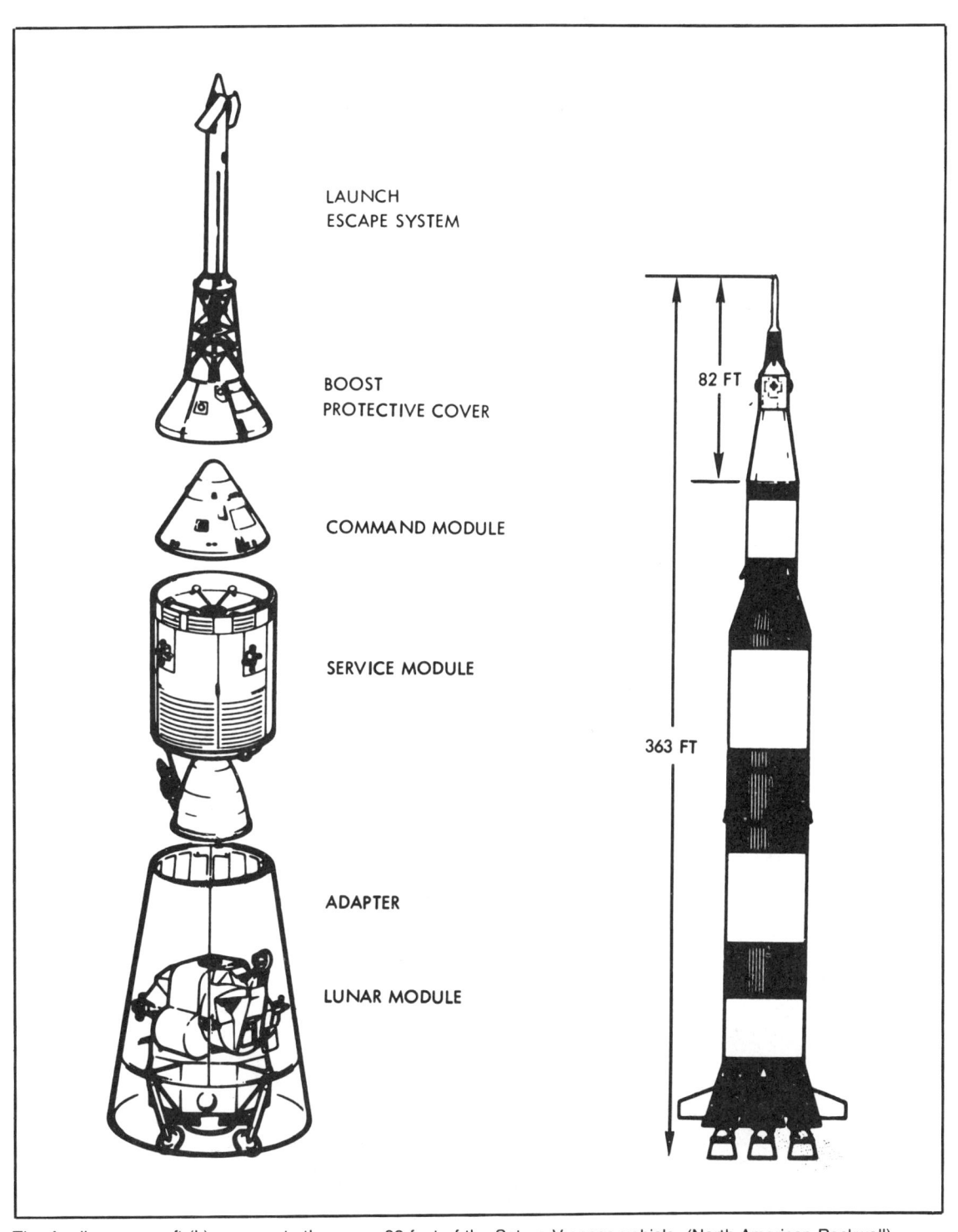

SERVICE MODULE

ADAPTER

LUNAR MODULE

82 FT

363 FT

The Apollo spacecraft (L) represents the upper 82-feet of the Saturn V space vehicle. (North American Rockwell)

The Apollo-Saturn vehicles weighed 6,200,000 pounds when fueled. Here an Apollo mission is transported toward its launch pad, 9 November 1967. (North American Rockwell)

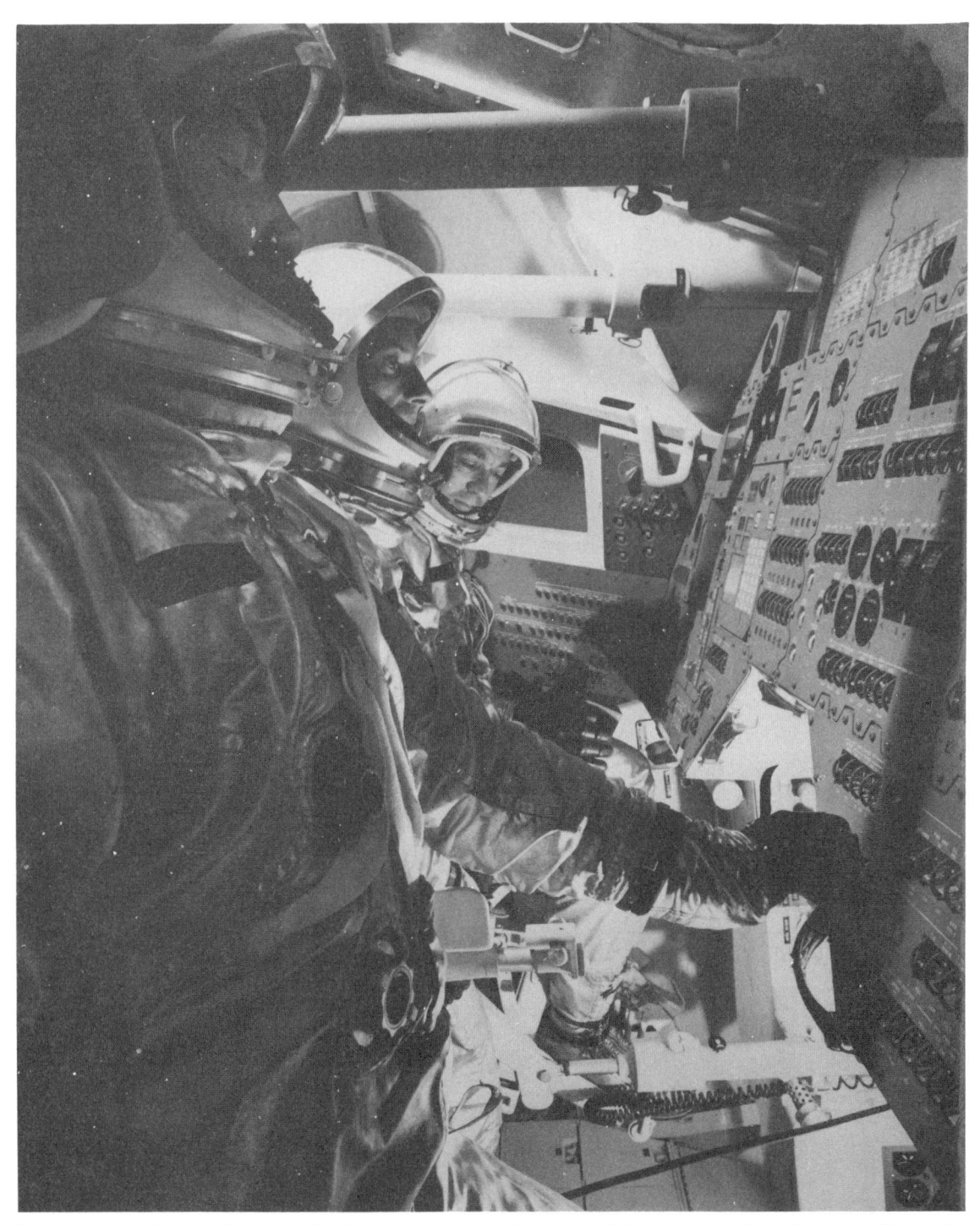

The control panel for the three-man Apollo command module had 250 switches and controllers, displayed 137 quantity measurements, and recorded 142 events. (North American Rockwell)

A Space Shuttle orbiter test vehicle in Lockheed-California Company's "reaction frame" which contains more than 350 hydraulic jacks that exert loads on the orbiter of up to one million pounds to prove its structural integrity. (Lockheed-California)

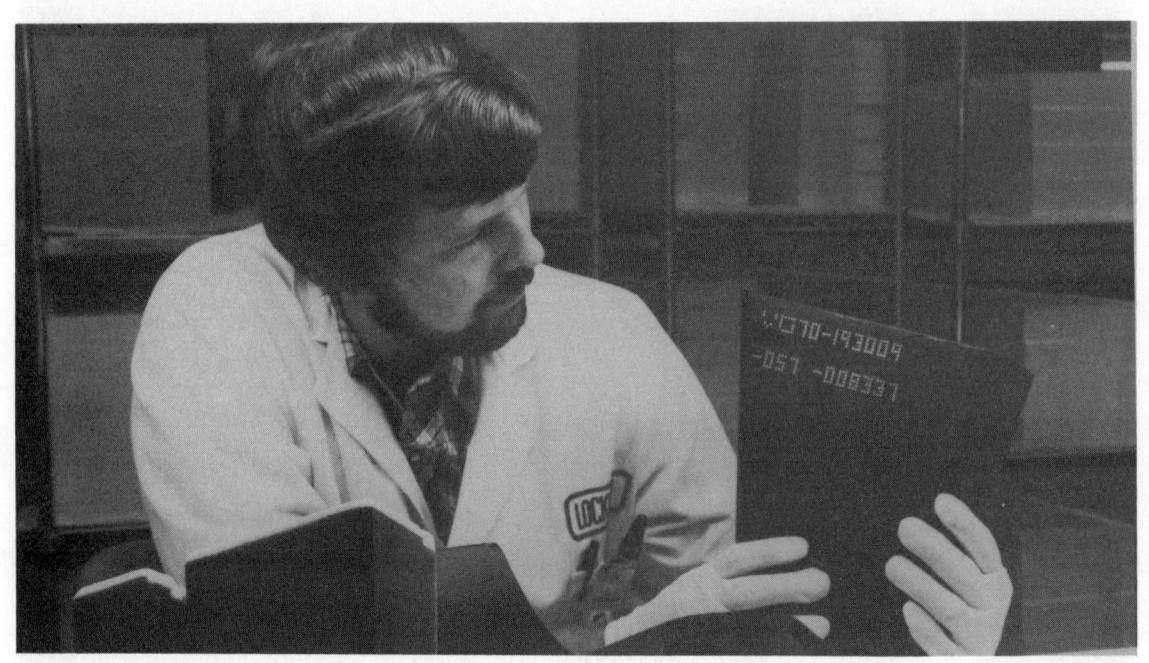

A Lockheed technician examines some of the silica tiles that shield the Space Shuttle from fiery temperatures when it returns from space to Earth's atmosphere. No two tiles are alike, and 24,000 are required for each of the four Shuttles. (Lockheed-California)

Astronauts Jack R. Lousma and C. Gordon Fullerton guide the Space Shuttle *Columbia* to a landing at the White Sands Missile Range, New Mexico, 30 March 1982, after the eight-day STS-3 mission. The Northrop T-38 chase planes seen here are also used by Shuttle astronauts to maintain proficiency in high-drag, low-thrust landings. (NASA)

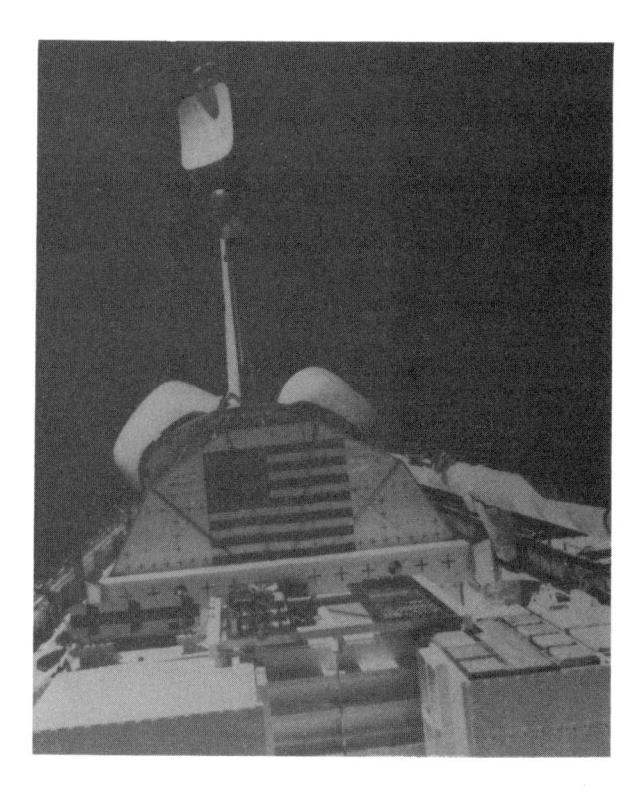

View of a Space Shuttle's cargo bay with doors open. (Lockheed-California)

Lockheed-built space telescope, scheduled for launch in 1986, will be able to see objects 50 times fainter and almost ten times farther away than the best Earth-bound telescopes. (Lockheed-California)

Index

A

A-1, vi
A-10, vi
A-20, 350
A-26, 167
A-4 Skyhawk, 428
A-6 Intruder, 414
A.B.C. Motors, Ltd. Dragonfly, 49
A.B.C. Motors, Ltd. Wasp, 49
Abrahamson, Lt. Gen. James, 440
Acheson, Dean, 409
Acosta, Bert, 261
A.C.E. Model La. 1, 54
A.D.C. Aircraft, Ltd., 55
AD Skyraider, vi
Adams & Farwell Company, 60
Adams, Ensign Charles E., 309
Advance Aircraft Company, 19, 22, 81
Aerial Experiment Association, xiii, 13, 47
Aero Commander, 196, 222
Aero Limited, 107
Aeromarine Sightseeing & Navigation Company, 107
Aeromarine West Indies Airways, Inc., 107
Aeromarine biplane, 30
Aeronautical Corporation of America (Aeronca), 83
Aeronca C-2, 83
Aeronca C-3, 83
Aeronca Champion, 83, 84, 194
Aeronca Chief, 84
Aeronca L-3, 84, 194
Aeronca Model K, 84
Aeronca 0-58, 84, 194
Aeronca Sedan, 38, 83
Air Carrier Economic Regulation Division, 121
Air Commerce Act of 1926, 75, 92, 110, 174
Air Corps Act of 1926, 268
Air Mail Act of 1934, (Kelly Bill), 56, 107
Air Mail Act of 1925 (H.R. 7064), 27
Air Mail, 24-27, 106, 107, 111, 112, 119
Air Transport Command (ATC), 355
Air racing, 154-173
Aircraft Production Board, 18, 48, 51, 108, 109, 232
Airline Pilots Association, 109
Airways Modernization Act, 196
Akagi, 348
Akers, Richard H., 121
Akron, 308
Alaskan Airways, 124
Albatros fighter, 244
Alcock, Capt. John, 260
Aldrin, Col. Edwin Jr., 439
Alexander Film Company, 76
Alexander, J.D., 76
Allied Aviation Industries, 83
Allied Control Council, 407
Allis-Chalmers, 108
Allison Engineering Company, 57
Allison, Ernest, 27
Allison, James, 57
Alon Aircoupe, 194
Alon Incorporated, 194
American Air Lines, 122
American Airlines, v, 56, 108, 120, 126
American Airlines, v, 56, 108, 120, 126
American Airmen's Association, 175
American Airplane and Engine Corporation, 55
American Cirrus Engine Company, 55
American Eagle Aircraft Corporation, 144
American Eagle, 53, 144
American Eagle Eaglet, 54, 144
American Export Airlines, Inc. (AMEX), 124
American Moth, 55
American Overseas, 190
American Volunteer Group (AVG), 354, 355
Ames Laboratory, 86
Ames Research Center, 441
AMEX, 125, 126
Anders, Major William, 438
Angle, Glenn D., 53, 83
Antique Airplane Association, 177
Anzani, Allesandro, 50
Apollo 8, 438-439
Apollo 9, 438-439
Apollo 10, 438-439
Apollo 11, 438-440
Apollo-Saturn, 450
AR-1 Avion Renault, 255
Arado 234, 408
Archdeacon, Ernest, 3
Aresti Aerocryptographic System, 178
Aresti, Count Jose L., 179
Argyll Company, 55
Arizona, 380
Armour, Lester, 108
Armstrong Siddeley, 49
Armstrong, Neil, 439
Arnold, Gen. Henry H. "Hap", 85, 268, 335, 344, 378, 408
Asp, M.B., 332
Atlantic Aircraft Manufacturing Company, 76
Atlantic Charter, 311
Atlantis, 441
Atlas missle, 438, 445
Auburn Automobile Company, 56
Aumann, H.H., 441
AVCO (Aviation Corporation), 56, 78, 113, 114, 191

Aviatak, 228
Aviation Corporation of the Americas, 113, 123
Aviation, 21
Avion Corporation, 112
Axelson Machine Company, 54

B

B&B Aviation, 195
B-1, vi
B-17, 334, 344, 346, 351, 353, 354, 356, 394, 407
B-18, 344
B-24 Liberator, 351, 354, 356
B-25 Mitchell, 183, 347, 350, 351, 355, 356, 407
B-29 Superfortress, 343, 358, 359, 407, 411
B-52 Stratofortress, 413, 414, 416, 417
B.F. Mahoney Aircraft Company, 105
Baden-Powell, Maj. B.F.S., xiii
Badoeng Strait, 410, 411, 425
Baldwin, Capt. Thomas S., 309
Ball, Clifford, 109
Battle of Britain, 311
Baumler, Maj. Albert "Ajax", 400
Bayerisch-Fleugzeugwerke, vi
Bayles, Lowell, 161
Bayley, Caro, 177
Beachey, Lincoln, 5, 6, 8, 11, 12
Becherau, Louis, 5
Beck, Don, 158, 172
Bee Line, Inc., 107
Beech, Olive Ann, 24, 206
Beech, Walter H., 21, 23, 39, 41, 43, 79, 80, 144, 154, 206, 260
Beech Aircraft Corporation, 24, 81
Beech AT-10, 341
Beech Baron, 217
Beech Bonanza, 191, 204, 217
Beech King Air, 198, 406
Beech Lightning, 216
Beech Model 17 Staggerwing, 54, 81, 99, 144, 152, 342
Beech Musketeer, 199
Beech 1900 Executive, 218
Beech Sierra, 199, 216
Beech Skipper, 191, 199, 215
Beech Starship I, 199, 220, 221
Beech Super King Air, 218
Behncke, David L., 109
Bell AH-1G Hueycobra, 432
Bell Aircraft, 408
Bell Jet Ranger, 213
Bell P-59 Airacomet, 409, 416
Bell P-63 Kingcobra, 156, 173, 376
Bell UH-1D Huey, 429
Bell X-1, 409, 417

455

Bell, Alexander Graham, viii, xiii
Bell, Lawrence D., 22, 283
Bellanca CF, 55, 79
Bellanca Viking, 194
Bellanca, Giuseppe, 6, 79, 144
Bellanca-Roos company, 79
Belleau Wood, 357
Bellinger, Lt. Cmdr. P.N.L., 259
Belmont Park Meet, 4
Bendix Aviation Corporation, 155
Bendix Trophy Race, 155-156
Berlin Airlift, 407-408
Berlin, Donovan, 83, 262, 378
Bettis, Lt. Cyrus, 261
Bingham, Hiram, 110, 267
Bingham-Parker-Merritt Bill, 110, 267
Birkigt, Marc, 15, 47
Bishop, Lester, 20
Bishop, Sam, 345
Bismark Sea, 358
Black, Hugo L., 118-121
Black-McKellar Bill (Air Mail Act of 1934), 121
Blair, Frank, 109
Bleriot, Louis, 3, 9
Bleriot, XI, 9
Blevins, Beeler, 37
BOAC, 190
Bock's Car, 359
Boeing 247, 119, 137
Boeing 314, 124
Boeing 40A, 111, 131
Boeing 707, 126, 190
Boeing 747, 440
Boeing 877 Stratocruiser, 190
Boeing Airplane Company, 15, 112
Boeing Airplane & Transport Company, 112
Boeing Air Transport, 109, 111-112, 120, 131
Boeing B-50, 407
Boeing C-97A, 74
Boeing KC-135, 427
Boeing Model C, 15
Boeing P-12, 303
Boeing P-26, 260, 306, 323, 346
Boeing PW-9, 48, 293
Boeing RB-47, vi
Boeing, William E., 15, 111-112, 124, 131, 301
Boeing-Vertol CH-47 Chinook, 431
Bogardus, George, 175
Bong, Richard, 146, 349
Borman, Col. Frank, 202, 438
Borrup, John, 52
Boston Harbor aerial meet, 4
Bowers Fly Baby, 181
Bowers, Pete, 181
Bowlus, Hawley, 42
Boxer, 411, 419
Boyington, Maj. Gregory "Pappy", 355, 402
Boyle, Lt. George Leroy, 25
Bradley, Gen. Omar, 352
Brandt, Gerald C., 268
Braniff Airways, 122, 126, 202-203
Braniff, Tom, 122
Breene, Lt. R.G., 155
Breese Penquin Trainer, 50, 174
Brennand, William, 156
Brereton, Gen. Lewis, 268, 346, 351, 354

Brewer, Griffith, xiv
Brewster F2A Buffalo, 306
Brewster F3A, vi
Brewster, Owen, 125
Bridgeport Machine Works, 79
Briggs, Lt. Benjamin, 426
Bristol Aeroplane Company, 49
Bristol Beaufighter, 390
Bristol, Capt. Mark L., 259
British Overseas Airways Corporation (BOAC), 125
Brittin, Col. L.H., 109, 118
Brodie, Otto, 262
Brookins, Walter, 4
Brow, H.J., 261
Brown, Capt. Roy, 247
Brown, D.S., viii
Brown, Donald, 52
Brown, Ensign E.W., 410
Brown, Harry, 345
Brown, Lt. Ben S., 383
Brown, Lt. Arthur Whitten, 260
Brown, Walter Folger, 114-115, 117-122, 134
Brown, Willis C., 145
Brownbach company, 81
Browning, Ed, 157
Bruce, David, 114
Bruckner, Clayton, 19, 38
Bruner, Donald L., 85
Buhl Aircraft Company, 76
Buhl Pup, 54, 76
Buhl, Lawrence, 76
Buhl-Verville, 76
Buhl-Verville Airster, 76, 111
Bulger, J.A., 332
Bull Pup, 54, 76
Bureau of Aeronautics, 110, 111, 307
Bureau of Lighthouses, 111
Bureau of Air Commerce, 117
Bureau of Aeronautics (BuAer), 50
Burgess, W. Starling, 17
Burgess-Dunn, 17
Burke, William A. "Billy", 19-21, 23, 39
Burleson, Albert, 25, 26
Burnelli, Vincent, 22, 41
Burnham, Walt, 80, 154

C

C-47, 352, 353, 354
Civil Aeronautics Authority, 146, 175
Civil Aeronautics Board, 126, 201
Cabot, 357
Caidin, Martin, 183
Caldwell, Frank W., 143
California, 380
Call, Ruell, 195
Call, Spencer, 195
CallAirs, 195
Caminez, Harold, 54, 69
Campbell, Capt. Douglas, 230, 255
Cannon, Gen. Howard, 197, 200, 352
Capital Airlines, 109
Caproni, 14, 26
Caproni-Campini N-1, 408
Carlstrom, Victor, 35
Carter, Jimmy, 197, 200
Case, Joyce, 177, 187
Castle, Col. Ben F., 81

Cayley, Sir George, viii, xiv
Central Airlines, 122, 138, 204
Central Intelligence Agency, 406
Central States Aero Company, 82, 83
Central States Aircraft Company, 83
Cessna 120, 191, 195
Cessna 140, 191, 195, 198
Cessna 150, 71, 198
Cessna 152, 208
Cessna 170, 191, 195
Cessna 172, 195, 209
Cessna 185, 208
Cessna 190/195, 54, 191, 205
Cessna 340, 211
Cessna A-37, 198, 430
Cessna Agwagon, 102
Cessna Aircraft Company, 79
Cessna AW, 55, 79, 95
Cessna C-16 Airmaster, 97
Cessna C-165 Airmaster, 97
Cessna C-34 Airmaster, 80
Cessna Centurion, 210
Cessna Citation, 198
Cessna Comet, 23, 43
Cessna Crusader, 211
Cessna DC-6, 79
Cessna EC-3, 97
Cessna FC-1, 96
Cessna GC-1, 80
Cessna Golden Eagle, 212
Cessna L-19, 195
Cessna Model A, 79
Cessna Model BW, 79
Cessna Silver Wings, 23, 42
Cessna Stationair, 209, 210
Cessna Type 3-120, 79
Cessna UC-78, 339
Cessna, Clyde V., 6, 9, 42, 70, 79, 80, 96,98
Cessna Conquest II, 198
Cessna, Eldon, 23, 29, 80, 96, 97
Cessna, Wanda, 23
Cessna-Roos Aircraft Company, 79
Chadbourne, Thomas, 15
Chaffee, Roger, 438
Challenger, 440
Chamberlain, Neville, 310
Chambers, Capt. Washington I., 259
Chambers, Reed, 109, 123
Champion Aircraft Company, 194
Champion Challenger, 194
Champion DXer, 194
Champion Decathlon, 195
Champion Lancer, 194
Champion Sky-Trac, 194
Champion Scout, 195
Champion Traveler, 194
Champion Tri-Traveler, 194
Champion Tri-Con, 194
Champlin, Doug, 194
Chance Vought Corporation, 112
Chance Vought Aeroplane Company, 17
Chandler, Harry, 108
Chanute, Octave, vii, 13
Chapman, Lt. C.G., 227
Chenango, 350
Chennault, Maj. Gen. Claire, 354, 355, 400
Chevalier, Lt. Cmdr. Godfrey, 264
Chiang Kai-shek, 355, 407, 409
Chicago Aircraft Show, 19, 20

456

Chicago and Southern Air Lines, 122, 126
China Air Task Force, 355, 400
China National Airways, 354
China National Aviation Corporation (CNAC), 124
Chitose, 357
Chiyoda, 357
Christianson, Hans, 345
Christofferson, Carl, 16
Christofferson, Simon "Chris", 17
Churchill, Winston, 311, 349, 350, 353, 406
Cierva Autogiro, 133
Cierva, Juan de la, 133
Cirrus Aero Engines, Ltd., 55
Citabria, 194
Civil Aeronautics Authority (CAA), 121, 196
Civil Aeronautics Board (CAB), 122, 126, 146, 175, 201
Civilian Pilot Training Program (CPTP), 82, 145
Clark, Homer, 21
Cleland, Cook, 156, 166
Clerget, 60
Cleveland National Air Races, 154
Cleveland Pneumatic Tool Company, 155
Coad, Dr., B.R., 84
Coanda, Dr. Henri, 87, 408
Cochran, Jackie, 163
Cody, Lt. j.g. Ernest D., 309
Coffin, Howard E., 17, 18, 109, 110, 232, 233, 267
Cohen, Marvin, 201
Cole, Duane, 176, 177, 188
Coleman, Lt. John, 395
Collier Trophy, 104
Collins, Lt. Col. Michael, 439
Colonial Air Lines, 107
Colonial Air Transport, 107, 108
Colonial Airways, 122
Columbia, 440, 453
Compania Mexicana de Aviacion, 123
Connecticut Aircraft Company, 308
Consolidated A-11, 49
Consolidated Aircraft, 17, 18
Consolidated B-24 Liberator, 382
Consolidated Commodore, 124, 140
Consolidated NY-2 biplane, 86
Consolidated N2Y, 308
Consolidated PBY, 382
Consolidated PT-1, 291
Consolidated XPT-4, 55
Constantinesco, George, 228
Constellation, 413
Continental Airlines, 122, 202
Continental engines, 143
Continental Motors Corporation, 55, 56, 156
Convair, 17, 190, 204, 240
Convair Atlas, 437
Convair B-36, 407, 415
Coolidge, Calvin, 14, 107, 110, 14, 118, 123, 267, 268
Coral Sea, Battle of, 347
Corben Baby Ace, 181
Corcoran, Thomas, 355
Cord, E.L., 56, 78
Cosmos Engineering Company, 49
Cote, Ray, 157
Craig, Malin, 310
Craven, Capt. Thomas, 263

Cronstedt, Val, 56
Crossfield, Scott, 417
Cullen, James K., 51
Culver Aircraft Company, 83
Culver Cadet, 83, 193
Culver Dart, 83
Culver V., 193
Culver, Lt. Paul, 25
Cunningham, Walter, 438
Cunningham, Lt. Alfred A., 226
Curtis-Robertson, 143
Curtiss 0-52 Owl, 337
Curtiss 18-T Wasp, 48, 282
Curtiss A-8, 49
Curtiss, AB-2, 241
Curtiss AB, 48
Curtiss Aeroplane and Motor Corporation, 13, 14, 17, 48, 49, 108
Curtiss AT-5A Hawk, 295
Curtiss AT-9 Jeep, 379
Curtiss B-2 Condor, 300
Curtiss B-2 bomber, 49
Curtiss Battleplane, 48
Curtiss BF2C-1, 327
Curtiss C-46 Commando, 355, 404
Curtiss Carrier Pigeon, 108, 112, 129
Curtiss CR-1, 261
Curtiss F Boat, 252
Curtiss F6C, 294, 296
Curtiss F7C-1 Seahawk, 297
Curtiss F9C-2 Sparrowhawk, 308, 324
Curtiss Falcons, 48
Curtiss Gold Bug, 4
Curtiss Golden Flyer, 4
Curtiss H5-2L, 15
Curtiss H81-A2, 355
Curtiss Hawk II, 298
Curtiss Hawk III, 328
Curtiss Hawk XFC6-6, 160
Curtiss Hawk, 76
Curtiss HA, vi, 44
Curtiss Hawk P-6, 49
Curtiss Hornet, 48
Curtiss HS-1L, 251
Curtiss HS2L, 32
Curtiss JN-2 Jenny, 226
Curtiss JN-4 Jenny, xiv, 14, 17, 250
Curtiss JN-6H, 25, 29, 279
Curtiss June Bug, xiii, 14, 47
Curtiss Kingbird, 81
Curtiss Model 0, 47
Curtiss Model L, 47
Curtiss N-9, 14, 251
Curtiss N2C-1, 304
Curtiss NC-4, 280
Curtiss Oriole, 20, 22
Curtiss P-1 Hawk, 294, 295, 296
Curtiss P-6 Hawk, 299, 321, 322
Curtiss P-6E, 322
Curtiss P-36, 260, 262, 336, 337, 350
Curtiss P-40, vi, 60, 262, 346, 376, 378, 398
Curtiss PN-1, 284
Curtiss PW-8, 48, 64, 261, 284, 287, 288
Curtiss R-4, 25, 35
Curtiss R2C-1, 261, 283
Curtiss R3C-1, 261
Curtiss Rheims Racer, 4
Curtiss Robin, 80
Curtiss S-3, 242

Curtiss SBC-4, 306
Curtiss SB2C, vi
Curtiss SOC-1 Seagull, 22, 325
Curtiss Teal, 81
Curtiss Triad, 225, 226, 240
Curtiss XF11C, 326, 327
Curtiss XP-6B, 298
Curtiss XP-10, 305
Curtiss XP-31 Swift, 325
Curtiss XP-37, 333
Curtiss XP-40, 334
Curtiss, Glen Hammond, xiii, xiv, xv, xvi, 3, 4, 5, 14, 17, 29, 35, 85, 225, 380
Curtiss-Robertson Company, 14
Curtiss-Wright Aeronautical Corporation, xvi, 13, 14, 49, 55, 57, 80, 84, 113

D

d'Arlandes, Marquis, vii
D-day, 353
Daines, John, 345
Daniels, Josephus, 259
Darwin, Charles, rr1
Davis, David R., 22
Davis, Doug, 98, 155
Davis, Dwight, 268
Davis, General Benjamin O. Jr., 389
Davis, William, 94
Davis-Douglas, 22
Davis-Douglas "Cloudster", 22
Day, Charles H., 17
Dayton-Wright Company, 17, 18, 56
Dayton-Wright XPS-1, 50
de Lesseps, Count Jaques, 4
de Rozier, Pilatre, vii
Deeds, Charles W., 52
Deeds, Edward A., 17, 18, 51, 52, 109, 232, 233
deHavilland Chipmunk, 186
deHavilland Comets, 163, 190
deHavilland DH-4, 18, 26, 45, 232, 282, 287
deHavilland Gipsy Moth, 55
deHavilland Gipsy, 55
Delta Air Corporation, 84
Delta Air Lines, 84, 115, 122
Delta Air Service, 84, 115
Denison, Arthur, 267
Department of Commerce, 110
Department of Defense, vi, 406
Department of Transportation, 196
Deperdussion control system, 5
Detroid Aircraft Engine Company, 53
Deutsche Luftschiffahrts Aktien Gesellshaft (DELAG), 24
Deutschland, 24
Dickinson, Charles, 19, 109
Dietz, Conrad, 83
Discovery, 440
Dollar, Robert, 56, 114
Doolittle Raid, 347
Doolittle, James Harold, 6, 86, 104, 176, 261, 262, 268, 305, 347, 348, 350, 252
Dornberger, Gen. Walter, 435
Dornier, Dr. Claude, 86
Douglas 0-38, 306, 321
Douglas 0-46, 324
Douglas 0-2H, 108
Douglas A-1E Skyraider, 427

457

Douglas A-4 Skyhawk, 431
Douglas A-20, 344, 392
Douglas AD Skyraider, 410
Douglas Aircraft Company, 22
Douglas B-18 Bolo, 334
Douglas B-19, 338
Douglas B-26 Invaders, 419
Douglas C-33, 139
Douglas C-34, 139
Douglas C-38, 139
Douglas C-39, 139
Douglas C-47, 391, 393
Douglas C-124, 74
Douglas Cruiser, 108
Douglas DC-4, 125, 190
Douglas DC-2, 137, 139
Douglas DC-1, 139
Douglas DC-3, 121, 141
Douglas DC-9, 203
Douglas DT-2, 289
Douglas DT, 22
Douglas M-2, 130
Douglas R2D, 139
Douglas SBD Dauntless, 339, 402
Douglas World Cruisers, 22, 264, 265, 282, 291
Douglas, Donald W., 22, 109, 191, 264, 283
Dowding, Sir Hugh, 311
Downey, Bob, 168
Downie, Don, 207
Driggs Dart, 56
Dunn, Maj. John W., 17
Durgin, Rear Adm., 358
Dutrieux, Helene, 4, 8

E

E.M. Laird Company, 20-22
Eaglerock Aircraft Company, 76
Eaglerock, Alexander, 90
Eaker, Gen. Ira., 352
Eareckson, Col. W.O., 356
Earhart, Amelia, 103
Eastern Air Lines, 14, 113, 120, 126, 139
Eastern Air Transport, 14, 113, 117, 120, 137
Eastern Airlines, 231
Eastern Solomons, Battle of, 349
Eaton, James M., 124
Eckener, Hugo, 307
Edgerton, Lt. James, 25
Edwards Air Force Base, 87, 200, 409, 441
Edwards, Capt. Glen W., 409
Egge, Carl F., 85
Eisele, Donn, 438
Eisenhower, Dwight David, 349, 350, 352, 354, 411, 412, 437
El Alamein, 350
Elder, Ruth, 76
Eleven-Day War, 413
Elias TA-1, 50
Elliot, A.G., 59
Ellyson, Lt. T.G., 225
Embry-Riddle Services, 114
Emmons, Harold 109
Engineering Research Corporation of Riverdale, 193
Enola Gay, 359
Enterprise, 269, 345, 347, 348, 349
Enterprise (nuclear), 420

Ercoupe, 191, 193
Escdrille Lafayette, 230
Esnault-Pelterie, Robert, xi, 3
Essex, 347, 358
Ely, Eugene, 240
European Space Agency, 440
Evans, Rowland, 197
Experimental Aircraft Association, 175
Explorer, 438

F

F-100 Super Sabre, 426
F-110, vi
F-111, 414
F-14, vi
F-4 Phantom, 428
F-8, vi, 424
F-16, vi
F-18, vi
F-51 Mustang, 411, 419
F-80, 411
F-82 Twin Mustang, 60, 72, 411
F-86 Sabre, 421
F4U Corsair, 349, 411
F8U Crusader, vi
F9F Panther, 410
Fairbank, Don, 171
Fairchild 6-390, 55
Fairchild 22, 56
Fairchild 24, 55, 195
Fairchild 71, 124
Fairchild Airplane Manufacturing Corporation, 114
Fairchild A-10 Thunderbolt II, 195
Fairchild PT-19, 55
Fairchild UC-61
Fairchild, Sherman, 69, 108
Fairchild, Walter L., 17
Fairchild-Caminez Engine Corporation, 54
Fairey Aviation Company, 59
Fairey Felix, 59
Fairey Fox, 59
Fairey, C.R., 59
Falk, Bill, 169
Farman, Henri, xii, 3, 4, 7, 228
Farwell, F.O., 60
Fedden, A.H. Roy, 49
Federal Aviation Act of 1958, 196
Federal Aviation Administration, 196
Federal Aviation Agency, 196
Federal Aeronautique Internationale (FAI), 177, 179
Ferris, Dick, 4
Field, Scott, 309
Fiske, Rear Adm. Bradley, 259, 266
Fitch, Rear Adm. A.W., 347
Fleet 2, 53
Fleet, Maj. Reuben H., 17, 25
Fletcher, Rear Adm. Frank J., 110, 267, 347, 348
Florida Airways Corporation, 109, 123
Flotorp, Ole, 6
Flyer III, xiii
Flying Tigers, 354, 355, 400
Focke-Wulf 190, 391, 392
Focke-Wulf Flugzeugbau A.G., 129
Fokker Aircraft Corporation, 77, 113
Fokker D-VII, 248
Fokker D. IX, 76

Fokker Dr-I Dreidecker, 60, 247
Fokker E-1 Eindeckers, 228
Fokker F-10 TriMotor, 90
Fokker F-14A, 133
Fokker F-32, 138
Fokker F. IV, 76
Fokker F-VIIa-3m, 77
Fokker Super Universal, 115
Fokker T-2, 76, 264
Fokker Trimotor, 77, 113, 123, 143
Fokker, Anthony Herman Gerald, 76, 77, 228
Folkerts, Clayton, 82, 83
Ford 4-AT, 77, 93
Ford 5-AT, 77, 93
Ford Air Transport, 108
Ford Model A, 174
Ford Reliability Tour, 1927, 78, 86
Ford Stout 2-AT, 92, 109
Ford Trimotor 3-AT, 92
Ford Trimotors, 86
Ford, Edsel, 77
Ford, Henry, 77, 78, 82, 108
Fornaire Aircoupe, 194
Forney Aircraft Company, 194
Foss, Joe, 146
Foulois, Capt. Benjamin, 225-227
Four-course transmitters, 86
Frank L. Odenbreidt Company, 55
Frank Robertson Aircraft Corporation, 108
Frank, K.G., 54
Frankfurt, 263
Franklin D. Roosevelt, 428, 431
Freidrichshafen bomber, 246
Frye, Jack, 115, 120
Fullerton, C. Gordon, 453
Funk, 195

G

Gabreski, Col. Francis S., 412
Gaffaney, Mary, 176, 177
Galileo Probe, 441
Galland, Lt. Gen. Adolf, 390
Gallaudet Scout, 17
Gallaudet Bullet, 17
Gallaudet, Edson F., 17
Gambier Bay, 357
Gardner, Edward, 25
Gardner, Trevor, 437
Garros, Roland, 243
Gates Flying Circus, 176
Gee Bee Model Z, 161
Gee Bee Model Y, 162
Gee Bee Q.E.D., 163
Gee Bee Sportster E, 54
Gemini, 438, 446
Gemini 6, 440
Gemini 7, 440
General Aircraft Corporation, 77, 113
General Dynamics F-111, 432
General Dynamics, 17
General Electric, 40, 262
General Motors Corporation, 14, 17, 57, 77, 113
Gibson, Harvey, 15
Gillett, Fred, 441
Gilman, Norman H., 57, 58
Glenn, John H. Jr., 438
Globe Aircraft Corporation, 196

Globe Swift, 196
Gloster Aircraft Company, 408
Glover, Warren, 114
Gnome Monosoupape, 60
Goddard, Dr. Robert H., 87, 408, 435, 436, 445
Goddard, Esther, 436
Goebel, Art, 94
Goering, Reichsmarschall Herman, 311
Goethals Engineering Company, 15
Goldwater, Barry Sr., 207
Golubev, I.N., viii
Goodyear F2G-1 Corsair, 156
Goodyear FG, vi
Goodyear Trophy Race, 156
Goodyear Zeppelin Company, 307
Gordon Bennett cup race, 4
Gorst, Vernon, 42, 108, 112
Gotha G-V, 249
Grace, W.R. & Company, 123
Graham Grosvenor, 113-114
Graham-White, Claude, 4
Great Lakes Aircraft Corporation, 19, 81
Great Lakes amphibian, 81
Great Lakes 2T-1 Sport-Trainer, 55, 81, 100
Green, Col. Carl, 85, 262
Greenamyer, Darryl, 157
Gregory, Ben, 129
Griffin, Lt. Cmdr. V.C., 264
Grissom, Virgil, 438
Grosvenor, G.B., 56
Grumman A-6A Intruder, 426
Grumman Ag Cat, 102
Grumman F4F Wildcat, vi, 306, 384, 385, 387
Grumman F6F Hellcat, vi, 403
Grumman F8F Bearcat, 157
Grumman F9F Panther, 420
Grumman J2F-1 Duck, 338
Grumman SF-1, 326
Grumman TBF Avenger, vi, 384, 401
Guadalcanal, 348, 349

H

Haigh, Henry, 178
Haizlip, James, 159
Haldeman, George, 76
Hall, Lt. Charles B., 389
Halle Trophy, 156
Hallett, Maj. George E.A., 85, 262
Halliburton, Erle, 116, 135
Halsey, Adm. W.F., 347, 349, 357
Hamilton Metalplane Company, 86
Hamilton Propeller, 112
Hamilton, W.A., 115
Hamilton-Standard Propeller Company, 143
Hammann, 348
Hammond, George, 145
Hanshue, Harris M. "Pop", 108, 109, 110, 115-116, 118, 122, 130
Harbord, Gen. James, 110, 267
Harding, John, 109, 123
Harding, Warren, 26, 258
Hargiss, M.T., 79
Harkom, Herb, 44
Harriman, W.A., 56
Harris, Lt. Harold R., 85, 262, 286
Hart, Adm. Thomas, 345, 346

Hawker Fury, 59
Hawker Sea Fury, 171
Hawks, Frank, 132, 159
Hawley Bowlus, 22
Hayden, Stone & Company, 14, 15, 113
Haynes, Colonel Caleb V., 354, 355
Hazen, R.M., 58
He 162, 408
He 178, 408
Hearst, William Randolph, 5, 117
Heath Parasol, 174
Heath, Ed, 174
Heinkel, Dr. Ernst, 408
Helms, J. Lynn, 201
Henderson, Clifford W., 155
Henderson, Phil T., 155
Henderson, Paul, 108
Herendeen, Bob, 177
Heron, Samuel D., 15, 49, 50-53, 56, 67, 155, 262
Herring, Augustus M., 13, 17
Herring-Curtiss Company, 13
Higginson, Lee & Company, 122
Hill, Maj. David "Tex", 400
Hill, Pete, 4
Hillard, Charles Jr., 176, 188
Hinckley, Robert H., 145
Hindenburg, 307
Hinton, Steve, 157
Hiryu, 348
Hitler, Adolf, 124, 146, 156, 307, 310, 311, 350, 353, 354, 390, 408, 436, 437
HMS Argus, 264
HMS Eagle, 264
HMS Hermes, 264
Ho Chi Minh, 412
Ho Chi Minh Trails, 413
Hockaday Comet, 195
Hockaday, Noel, 144
Hoffman, Maj. E.L., 85
Holman, Charles "Speed", 37, 109
Hoover, Herbert, 110, 114
Hoover-Owens-Rentschler Tool Company, 51
Hornet, 269
Hotchkiss machine gun, 228, 243
Houser, J.S., 84
Howard UC-70, 54
Howard, Ben O., 162
Howard, Beverly "Bevo", 176, 187
Hoyt, Capt. Ross G., 298
Hoyt, Fred, 78
Hoyt, Richard, 15
Hubbard Air Transport, 112
Hubbard, Edward, 107, 109, 111
Hudson, William G., 410
Huff-Daland Dusters, Incorporated, 84
Huff-Daland Petrels, 84
Hughes H-1, 165
Hughes, Howard, 165
Hyatt Corp., 202

I

Illinois Aero Club, 77
Immelmann, Max, 176, 245
Indiana, 262
Infrared Astronomical Satellite, 441
Ingalls, Lt., David S., 232
International Aerobatic Club, 175, 176, 179

International Air Transportation Assosciation (IATA), 125
International Schneider Cup, 261
Intrepid, 309, 357
Iowa, 263
Irving, Leslie, 286
Iwo Jima, 358

J

Jacobson, Joe, 164
Jameson, Dave, 101
Jannus, Tony, 107
JATO, 408
Jay, Ken, 16
Jet Propulsion Laboratory, 441
Jocelyn, Rodney, 176
Johnson Rocket, 195
Johnson, Capt. Robert S., 396
Johnson, Lyndon, 196, 413
Johnson, Robert, 146
Johnstone, Ralph, 4
Jones "Light Six", 23
Jones, John Paul, 156
Jones, Lt. Edward T., 15, 51, 52, 56, 67, 85, 155
Joyce, Temple, 297
Junkers Ju 88, 376, 395
Junkers Ju 87 Stuka, 376, 379
Junkers Ju 52, 183, 351
Junkers-Larson JL-6, 26
Junkin, Edwood "Sam", 19, 38
Jupiter C., 438
Jurney, Chesley, 118

K

Kaga, 348
Kahn, Alfred, 197, 200, 201
Kalinin Bay, 357
kamikaze, 357, 358, 377
Kari Keen, 144
Kawabe, Lt. Gen. Masakaza, 343
Kearny, 311
Keefe, Howie, 167
Keeler, F.S., 16
Keitel, Field Marshal Wilhelm, 390
Keller, Fred, 199
Keller, Pop, 6, 11
Kellett Autogiro, 55
Kelly Bill, 1925, 27, 75, 106
Kelly, Clyde, 106
Kelly, Oakley, 264
Kelsey, Lt. Ben, 86, 335
Kendall Oil Trophy, 156
Kennedy Space Center, 440
Kennedy, John, F., 412, 438
Kennedy, Ted, 197, 200
Kenney, Gen. George, 40, 262, 349, 357
Kepner, W.E., 332
Kesselring, 352
Kettering, Charles F., 17, 77, 108, 109, 233
Keys, Clement Melville, 13, 14, 48, 108, 113, 131
Keystone Aircraft Manufacturing Company, 84
Kindelberger, James Howard "Dutch", 77, 146
Kinkaid, Vice Adm. T.C., 357
Kinner Airplane and Motor Corporation, 43
Kinner, C.M., 53

459

Kinner, W.B., 53
Kirkham, Charles, xiii, 38, 47, 48, 63
Kitkun Bay, 347
Kitt Peak Observatory, 441-442
Klecker, Henry, 47
Klingensmith, Florence, 162
Knabenshue, Roy, 5, 8
Knight Twister, 171
Knight, James H. "Jack", 26, 27
Knox, Frank, 355
Koppen, Otto, 77
Korean War, 410, 412
Kraus, Lt. Cmdr, 50
Krier, Harold, 176, 178, 186

L

L-8, 309
Lacy, Clay, 157, 167
LaGrone, Tex, 37
Lahm, Lt. Frank, 225
Laird Boneshaker, 20
Laird Commercial, 17, 20, 24, 103
Laird LC-RW300 Speedwing, 37, 103
Laird Model S, 16, 20
Laird Solution, 159
Laird Swallow, 21, 22, 39, 40, 144
Laird Transport, 21
Laird, Charles, 6, 20, 40
Laird, Emil Matthew "Matty", 6, 11, 12, 16-18, 20, 21, 23, 24, 34-35, 56, 109, 144
Laird-Turner Special, 260
Lambert, Lt. Col. William C., 230
Lampert-Perkins Committee, 266
Langley Aerodrome, viii, ix, xiv, xv, xvi
Langley, 264
Langley, Edward, 25
Langley, Samuel P., viii, x
Lanphier, Capt. Thomas, 349
Lansdowne, Lt. Cmdr. Zachary, 267
LaPorte, Capt. Arthur, 124
Larson, Agnew, 76
Lassiter, Gen. William, 266
Latham, Hubert, 4
Law, Ruth, 4
Lawrance Aero Engines Company, 50
Lawrence, Charles Lanier, 50, 52, 108
Lawson-Willard-Fowler, 17
Le Blond Aircraft Engine Company, 53
Le Blond Machine Tool Company, 53
Lea, Clarence, 121
Lear, Bill, 103
Learjet 55, 222
Learjet, 198
Leblanc, Alfred, 4
Lee, J.D., 332
Lee, John, 77-78
Leef-Robinson, Lt. W., 229
Lehman, Robert, 56, 114
Leighton, Lt. Cmdr. Bruce, 50
Leisy, Cliff, 81, 100
Leland, Henry M., 17
LeMay, Gen. Curtis E., 343, 344, 358, 359, 410
Lend-Lease Act, 310, 376
Levavasseur, Leon, 3
Lewis, Drew, 201
Lewis, Fulton, 119
Lexington, 264, 267, 301, 345, 347, 357

Leyte Gulf, Battles of, 357
Lilienthal, Otto, vii, viii, x
Lillie, Max, 6
Lincoln Aircraft Company, 22
Lincoln-Standard, 22
Spirit of St. Louis, 105
Lindbergh, Anne, 149
Lindbergh, Charles A., 14, 52, 75, 89, 99, 108, 110, 123, 149, 260, 267, 436
Lippisch, Dr. Alexander, 408
Lipsner, Captain Benjamin B., 25
Little, Maj. James W., 410
Livingston, John, 163
Lloyd, Ralph, 81
Locke, Walter, 42
Lockeed, Allan, 16
Lockheed Constellation, 125, 142, 190
Lockheed Agena Target Vehicle, 446
Lockheed Aircraft Company, 16
Lockheed Air Express, 132
Lockheed Altair, 149
Lockheed C-5A Galaxy, 433
Lockheed C-130 Hercules, 434
Lockheed F-1, 31
Lockheed L-1011 TriStar, 221
Lockheed Model 12, 153
Lockheed Model G, 31
Lockheed Orion, 16, 132, 153
Lockheed F-80/P-80 Shooting Star, 409, 418, 423
Lockheed P-38 Lightning, 156, 335, 397, 399, 409
Lockheed SR-71 Blackbird, 418
Lockheed Sirius, 149
Lockheed U-2, 433
Lockheed Vega, 16, 33, 103, 155
Lockheed XC-35, 104
Lockheed YP-24, 322
Lockheed, Allan, 6
Lockheed, Dorothy, 16
Loening amphibians, 57
Loening PW-2A, 285
Loening R-4, 281
Loening, Grover C., 14, 22, 23, 304
Long Island Airways, 122
Longren, A.L., 10
Longren, A.K., 20
Lorenzo, Frank, 202
Los Angeles, 260, 267
Los Angeles-San Diego Air Lines, 22
Lott, E.P., 18, 19
Loudenslager, Leo, 178
Loughead Aircraft Manufacturing Company, 16, 32
Loughead Model S-1, 16
Loughead S-1, 33
Loughead Victor, 15
Loughead, see also Lockheed
Loughead, Flora Haines, 15
Loughead, Allan, 15, 31
Loughead, Malcolm, 15, 31
Loughead Model G, 16
Loughead Model F-1, 16
Louhead, Victor, 15
Lousma, Jack R., 453
Lovell, Captain James, 438
Lowe, Prof. Thaddeus, 309
Ludington Lines, 117, 124, 136
Ludington, Charles, 117

Ludington, Nicholas, 117
Luftschiffbau-Zeppelin, 246
Luke, Lt. Frank, Jr., 230, 252-253
Lunar Excursion Module, 438, 439
Lund, "Fearless" Freddie, 104, 176
Luscombe Silvair, 196, 213
Luscombe, Don, 82, 196
Lycoming Division of AVCO, 56

M

MacArthur, Gen. Douglas, 268, 346, 349, 357, 410, 411
Macchi MC 202, 352
Macchi MC 205, 352
MacCracken, William P., Jr., 110, 118
Mace R-2 Shark, 170
Mace, Harvey, 170
Machado, Gerardo, 123
MacNair, G.H., 332
Macon, 412
Macready, Lt. John, 262, 264
Maddox, 412
Mahoney, B. Franklin, 22
Malcolm, Allan, 15
Malcolm, Victor, 15
Manley, Charles, viii, ix, x
Mantz, Paul, 156
Mao Tse-Tung, 406-407
Mariner Nine, 441
Mariner Two, 441
Marks, Charles, 52
Marshall, Capt. A., 269
Marshall, Gen. George C., 344
Martin 202, 190
Martin B-10, 306
Martin B-26 Marauder, 391
Martin GMB, 283-284
Martin M-130, 124
Martin MB-2, 290
Martin Model TT, 14
Martin T4M-1, 81
Martin TA, 15
Martin TT, 15
Martin, Glenn Luther, 6, 14, 15, 22, 24, 81, 85, 124
Maryland, 380
Maughan, Lt. Russell, 261
Maule, 194
McCain, Vice Adm. J.S., 358
McCarran, Pat, 118, 119, 121, 125
McCormick, Willis, 23
McDonald, Capt. W.D., 400
McDonald, Wally, 167
McDonnell F-4 Phantom, 427
McDonnell XP-67 Tornado, 58
McDonnell, James S., 109
McDonnell-Douglas, 109
McElroy, Col. Ivan, 398
McGuire, Thomas, 349
NcKellar, Kenneth, 121
McKinley, William, viii
McNamara, Robert, 413
McNamara, Ens. John F., 232
McNary, Charles, 114
McNary-Watres Bill, 114, 115, 118
McWhorter, Rear Ad. E.D., 350
Mead, George, 15, 49, 50, 51, 52
Mellon, Andrew, 114

460

Mellor, Olive Ann, 24, 41, 206, 412
Melvill, Mike, 200
Melvill, Sally, 200
Menasco Motor Company, 53
Menasco, Al, 53
Menoher, Gen. Charles T., 259, 262, 263
Merritt, Schuyler, 110
Messerschmitt Me 109, vi, 351, 376, 393
Messerschmitt, Willy, vi
Metcalf, Bob, 168, 171
Meyers Midget, 19
Meyers, Charles W., 16, 18, 19, 54, 81, 100, 131
Meyers, Linda, 178
Michigan Aero Engine Company, 56
Mickley, Nyle S., 410
Mid-Continent Company, 144
Midway, Battle of, 348, 356
MiG-15, 411, 422
Mikuma, 348
Milbank, Jeremiah, 108
Milch, Field Marshal Erhard, 390
Miles, Lee, 164
Miller, Jim, 170
Miller, Max, 25
Mills, Lt. H.H., 261
Mitchell P-38, 214
Mitchell, Brig. Gen. William "Billy", 99, 110, 231, 233, 258, 259, 261, 266-268, 279, 310
Mitchell, Lt. Mack, 400
Mitscher, Vice Adm. M.A., 357, 358
Mitsubishi A6M5, 403
Mitsubishi Ki-21, 404
Moellendick, Jacob Melvin, 19-21, 23, 40, 94
Moffett, Rear Adm. William A., 50, 51, 259, 264, 269, 307, 308
Mogami, 348
Moisant, 5
Moisant, John B., 4, 86
Moisant, Mathilde, 4, 86
Mono Aircraft Company, 83
Monocoach, 83
Monocoupe 113, 54
Monocoupe Model 113, 83
Monocoupe Monoprep, 83
Monocoupe Monosport, 83
Monocoupe 90, 83
Monocoupe Corporation, 54, 82, 83, 163
Monocoupe 110, 101
Monteith, Lt. Charles N., 85, 265
Montgolfier, Etienne, viii
Montgolfier, Joseph, vii
Montgomery, Gen. Bernard, 350
Montgomery, John K., 123, 124
Montieth, C.N., 111
Mooney Aircraft, Incorporated, 193, 199
Mooney Cadet, 194
Mooney M20, 193
Mooney Mite, 193
Mooney, Al, 76, 83, 193
Moore, Mike, 345
Moran, Charles B., 410
Morane-Saulnier Type L, 243
Morehouse, Harold E., 56
Morgan, J.P., 110, 118, 267
Morgenthau, Henry, 355
Morrow Board, 99, 110, 267, 268

Morrow, Anne, 99
Morrow, Dwight, 99, 110, 267
Morse, Thomas, 22
Moss, Dr. Sanford, 262, 299
Moth Aircraft, 14
Mountbatten, Lord, 356
Mozhaisky, Alexander, viii
Murphy, Capt. W.H., 86
Murray, Rear Adm. G.D., 349
Musick, Capt. Ed, 124
Mussolini, Benito, 307, 352
Mustin, Lt. Cmdr. H.C,, 241

N

Nagumo, Adm. Chuichi, 348
Nakajima L2D3, 405
Nassau, 356
National Advisory, Committee for Aeronautics (NACA), 86, 110, 118
National Aeronautics and Space Administration (NASA), 87, 441, 438
National Aeronautic Association, 178, 437
National Air and Space Museum, 308
National Air Lines, 126
National Air Transport (NAT), 14, 93, 107, 108, 112, 129
National Security Act, 406
National Security Council, 406
Naval Academy, 226
Naval Aircraft Factory, 306, 307
Navion Rangemaster, 191
Navion, 38, 193
Neale, Robert H., 355
Neillie, C.R., 84
Nelson, Eric, 265
Nelson, George, 440
Neosho, 380
Netherlands Aircraft Manufacturing Company, 76
Neutrality Act of 1937, 311
Nevada, 380
Nevill, John, 23
New Standard Aircraft Corporation, 17
New York Air Terminals, Inc, 14
New York, Rio & Buenos Aires Line, 124
New, Gen. Harry S., 56, 107, 111, 114, 123
Newell, Joe, 85, 262
Newton, Maurice, 25
Newton, Sir Isaac, 87
Nicholas-Beasley NB3, 54
Nieuport 11, 247
Nike-Ajax, 437
Niles, Alfred, 85, 262
Niles-Bement-Pond, 51
Nimitz, Adm. Chester, 348, 357, 358
Nixon, Richard, 413, 414, 440
Noonan, Dan, 76
Noorduyn, Bob, 76
North American AT-6, 67, 341
North American Aviation, 77, 113, 114, 124, 131, 145, 191
North American B-25 Mitchell, 386
North American F-51/P-51 Mustang, 157, 191, 422
North American F-86 Sabre, 411
North American F-100 Super Sabre, 409
North American Navion, 191, 206
North American O-47, 310, 336
North American Rockwell, 113

North American X-15, 417
North Carolina, 241
Northeast Airlines, 117
Northrop A-17, 310
Northrop Aircraft Corporation, 112
Northrop Alpha, 16, 112, 134
Northrop Beta, 16
Northrop F-5, vi
Northrop Gamma, 16
Northrop P-61 Black Widow, 398
Northrop T-38, 453
Northrop Vega, 32
Northrop XB-35 Flying Wing, 415
Northrop, John K. "Jack", 16, 32, 42, 112, 134, 143, 407
Northwest Airlines, 109, 126
Northwest Airways, 109, 115, 118
Northwest Orient Airlines, 37, 109, 126
Nourse Oil Company, 21
Novak, Robert, 197
Nutt, Arthur, 48, 49, 59
NYRBA do Brasil, 124

O

O'Hare, Lt. Edward "Butch", 384
O'Neill, Ralph, 124
O.E. Szekely Corporation, 54
O.W. Timm Airplane Corporation, 53
Oberth, Hermann, 437
Odom, Bill, 156
Ogden, Henry, 265-266
Ohio Transport, 116
Oklahoma, 380
Oligvie, Alec, 4
On Aerial Navigation, viii
Operation Torch, 350
Operation Vittles, 408
Ostfriesland, 263
Ovington, Earle, 24

P

P-35, 344
P-36 Hawk, 68, 334, 344, 345
P-38 Lightning, 58, 60, 156, 169, 349, 350
P-39 Airacobra, 156, 349, 356
P-40 Warhawks, 311, 334, 344, 345, 350, 351, 356, 381, 383, 388
P-47 Thunderbolt, 351, 353
P-51 Mustang, 60, 73, 156, 157, 353, 356, 358
Pacific Aero Products Company, 15
Pacific Air Transport, 42, 108, 126
Page, Capt. Arthur, 160
Page, Ray, 22
Palmer, Pat, 173
Pan American Airways, 108-09, 118, 122, 123-126, 140, 190, 202, 354
Pan American Grace Airways (PANAGRA), 124
Panair do Brasil, 124
Panama Pacific Exposition, 6
Pangborn, Clyde, 163
Papana, Count Alex, 187
Parker, James, 118, 267
Parker, John, 157
Parkins, W.A., 46
Parsons, Lindsey, 176
Partridge, Elmer, 6, 11, 109
Patrick, Maj. Gen. Mason M., 56, 114, 231, 236

461

Patterson, William A., 109, 126
Patton, Gen. George, 353
Paulhan, Louis, 4
Pegoud, Adolphe, 6, 176, 185
Pennsylvania Airlines, 109, 122
Pennsylvania Central Airlines, 109, 122
Pennsylvania Railroad, 113
Pershing, Brig. Gen. John "Blackjack", 110, 227, 230, 231, 258
Pfalz, Reinhold, 77, 248
Philippine Sea, 410
Pickford, Mary, 85
Pickins, Bill, 5, 6, 12
Pierson, Lewis, 123, 124
Pietenpol Air Camper, 174, 180
Pietenpol, Bernard, 174
Pioneer 10, 441
Pioneer Pole & Shaft Company, 145
Piper 0-59, 82
Piper Aircraft Corporation, 82
Piper Caribbean, 207
Piper Cherokee, 192, 193
Piper Clipper, 192
Piper Colt, 193
Piper Commanche, 193
Piper Cub Special, 192
Piper Cub, 143
Piper Family Cruiser, 192
Piper J-2, 82
Piper J-3 Cub, 56, 81, 82
Piper J-4 Cub, 82
Piper J-4 Cub Coupe, 82
Piper L-4, 82
Piper Pacer, 192
Piper Super Cruiser, 191, 192, 205
Piper Tomahawk, 191
Piper TriPacer, 192, 207
Piper Vagabond, 192
Piper, William T. Sr. 81, 193
Pitcairn Arrow, 76
Pitcairn Aviation, Incorporated, 113
Pitcairn Fleetwing, 76
Pitcairn Orowing, 76
Pitcairn PA-5 Mailwing, 76
Pitcairn, Harold, 76, 113
Pittman, F. Don, 189
Pitts Special, 172, 177
Pittsburgh Airways, 116
Pittsburgh Aviation Industries Corporation, 109, 113, 117
Pittsburgh Plate Glass Company, 76
Plew, James E., 15
Plog, Lt. j.g. L.H., 410
Poberezny, Tom, 177
Poberezny, Paul, 175
Pogue, L. Welch, 126
Pollard, George, 150
Porter, Finlay R., 48
Porterfield, Edward E., 144
Porterfield Flying School, 144
Post, Wiley, 153
Povey, Leonard, 176
Power Jets, Ltd., 408
Praeger, Otto, 25-26
Pratt & Whitney Tool Company, 51-52, 112
Pratt truss system, vii, x
Pratt, Francis, 51
President's Aircraft Board, 110
Price, Frank, 176

Prince of Wales, 346
Princeton, 357, 420
Prix de la Vitesse, 4
Professional Air Traffic Controllers Organization (PATCO), 201
Project Gemini, 440
Project Mercury, 438
Prudden, George, 77
Pulitzer Trophy, 22, 260
Putt, Maj. Gen. Donald L., 437-438

Q

Queen Monoplane, 23
Quimby, Harriet, 4

R

Rader, Lt. Ira, 227
Railway Clerks Union, 106
Raleigh, 380
Ralph, Spec. 4 James, 429
Rand Corp., 124
Rand, James, 124
Rangemaster Navion, 206
Rankin, Tex, 176
Rasmussen, P., 345
Rawdon, Herb, 80, 81, 154
Raytheon Company, 24
Reaction Motors, Inc., 409
Read, Lt. Cmdr. Albert C., 259, 280
Reagan, Ronald, 197, 200-201
Redstone Intercontinental Ballistics Missile, 437-438
Reed, Albert C., 37
Reed, Dr. Sylvanus Albert, 261
Reed, Lt. Joe R., 279, 286
Reimer, Dwight, 167
Remington-Rand, 124
Reno National Air Races, 158
Rentschler, Capt. Frederick Brant, 15, 49, 50, 51, 111, 112, 131, 301
Rentschler, George, 51
Republic F-105 Thunderchief, 429
Republic P-47 Thunderbolt, 351, 353, 396
Republic Seabee, 191, 206
Republic Steel, 199
Repulse, 346
Reynolds, Earle, 113
Rheims Meet, 1909, 4, 7, 13
Richardson, Capt. H.C., 81
Richter, J.P., 264
Rickenbacker, Edward V., 120, 123, 230, 255
Riddick, Merrill, 107
Ride, Sally, 440
Ritchie, Margaret, 176
Robbins, Gen. A.W., 335
Robertson Aircraft Corporation, 107
Robertson, Frank, 108
Robertson, William, 108
Robinson, Sgt. Robert G., 232
Roche, Jean A., 83
Roche-Dohse "Flivver" plane, 83
Rockefeller, William A., 108
Rodgers, Calbraith P., 5, 9
Rodman, Burton, 16
Rogers, Bob, 345
Rogers, P.B., 81
Rogers, Will, 153
Rolan Garros, 228

Rolls, Charles, 59
Rolls-Royce, Ltd., 59
Roma, 309
Rommel, Field Marshal Erwin, 350, 351, 352, 356, 437
Roos, Victor, 79
Roosevelt, Franklin Delano, 25, 35, 106, 117, 119, 121, 122, 125, 146, 259, 310, 311, 344, 350, 353, 407
Roosevelt, Elliot, 121
Royal Aeronautical Society, xiv
Royal Aircraft Factory, 49
Royce, Henry, 59
Rutan Aircraft Factory, 200
Rutan Defiant, 199, 219
Rutan VariEze, 199
Rutan VariViggen, 200
Rutan, Elbert L. "Burt", 199, 219
Ryan Airlines, 22
Ryan Brougham, 105
Ryan Flying Company, 22
Ryan M-1, 42, 109
Ryan M-2 Bluebird, 42
Ryan Navion, 191
Ryan NYP, 89
Ryan PT-22, 53, 183
Ryan SC, 54
Ryan ST, 54
Ryan, Tubal Claude, 22
Ryujo, 349

S

Sachsen, 24
Sagan, Carl, 441
de Sait Phalle, Brigitte, 178
Salina-Hoffman biplane racer, 172
Sanders, Frank, 171
Sanders, Lewis, 345
Sangamon, 350, 357
Santee, 350, 357
Santos-Dumont, Alberto, 3-4
Santos-Dumont Demoiselle, 3
Saratoga, 264, 307, 345, 349, 358
SAS, 190
Saturn V, 438, 447-449
Sayre, Daniel C., 24
Schirra, Walter, 438
Schneider, Franz, 228
Scholl, Art, 185-186
Schroeder, Maj. Rudolph "Shorty", 77, 262
Schweizer glider, 185
Scott, Blanche, 4
Scott, Maj. G.H., 260
SEATO Treaty, 413
Seguin, Laurent, 60
Selfridge, Lt. Thomas E., xii, 225
Sessums, Brig. Gen. John W. Jr., 437-438
Seversky P-35, 68, 260, 331, 346
Shank, Robert, 25
Shell Oil Company, 268
Shelton, Lyle, 171
Shenandoah, 267
Sheppard, Morris, 24-25
Shoho, 347
Short, Mac, 78
Shutte-Lanze, 246
Sicily, 410, 411
Siddeley Deasy Company, 49
Siedhoff, George, 20, 79

Siemens & Halske, 54, 60, 69
Sikorsky Manufacturing Corporation, 112
Sikorsky S-42, 141
Silvaire Aircraft Corporation, 196
Simplex Automobile Company, 15
Simplex Red Arrow, 53
Sims, Rear Adm. William S., 259, 263, 266
Skeel, Capt. Burt, 261
Skelly, William G., 144, 145
Skelton, Betty, 177, 186
Sky Gem, 195
Skylab, 440
Slater, John E., 124
Sloan Aircraft, 17
Slocum, Col. H.J., 227
Slovak, Mira, 157
Smith, Art, 6, 35
Smith, Floyd, 85, 262, 286
Smith, Lowell H., 264-265
Smith, Mike, 173
Smith, Ted, 222
Smithsonian Institution, viii, xiv, xv, 436
Snook, William, 21, 24
Societe des Moteurs Gnome, 60
Societe Pour Aviation et Derives, 5
Sohio Trophy, 156
Solace, 380
Solar Max, 440
Sopwith Camel, 60, 232, 249
Sopwith "One and-a-Half Strutter", 228
Sopwith Snipe, 254
Sopwith Tabloid, 228
Soryu, 348
Soucy, Gene, 177
Southwest Air Fast Express (SAFEway), 116, 135
Spaatz, Gen. Carl A. "Tooey", 268, 350-352, 407
Space Shuttles, 440, 452
Space Transportation Systems (STS), 440
Spartan 7W executive, 145, 152
Spartan Aircraft Company, 144
Spartan C2, 145
Spartan C3, 145
Spartan C4, 145
Speer, Albert, 343
Spencer, Ivan, 195
Sperry Directional Gyro, 86
Sperry Gyroscope, 14
Sperry Gyro Horizon, 86
Sperry, Lawrence, 86
Sperry-Rand, 124
Spezio Tuholer, 182
Spezio, Tony, 182
Spirit of St. Louis, 65
Supermarine Spitfire, 311, 350-351
Sprague, Rear Adm. T.L., 357
Spruance, Rear Adm. Raymond, 348
Sputnik I, 438
Sputnik II, 438
St. Lo, 357
St. Louis Cardinal, 82
St. Petersburg-Tampa Airboat Line, 107
Standard Aircraft Corporation, 17
Standard Airlines, 115, 117
Standard JR-1B, 17, 25
Standard Steel Propeller, 112
Stanley, Robert M., 409
Stanley, Taylor, 83

Stanton, Stan, 80
Star Cavalier, 54, 82
Stead, Bill, 156-158
Stearman Aircraft Company, 78, 112
Stearman C-3, 79
Stearman C-1, 78
Stearman PT-17, 79
Stearman, Lloyd C., 21, 23, 39, 40, 78, 79, 144
Stearman, Waverly, 21, 144
Sterling Oil Company, 21
Sterling, Gordon, 345
Stewart, Betty, 178
Stewart, Earl, 81
Stewart, Maj. Everett, 395
Stinson 105, 143
Stinson 6000, 136, 137
Stinson Aircraft Corporation, 78
Stinson Detroiter, 78, 109, 191
Stinson Model A, 138
Stinson Model U, 135, 136
Stinson Reliant, 144, 150
Stinson SM-1, 78
Stinson Station Wagon, 191
Stinson Trimotor, 117
Stinson Voyager, 191
Stinson, Eddie, 6, 78, 94, 176, 191
Stinson, Emma, 6
Stinson, Katherine, 4-6, 11-12, 30, 35, 78
Stinson, Marjorie, 4, 6, 8, 12, 78
Stone, Lt. Elmer F., 226
Storch, Pat, 200
Stout Air Lines, 113
Stout Air Services, 108
Stout Air Pullman, 77
Stout Air Sedan, 77, 91
Stout Batwing, 77, 91
Stout Metal Plane Company, 77
Stout Model 2-AT, 77
Stout, William B., 77, 109
Street, Capt. St. Clair, 262
Strobel, Walter "Pop", 21
Sugden, Lt. Charles E., 226
Suwannee, 350, 357, 387
Swallow Airplane Manufacturing Company, 23
Swallow Sport, 55
Swallow TP-K, 53

T

Talbot, Lt. Ralph, 232
Talbott, Harold E., Jr., 233, 437
Tangier, 380
Taylor Aircraft Company, 82
Taylor Brothers Aircraft Corporation, 81
Taylor Chummy, 81
Taylor Cub, 56, 81, 143
Taylor, Charles, 46, 81, 82, 146
Taylor, E. Gilbert, 82, 193
Taylor, Gordon, 81
Taylor, Kenneth, 345
Taylor-Young Airplane Company, 82
Taylorcraft, 191
Taylorcraft B-12, 82, 193
Taylorcraft BC-12D, 82, 193
Taylorcraft 0-57, 82, 193
Taylorcraft L-2, 82, 193
Taylorcraft Sportsman, 193
Taylorcraft F-19 Sportsman, 82

Tedder, Air Marshal A.W., 350, 351
Temco Swift, 196, 212
Tennessee, 380
Texas Engineering and Manufacturing Company (TEMCO), 196
Thacker, Tony, 345
Thomas-Morse company, 77
Thomas-Morse MB-3, 279
Thomas-Morse S-4C, 60
Thomas-Morse Scout, 22, 150, 257
Thomas, Douglas, xiv
Thompson Products Company, 155
Thompson Trophy Race, 80, 155, 156
Thompson, DeLloyd, 6
Thompson, W.B., 15
Thurston, Mike, 261
Tibbets, Col. Paul, 359
Ticonderoga, 413
Timm Biplane, 53
Timm, Roy, 176
Tinker, Clarence, 296
Tinnerman Trophy, 156
Tips and Smith, Incorporated, 61
Tissandier, Maurice, 4
Titan, 438
TLR Flying Circus, 37
Todd, Lt. William A., 421
Torrance, Merle, 176
Tour de liste, 4
Towers, Cmdr. John H., 259
Towle, Tom, 78
Trans-World, 113
Transcontinental Air Transport (TAT), 14, 113
Transcontinental & Western Air, Incorporated, 14, 113, 116
Travel Air 2000, 24
Travel Air 4000, 44, 48
Travel Air 5000, 79, 94
Travel Air 6000, 76, 79, 95, 99, 145
Travel Air Manufacturing Company, 14, 24, 55, 78, 144
Travel Air Model R Mystery Ship, 80, 98, 154, 159, 260
Treaty of Versailles, 307
Trenchard, Maj. Gen. Sir Hugh "Boom", 231
Trippe & Company, 123
Trippe, Juan Terry, 108, 122, 123, 125, 126, 140
Truman, Harry S, 122, 125, 359, 406, 409, 410, 412
Trumbull John, 107
TRW, 441
Turner, J.H., 20
Turner, Roscoe, 144, 155, 161, 163, 260
Tuxhorn, Blaine, 144
TWA, 108, 113, 115, 116, 117, 120, 121, 130, 202
Twining, Gen. Nathan, 352
Twomby, Ralph, 170

U

Ultralights, 198, 215
Union Switch & Signal Company, 68
Union, 309
United Air Lines, 93, 108, 109, 112, 113, 120, 131
United Airlines, 88

463

United Aircraft & Transport Corporation, 52, 79, 112, 301
United Avigation Company, 116
United Motors, 17, 18
United States Aircraft Company, 17
United States Airways, 116, 117
United Technologies, xiv
Universal Moulded Products Corporation, 83
Unmanned Space Vehicles, 441
Utah, 380

V

V-1, 353, 397
V-2, 353, 435, 437
Valley Forge,, 410
Van Allen Radiation Belt, 438
van Hoften, James, 440
van Sicklen, Charles, 81
Vance, Cyrus, 413
Vandenberg, Gen. Hoyt, 352, 408
Varney Air Lines, 108, 113
Varney Air Transport, 122
Varney Speed Lines, 107, 113, 122
Varney, Walter T., 108, 132
Velie L-9, 54
Velie M-5, 54
Velie Motors Corporation, 54
Velie, W.L., 83
Vergeltungswaffe, 353
Verville, Alfred, 35
Verville, Fred, 76
Verville-Sperry R-3, 260, 261, 281
Vestal, 380
Vickers Gunbus, 228
Vickers-Challenger, 228
Vidal, Gene, 117
Vidal, Gore, 117
Viet Cong, 413, 414
Viet Minh, 412
Vietnam, 412-413
Viktoria Luise, 24
Villa, Pancho, 227
Vincent, Jesse G., 17, 18
Vinson, Carl, 267
Vo Nguyen Giap, 413
Voisin 10, 244
Voisin, Gabriel, 3
Voisin, Henry, 3
von Braun Dornberger, 437
von Braun, Werner, 435
von Chamier-Glisezinski, Maj. Gen. Wolfgang, 435, 437
von Ohain, Hans, 408
von Richthofen, Manfred, 245, 247
von Zeppelin, Count Ferdinand, vii
VOR, 86
Vought 02U-1 Corsair, 112
Vought AU-1, 401
Vought Corsairs, 301
Vought F4U Corsair, vi, 401, 425
Vought SU2, 302

Vought UO-1, 308
Vought VE-7, 264, 280
Vought, Chance, xiv, 17, 131, 301
Voyager Two, 441
Vultee BT-13 Valiant, 340

W

WACO "Cooties", 16
WACO 3, 19
WACO 5, 19
WACO 6, 19
WACO 9, 19, 24, 109
WACO 10, 19, 54, 69, 130
WACO Aircraft Corporation, 6, 19
WACO GC-4A, 144, 391
WACO Model E Aristocrat, 151
WACO RNF, 105
WACO S3HD, 151
WACO XJ1-W, 308
Wade, Lt. Leigh, 265-266, 268
Wadsworth, James W., 110
Wagner, Judy (Mrs. Les), 158
Wainwright, Gen. Johnathan, 346
Walcott, Dr. Charles D., xiv
Walker, Thomas, viii
Wallace, Dwane, 80, 97, 98
War Training Service, 82, 145
Ward, Effie, 436
Warren, Dr. E. Francis, 85
Washington Treaty, 269
Watres, Laurence, 114
Weaver Aircraft Company (WACO), 16, 19
Weaver, George "Buck", 6, 16, 18, 19, 21, 38, 43
Webb, Lt. Torrey, 25
Weber, Walter, 20-21
Webster, J.D., 27
Webster, Lt. John, 345
Wedell, Jimmy, 160
Wedell-Williams Number 44, 160-161
Weeks, John, 264, 266
Weeks, Kermit, 178
Weick, Fred, 193
Welch OW5-M, 101
Welch, George, 345
Wells Fargo Bank, 109
Wells, Capt. Hugh, 123
Wells, Ted, 80, 81
West Indian Aerial Express, 123
West Indies Airways, 107
West Virginia, 380
Western Air Express, 14, 90, 107-109, 113, 116, 122, 130, 138
Western Air Lines, 122
Westervelt, Conrad, 15
Whirlwind-Cessna, 79
White Sands Missile Test Center, 357, 435
White, Edward, 438
White, Sidney, 158
Whiting, Cmdr. Kenneth, 264
Whitnan, George, 345
Whitney, Amos, 51

Wichita Aircraft Company, 20, 21
Wichita Eagle, 24
Wiggen, Albert, 15
Wild, Horace, 22
Willgoos, Andrew, 15, 49-52
Williams, Alford, 261, 283
Williams, Errett, 21
Williamson, Lt. Col. John H., 400
Willys Car Company, 13, 14
Willys, John North, 13, 48
Willys-Overland, 13
Wilson, Field Marshal Sir Henry, 352
Wilson, Woodrow, 18, 25, 227, 232
Winslow, Alan, 230
Winslow, Samuel F., 110
Wise, John, 24
Wise, W.H., 332
Witney, Cornelius Vanderbilt, 108
Witt, Joe, 20
Witteman-Lewis XNBL-1, 290
Wittman, Steve, 156
Wittermore, Harris, 107
Woolman, C.E., 84, 115, 122
Wright Aeronautical Corporation, 14, 15, 49, 52, 56, 57, 59, 89, 108
Wright Flyers, xi, xii, xv, 46
Wright glider, 1902, xi, 3
Wright Military Flyer, 225
Wright Model A, 49
Wright Model B, 5, 6, 9
Wright Model C, 4
Wright Model E, 49
Wright Model H, 49
Wright Model R-1, 49
Wright, Burdette, 378
Wright, Orville, vii, viii, ix, x, xi, xii, xiii, xv, 3, 5, 8, 14, 15, 17, 46, 47, 225, 309
Wright, Wilbur, vii, viii, ix, x, xi, xii, xiii, xv, 3, 5, 8, 46, 47, 176
Wright-Martin Aircraft Corporation, 13, 15, 48, 49
Wrigley, Philip, 108

Y

Yak-18, 177, 189
Yalta Conference, 407
Yamamoto, Adm. Isoroku, 348, 349
Yamato, 358
Yankee-Siemens, 54
Yarnell, Capt. H.E., 269
Yeager, Capt. Charles "Chuck", 409, 417
Yeager, Frank, 26
Yorktown, 267, 269, 347, 348, 386, 439
Young, Maj. Clarence, 111

Z

Zahm, Dr. Albert P., xiv, xv
Zlinn, 177
ZR-1 Shenandoah, 307
ZR.3 *Los Angeles*, 300, 307, 323
ZRS-4 *Akron*, 307
ZRS-5 *Macon*, 307
Zuikaku, 357